Sociology

Sociology

CONCEPTS AND CHARACTERISTICS

Ninth Edition

JUDSON R. LANDIS
California State University, Sacramento

Wadsworth Publishing Company
I(T)P™ An International Thomson Publishing Company

Belmont • Albany • Bonn • Boston • Cincinatti • Detroit • London • Madrid • Melbourne
Mexico City • New York • Paris • San Francisco • Singapore • Tokyo • Toronto • Washington

Assistant Editor: *Susan Shook*
Production Editor: *Karen Garrison*
Managing Designer: *Ann Butler*
Print Buyer: *Barbara Britton*
Art Editor: *Bobby Broyer*
Permissions Editor: *Jeanne Bosschart*
Designer: *Donna Davis*

Copy Editor: *Karie Youngdahl*
Photo Researcher: *Stephen Forsling*
Cartoonist: *Tony Hall*
Compositor: *Graphic Composition, Inc.*
Cover: *Ellen Pettengel Design*
Cover Illustrator: *Helen B. O'Rourke*
Printer: *Quebecor/Fairfield*

*This book is printed on
acid-free recycled paper.*

Printed in the United States of America
4 5 6 7 8 9 10—01 00 99 98 97

For more information, contact Wadsworth Publishing Company:

Wadsworth Publishing Company
10 Davis Drive
Belmont, California 94002, USA

International Thomson Publishing Europe
Berkshire House 168-173
High Holborn
London, WC1V 7AA, England

Thomas Nelson Australia
102 Dodds Street
South Melbourne 3205
Victoria, Australia

Nelson Canada
1120 Birchmount Road
Scarborough, Ontario
Canada M1K 5G4

International Thomson Editores
Campos Eliseos 385, Piso 7
Col. Polanco
11560 México D.F. México

International Thomson Publishing GmbH
Königswinterer Strasse 418
53227 Bonn, Germany

International Thomson Publishing Asia
221 Henderson Road
#05–10 Henderson Building
Singapore 0315

International Thomson Publishing Japan
Hirakawacho Kyowa Building, 3F
2-2-1 Hirakawacho
Chiyoda-ku, Tokyo 102, Japan

Library of Congress Cataloging-in-Publication Data

Landis, Judson R.
    Sociology : concepts and characteristics / Judson R. Landis. —
9th ed.
        p. cm.
    Includes bibliographical references (p.) and index.
    ISBN 0-534-23754-1
    1. Sociology.   I. Title.
HM51.L255 1995
301—dc 20   94-35558

Contents

3   NORMS, ROLES, CULTURE, SOCIETY   72

Part 2   Social Organization   107

4   GROUPS   110

5 SOCIAL INEQUALITY: SOCIAL CLASS   146

6 SOCIAL INEQUALITY: RACE, ETHNICITY, GENDER, AGE   180

7   INSTITUTIONS: FAMILY   220

11 POPULATION AND ECOLOGY 328

Part 3   Social Change and Social Deviance   363

12 SOCIAL CHANGE AND SOCIAL DISORGANIZATION 366

Contents

Preface

THE PURPOSES of a first course in a discipline usually are several—to introduce an area of study, to communicate its unique perspective or way of looking at the world, and to offer the promise of secrets yet to be discovered—all in the hope that interested students will come back for more (maybe even thirty units more). That there are many different views of the best way to achieve these ambitions is apparent in sociology from some of the introductory textbooks available.

Many books provide an encyclopedic analysis of what a discipline is and does—all its tools, techniques, and substantive areas. I have decided not to try to do that in this book. It seems to me that a feeling for an area of study and its unique perspective may be gained by sampling the concepts of that field. Concepts are the building blocks—the language—of most disciplines. Once one has this language, the rest comes more easily and makes more sense. This book approaches the task of helping students develop a general understanding of the sociological perspective by focusing on the basic concepts of sociology. Through examining several substantive areas and the research connected with them, students can begin to see how the concepts contribute to sociological analysis.

At the end of most of the chapters, I have included readings to further illustrate the chapter's concepts. These readings are drawn from a variety of sources. Included are descriptions of research and excerpts from novels, autobiographies, and nonfiction works. Some of the selections were written by professional social scientists; some were not. For example, writings of both Helen Keller and Malcolm X are used to illustrate socialization and development of self. In addition, there are selections by William Domhoff on where the powerful play, Christopher Edwards on joining the "moonies," Dee Brown on the history of the American Indian, and Barbara Garson on the evils of automation, as well as William Kephart's analysis of the Gypsies, Harry Gracey's comparison of a day in kindergarten to a day in boot camp, and *The New Republic's* interesting debate on the "Jeweler's Dilemma." The readings, I think, illustrate the concepts and ideas in an unusual way. It's always a challenge to find good readings, readings that are both illustrative and interesting. Four of the readings (those written by Helen Keller, Malcolm X, Ralph Linton, and Harper Lee) now have survived all nine editions of this book.

This edition reflects the following: Most chapters have been revised at least slightly and all statistical information including graphs and tables has been updated to reflect the most recent data available. I have added eight new readings and dropped four old ones.

Many people have helped me on this and earlier editions with ideas and criticism. These include especially my colleagues at CSU, Sacramento, from whom I have borrowed books and stolen ideas. The following reviewers were also helpful: Debra J. Berger, Oakton Community College; Paula M. Fahy, Monroe Community College; Douglas B. Gutknecht, San Bernardino Valley College; Christine Lovizio, SUNY Farmingdale; Kooras Mahmoudi, Northern Arizona University; and Paul Renger, St. Philip's College. I'd also like to thank Lois Hill, Verica Dering, Steve Rutter, Sheryl Fullerton, and Serina Beauparlant and the people at Wadsworth—Karen Garrison, Karie Youngdahl, Susan Shook, and Jason Moore.

This is dedicated to Sheron, who helped whenever I asked, and to Jeffrey, Brian, and Kevin, who have continued to be patient with me.

Judson R. Landis

Sociology

CHAPTER ONE

Introduction: Knowledge, Science, and Sociology

THIS BOOK IS designed to acquaint you with a field of study called *sociology*. Generally speaking, sociology is the study of human society; it is the study of social behavior and the interaction of people in groups. Sociologists study various aspects of social life, including behavior in large organizations and small groups, deviant behavior, and the characteristics of political and religious institutions and social movements; sociologists also examine the social-psychological explanations for behavior. Our view is that you understand human behavior by looking at the groups people belong to and the culture in which they live. We focus on social structure—that complicated network of ways that people are connected. In this chapter, we look first at the idea of *knowledge*, then at a particular way of gaining knowledge called *science*, and finally at one of the social sciences called *sociology*.

However, before I get into these topics, I'd like to get you accustomed to the way social scientists look at the world, so let's consider several situations.

1. A white sports fan is fascinated by the fact that while the majority of our society is white, the majority of the athletes playing professional basketball are black. He concludes that because of his racial, genetic, and hereditary background factors, he had better take up tennis. How good is his logic?

2. A famous person (Marilyn Monroe, Ernest Hemingway, Freddie Prinz, and so on) commits suicide. Or a murder-suicide takes place in which a person kills

David Harding/Tony Stone Images

at least two other people and then kills himself or herself. What, if anything, will happen among those in the rest of the society who know about these events?

3. A statistics fanatic finds that in a particular year, Mexico and Miami had a lot of murder but little suicide and that Hungary, Denmark, and western states like New Mexico, Nevada, and Wyoming had many suicides. Our numbers freak wonders if this is a one-time event or typical, and if typical, why?

4. A poll tells us that most people (80–85 percent) favor using the death penalty as punishment for serious crimes like murder. That same poll mentions that a good portion of those who favor the death penalty do so because they believe that it deters others from committing crime. Is that really so, and is the poll accurate?

5. We're loaded with crime these days—burglary, robbery, auto theft, you name it. But what about 300 years ago: Were there crime waves then like now? If so, what were the criminals doing—burglary, robbery, auto theft?

6. A sane individual goes to the admitting officer of a mental hospital and gives that person an accurate case history (which in general states that he is as sane as you and I) and also says one thing that is not true: that he occasionally hears voices, unclear voices that seem to be saying words like "empty," "hollow," and "thud." What will happen to our sane person?

3

7. A teacher is led to believe (falsely) that several of his or her students are very sharp and are bound to do well this semester. How does the teacher behave toward the "sharp" students, and how do the students perform?

These are the types of situations that social scientists study and the types of questions that we try to answer. In fact, the situations listed have been studied, and some fairly interesting answers have emerged. Later in this chapter and in following chapters, I'll return to each of these and describe to you how we know what we know.

KNOWLEDGE

Why do we want to study sociology? For ages people have been fascinated with their own existence, and they have persistently tried to find out more about themselves and the world in which they live. Their attempts have occasionally been bumbling and crude, sometimes amazingly sophisticated. For example, around 1900 B.C., a group of immigrants from what is now Holland and the Rhineland arrived in southern England. (They eventually became known as Beaker people because pottery drinking vessels were found in their graves.) Upon their arrival, they began building a structure of posts, mounds, holes, and stones on the Salisbury plain southwest of London that later became known as Stonehenge. Some of the stones they used were as long as 30 feet and weighed 50 tons. The stones were placed on end in holes in an intricate pattern of circles and alleys. It took the Beaker people and the Wessex people, who continued the work, some 300 years to complete what amounted to a giant calendar, a sort of Stone Age computer. The stones were set up so that sighting through cracks between stones would show sunrise, sunset, moonrise, and moonset on certain important days (for example, midsummer day, midwinter day, first day of summer, first day of winter). Other sightings allowed the Wessex people to predict eclipses of the moon.

Why was this elaborate structure built? As a calendar it was useful for judging the proper time to plant and harvest crops. It probably also proved helpful to priests, who could impress people with their apparent power to make the sun and moon rise and set wherever they wanted. And possibly the structure was built just because the Beaker and Wessex people were curious to see if it could be done.

Aside from the enormous physical work involved, the builders of Stonehenge had to collect and correlate a great deal of information: planting and harvesting times, times of sunrises and sunsets, and phases of the moon. They had to pass this information on from generation to generation. Archaeologists who have studied Stonehenge tell us that what looks like a single monument is actually several. The first structure didn't work as well as the builders wanted, and a second model was built over the first. It likewise needed improvement, and a third was constructed.

Unlike most primitive structures, Stonehenge has lasted to modern times. Some stones have weathered and fallen from their positions; others have been taken by farmers for stone fences. But much of the structure remains intact.

For years, its presence and purpose had confounded people. And although some who studied Stonehenge felt that its existence had something to do with the positions of the sun and moon, the true complexity of Stonehenge was first revealed only in 1961, when astronomy professor Gerald Hawkins fed data on the positions and relationships of the stones, posts, and holes into a computer for analysis. Then modern people learned the significance of what the Beaker and Wessex people had learned 3,500 years earlier.[1]

In the late 1600s, some chemists who were attempting to explain what happens when substances are heated or burned came up with the theory of *phlogiston.* The theory stated that phlogiston is an invisible substance that exists in all combustible bodies and is released during combustion, the act of burning. Substances rich in phlogiston burned easily, and perhaps fire was a manifestation of phlogiston. The new theory was immediately accepted: It helped explain various phenomena, and it guided the famous scientists of the day. The problem, of course, is that it was totally wrong. Even when confronted with evidence that discredited the phlogiston theory, scientists in the 1780s were very slow to give up on it. Joseph Priestley, a famous scientist of the time, died in 1804 still believing in phlogiston.[2]

These are just two of countless examples of people's attempts to describe and understand their world, to obtain knowledge about their existence. Regardless of how it is collected and sometimes even regardless of its accuracy, knowledge tends to accumulate. Explanations seem to last if they are convincing; that is, they are believed as long as they seem to explain a part of human experience. If the explanations leave out important elements, they tend to be modified or even abandoned to be replaced by a better theory.

Exchanging an inaccurate explanation for a better one, however, isn't always easy. Some explanations are very dearly held. Their backers (even fans) may believe in them so fiercely that the ideas remain in vogue even when other theories work better. Scientists have been known to compete with and to conceal findings from each other to arrive at a discovery first and to make certain the discovery stands. This situation, unfortunately, can hinder the growth of knowledge.

But one way or another, knowledge accumulates, and it usually does so in one of three general ways: mysticism, rationalism, or empiricism. **Mysticism** refers to knowledge gained by intuition, revelation, inspiration, magic, visions, or spells. Societies often hold special places for those individuals who seem to possess unusual powers: the magician, the spiritualist, the priest, the witch doctor. The ceremonial use of drugs to produce visions is also highly valued among some peoples. In the 17th century, the Puritans of the Massachusetts Bay Colony were heartily convinced of the presence of witchcraft in their midst. Their evidence came from several young girls, who, between dramatic seizures, pointed out fellow villagers who were aiding the devil in his work. The girls became so energetic in their attempts to cast out evil that 350 people were taken into custody and accused of witchcraft.[3]

Knowledge arrived at through mysticism tends to be private. Only the person experiencing the revelation or vision has it, and others must take the mystic's word for it. Even though this may enhance the appeal of the mystic, it tends to compromise the quality of the product. If no one else sees or feels it, can it really be true?

Rationalism refers to knowledge gained through common sense, logic, and reason. The writings of Aristotle and the dialogues of Plato in ancient times, and the ideas of Descartes, Spinoza, Voltaire, and Thomas Paine in the 17th and 18th centuries, provide examples of knowledge emerging through the careful use of logic and reason. In the 20th century, many continue to find the use of common sense, logic, and reason the most appropriate way to arrive at knowledge. As with mysticism, however, there are limitations. For example, at one time knowledge emerging from a commonsense analysis told us that the earth was flat ("Couldn't be round; people on the other side would fall off...."), and that if two metal objects—a heavy one and a light one—were dropped from a height, the heavy object would fall much faster. We see that the intellectual approach alone can be as inaccurate as the mystical one.

Empiricism, on the other hand, refers to knowledge that is gained by sense observation—by observing or experiencing phenomena with the senses of touch, sight, hearing, smell, or taste. This is the basis of science. Instead of simply thinking about the heavy and light objects, we actually drop them from a tall building and observe whether the heavy one falls faster. Public knowledge then grows as we invite a crowd of observers to help us watch the fall of the objects, or we suggest that the experiment be tried by others in other places.

SCIENCE

We have said that empiricism is the basis of science. A one-sentence definition of what scientists are up to could go something like this: **Science** is *understanding* and explanation through *description* by means of *measurement*, making possible *prediction* and thus *adjustment* to or *control* of the environment.

For example, violent windstorms occasionally spring out of the Caribbean and the western North Atlantic. Huge waves caused by the storms are hazardous to shipping, and tremendous damage often results if a storm crosses land. Scientists try to understand these storms by collecting information about them and listing their common traits until a fairly consistent description emerges: Certain conditions of wind and atmospheric pressure lead to a situation in which a low-pressure cell is surrounded by winds circulating counterclockwise at speeds over 75 miles per hour. Then predictions about these storms, called hurricanes, become possible. The conditions that produce them, the time of year when they are most likely to occur (June to November, most frequent in September), their direction (west, then north, finally east), their speed (10–30 miles per hour), and their destructive power can all be predicted, allowing societies to control their impact. Advance knowledge of the hurricane's arrival means that clouds can be seeded with chemicals to reduce the force of the hurricane, and measures can be taken to move people and objects out of its way.

In a similar way, medical researchers observe a disease. Careful description of its characteristics allows predictions about its future occurrence and suggestions for its control through antibiotics, isolation, rest, and so on.

Science (or empiricism) as a way of obtaining knowledge involves specific characteristics and assumptions. One such assumption is that the best way to know about the world is through the senses, aided when possible by mechanical means such as microscopes, scales, telescopes, and other such extensions of the senses. Scientists assume that knowledge obtained in this way is better than that of the rationalists and mystics before them. Knowledge gained through science seems to explain and predict phenomena more adequately than does that obtained by other methods.

Another basic assumption of most scientific endeavors is that phenomena are related causally. Science maintains that a **cause-effect relationship** exists in that each event has a prior cause or causes that can be discovered. Gravity (cause) makes an object fall (effect) from a building. If a heavy object falls faster or if objects fall at the same speed, these effects are also caused by phenomena that we can try to discover.

However, an event is usually caused by more than one factor. This is referred to as **multiple causation.** Someone runs into my car and crunches it. Why—what is the cause? It happened while I was trying to turn left at Watt and Arden, which happens to be the busiest and most poorly designed intersection in the county (top of a hill, a drop-off in three directions, poorly lit, many dogs), and traffic was backed up in all directions. It was near dusk and wet. The person who hit me coming the other way had just been in a fight with her boss and was a little bit tightened up about life. She had to pull around a drunk who had gone to sleep at the wheel; as she turned to talk to a guy about it, she hit me. It's possible, although unlikely, that I was slightly distracted by the exciting Lakers game that I was listening to. Now, what caused the accident?

Because isolating a single cause or even several causes becomes enormously difficult when working on a complex problem, scientists search for relationships and correlations. Look at **Situation 1** at the beginning of the chapter. Blacks, while a small percentage of the population, represent a high percentage of those in professional sports, especially in professional basketball. Our sports fan who is about to turn to tennis thinks that blacks have inherited a predisposition toward basketball and sports. How's his logic— what's the answer? Sociologists look for a structural explanation and see a number of influences or relationships. Basketball is an inner-city playground game in the United States and is more likely to be played by blacks. It doesn't take any special equipment or club memberships. Making it in professional sports is seen as a good path to upward mobility, especially if other ways (like having a middle-class or wealthy family background, inheriting a family business, or attending and graduating from a high-level university) are blocked. Also look at role models—the inner-city kid's hero, the person he models his behavior after, is probably a sports star, someone who has made big bucks in professional sports. Being a basketball star probably seems a lot more possible than does being a doctor, politician, or professor. Does genetics explain basketball success? I doubt it. The answer lies in a series of factors in the social structure working together in complicated ways.

The concept of **control**, essential to science, has several meanings. As used earlier, it describes a product of knowledge. For example, if we know specific facts, such as when a hurricane will hit, then we can regulate or control events to a certain extent, such as moving people to safety. Or, as we learn about disease we can often control its effects with antibiotics. The second meaning of control refers to its use as an aid in the process of obtaining knowledge. Scientists attempt to control those factors in the environment that might have an effect on what they are studying. Our friends who are dropping objects from tall buildings may decide to perform their experiment only on calm, clear days. They are not studying wind and rain, but these factors might affect how the objects drop, and so they would want to make allowances for them. (Ideally, they would probably want to perform the experiment in a vacuum, if that were possible.)

As problems or experiments become more complicated, making allowances for all the relevant variables becomes more difficult. Important control variables may be left out, either because they are overlooked or because they are too difficult to control. For example, two psychologists decided to test the idea that hospitalization for psychosis, a serious mental illness, had increased over the century between 1845 and 1945. They found that it had indeed increased substantially, and they had many good reasons to offer in explanation, among them increased stress, shift from farm to city, breakdown of family and religion, and more complex life. The information and conclusion make sense. But they were wrong. The results and interpretation changed completely when the psychologists controlled for age. When they looked at hospitalization for each age group, instead of an increase in hospitalization they found virtually no change over the hundred-year period. People lived longer in 1945 than in 1845. That fact meant that the population contained more old people, who are more susceptible to the senility psychoses related to physiological degeneration of the brain and central nervous system. The conclusions follow: (1) The percentage of hospitalization in each age group remained the same from 1845 to 1945; (2) 1945 showed more total hospitalizations only because there were more older people (over 50 years old) in the population; and (3) the variations observed have nothing or very little to do with societal stress.[4]

Careful definitions of terms, often called **operational definitions** or *working definitions,* are part of science. They allow others to know precisely what the scientists studied and how they carried out their experiments. Look back at **Situation 2,** the study of what happens when a famous person commits suicide. What's a "famous person"? For David Phillips, who studied this issue, a famous person was one whose suicide was reported on the front page of the *New York Times.* In another of his studies dealing with murder-suicide, his operational definition of an acceptable case was that the story had to appear on the front page of the *New York Times* or the *Los Angeles Times* or be on the network TV nightly news.

Scientists are assumed to be *objective.* They are expected to record what they actually see, not what they hope to see or wish they had seen. They don't select or choose only those data that will fit or prove the hypothesis on which they are working while ignoring contrary evidence. One doesn't test the idea that sociology professors are absentminded by looking for a sociology professor who is thus afflicted and then saying, "I told you so." If the light object falls faster, or if both objects *rise* when dropped, then that is what the scientist reports.

Of course, absolute objectivity is impossible, and scientists who suggest that they are totally objective are misleading themselves and everyone else. Scientists are humans with beliefs and values that are bound to affect their work. The suggestion that when scientists put on their white coats they leave behind all nonscientific human emotions and biases is unreasonable. Objectivity requires rather that scientists try to be aware of their biases and make them public so that they may be considered when their work is evaluated.

Finally, **replication** is an essential characteristic of science. Studies are repeated by others in similar and in unique circumstances to see if the results are consistent. The objects would be dropped numerous times from the same height. Then the height might be changed from 30 feet to 60 feet, then to 90 feet. Then the weight of the objects might be varied, and so on. It is through replication that one develops confidence in the results.

Types of Science

The range of matters in which scientists are involved is probably far wider than most people think. There are numerous types or categories of science, and they involve widely differing methods. *Pure* and *applied* science represent two complementary approaches. **Pure science** attempts to discover facts and principles about the universe without regard to the possible practical uses the knowledge may have. **Applied science** concerns itself with making knowledge useful to people. The applied scientist devises practical and utilitarian uses of knowledge obtained through pure scientific endeavors. Splitting the atom was for the most part a product of pure science. Harnessing the atom to make bombs, to power submarines, and to fuel power plants is a product of applied science. The two—pure and applied—are closely linked, and the distinction between them is sometimes more artificial than real.

Science is also categorized by subject matter. Hence, you have the *physical* sciences like chemistry that study the physical universe; the *biological* sciences like zoology that study living organisms; and the *social* sciences like sociology that study various aspects of people and societies.

Social Science

Social science in general and sociology in particular have developed as a result of studying certain characteristics observed in human beings. For example, humans tend to group together and to cooperate. People discovered long ago that they could accomplish much more working together than they could separately. Cooperation has obvious rewards. But we know that not all human behavior is cooperative: Conflict is also an important part of the human condition. In fact, one of the first cooperative group efforts probably involved a battle or conflict with another group. Conflict arises when people compete for the same resources, such as wealth or power. Very often when one party or group benefits, another party or group is deprived; this conflict is a major focal point of social science.

Another observable characteristic of human beings is that they tend in similar situations to behave in the same ways, time after time. We all know this. When a car comes toward us, we have a pretty good idea of which side of the road it will be on. Students can correctly predict the behavior of their instructor on the first day of class, just as the instructor can usually anticipate the questions and concerns the students will have about the class. The behavior of most people on almost any elevator is also predictable. Generally they will get in, push the button for their floor, stand as far as possible from others, face the front, stare at the numbers flashing, and be quiet and subdued. Amazing? Not really, for human behavior in its simpler aspects is not difficult to predict.

The complex aspects of behavior, however, are more difficult to predict. As people get together in groups, organizations, and societies, factors multiply, and behavior becomes more complicated. But behavior can be understood. Social scientists try to find the regularities and consistent patterns in human activities. They work through description, understanding, and explanation to prediction and perhaps control. Let's say, for instance, that we're worried about airplane hijacking. We first try to describe the condition. What sort of people are involved—first offenders or people with prior criminal histories? What motivates them—psychological problems, financial needs, a political viewpoint? Adequate description of the event may allow us to predict its occurrence. Careful screening of passengers to point out the hijacker type may, if we're very good at it, allow us to control the occurrence of the event. The experts aren't that good at it yet, as you know; there are fewer airplane hijackings than there used to be, but they still occur. Maybe we'll learn to control them better. But the point is that knowledge can be gained about human behavior using the general principles of science. This is the work of the social scientist.

The various social sciences approach the task of studying and recording human behavior in diverse ways. *Anthropologists* traditionally have gone about the task by examining artifacts and remains of long-extinct communities, or by living with and studying nonliterate tribes and societies. The study of culture is of central importance in anthropological analysis. *Historians* record an accurate chronology of past events. As events are placed in perspective, analysis of emerging trends may lead historians to make predictions about the future. *Political scientists* study the characteristics and patterns of political systems and the principles and conduct of government. Their topics include political parties, elections, systems of government, foreign policy, and the comparative structure of governments. *Economists* are interested in patterns of production, distribution, and consumption of goods and services. They study such topics as price and market theories; consumer behavior; merchandising and selling practices; money, banking, and credit; and economic growth and development. *Psychologists* deal with individuals; their adjustment and personalities; and their patterns of learning, motivation, and perception. *Sociologists* focus on groups, patterns of interaction, and descriptions of the institutions and social organization of society.

SOCIOLOGY

A sociologist is a person who goes to a football game and watches the crowd.

The dictionary tells us that **sociology** is the "science or study of the origin, development, organization, and functioning of human society." Sociologists study human society and social behavior, and they focus on groups, institutions, and social organizations. The discipline of sociology includes several areas of study. For example, sociologists specialize in such areas as small groups, large-scale organizations, race relations, religion, marriage and the family, social problems, collective behavior, criminology and delinquency, social class, urban and rural sociology, population and demography, age and gender roles, political sociology, and the sociology of medicine and law.

In sociology, as in other disciplines, the process of discovery centers on the interplay of two elements: *theory* and *research methods*. A *theory*, which provides a framework for understanding phenomena, suggests specific hypotheses or predictions. Then, by using appropriate research methods, one can test the hypotheses empirically through controlled sense observation. The *research findings* often suggest modifications of the theory that in turn will make it more complete. This is the interplay of theory and research. For example, Einstein's theory of relativity provides a framework for making predictions about the nature of space and time under various conditions, assuming the speed of light is a constant. Physicists and astronomers test these predictions using sophisticated research methods that involve laser beams, high-speed aircraft, and complicated mathematics.

Data on cause of death do indeed tell us that year after year Miami has many homicides and that certain western states have high suicide rates as described in **Situation 3.** Explanations for Miami's murders easily come to mind—high percentage of gun ownership, high crime rate generally, and especially the high volume of narcotics traffic and the violent crime associated with it. But why so many suicides in New Mexico, Nevada, Wyoming, Arizona, and Colorado? It turns out that years ago, when he was studying suicide in western Europe, Emile Durkheim theorized that one type of suicide (egoistic) would be related to lack of social integration. Lack of social integration may be seen in such factors as high divorce rates, rapid population growth, rapid population turnover, and low church membership. Durkheim's theory presents us with a possible explanation for high suicide rates in the western states. Now we can go back to those states and see if they have the other characteristics that the theory suggests they should. Ideally, there is an interplay like this one between theory and method.

In the next few pages we will examine more closely the theory and research methods sociologists use in their efforts to understand the human condition. First, let's look at sociological theories.

Sociological Theories

A *theory* is a coherent set of propositions used to explain a class of phenomena. Scientists develop theories to explain that aspect of the world they are studying. Next they attempt to test the theories with the information or data they subsequently collect. Theories are crucially important, for scientific research is organized around them. If the theory is reasonably accurate, it will explain many of the phenomena with which it is concerned. If it is not, then it will not be very helpful, and eventually a better theory—one that explains more about the phenomena or subject—will replace it.

In the second century A.D., the Greek scholar Ptolemy theorized that the earth was the center of the universe and the other planets rotated in complicated ways around it. This was the accepted theory, and it had strong support in the scientific and religious communities. But it did not explain the movements of the planets as well as it should have. Some 1,300 years later, the Polish astronomer Copernicus began publishing papers that outlined a new theory—that the sun was the center of the universe and the earth and other planets rotated around the sun. It was a revolutionary idea. In the years after Copernicus died in 1543, Kepler, Galileo, and others made continued observations enabling them to refine the theory that Copernicus had suggested. Substantial opposition arose to the new theory because it challenged some beliefs that were very important to the Catholic Church. In fact, Galileo was imprisoned for a time because of his support of the theory. It slowly became accepted, however, because it was a better explanation for astronomical phenomena than the Ptolemaic theory before it had been.

All disciplines have competing theories that attempt to explain the phenomena that are important to that discipline, and sociology is no exception. In the following paragraphs, we will briefly examine some of the major theories in sociology: *functional analysis, conflict theory,* and *symbolic interaction.*

Functional Analysis   The key terms in **functional analysis** are *structure* and *function.* A good way to understand functional analysis is to start with a biological analogy. If one looks at the human body, one sees that numerous functions must be performed—breathing and eating, for example—for the organism to survive. To perform these functions, structures like the nose, lungs, digestive tract, and so on have developed. Functional analysis holds that society can be analyzed in the same manner. Societies (and groups) need to perform certain functions to maintain their existence. They must populate themselves, they must care for the sick, the young must be socialized, goods and services must be distributed. Structures develop to perform these functions: a family system to control reproduction, an educational system to train the young, and economic and medical institutions to carry out other functions. This theory holds that structure and function are closely linked and that they are the crucial factors in understanding and explaining society. Again, functional analysis explains a given pattern of activity by defining its contribution, or function, to the group or society of which it is a part. Functional analysis tends to emphasize social equilibrium, stability, and the integration of the elements of society. One of the criticisms of this theory is that it has a conservative bias; with its emphasis on social order, functional analysis tends to ignore conflict and social change.

Several terms commonly connected with functional analysis should be clarified. **Function** refers to an act that contributes to the existence of a unit, such as breathing and eating for the life of the organism. **Dysfunction** is the opposite—an act that leads to the change or destruction of a unit, like cancer in an organism. **Manifest function** is intended and recognized; **latent function** is unintended and unrecognized. According to Robert Merton, the Hopi Indians continue their ceremonial rain dancing even though the dances do not produce much rain. The latent function—reinforcing group identity by providing a periodic reason for getting together and engaging in a common activity—has become more important than the manifest function of producing rain.[5]

Try a functional analysis of a sociology class. Why *is* there such a class? It functions to pass on knowledge, viewpoints, and information. A structure has developed consisting of an instructor, some number of students, and a set of rules to facilitate this transmission of knowledge. It also functions to provide the teacher with a job. Some feel a need to find out how well the knowledge is reaching students, and so the structure of tests and a grading system develops. Attending class is functional if the student wants to get a good grade; cutting class often is dysfunctional. The first big test given in a class may be dysfunctional if it changes the mood of the class. The rapport and mutual understanding that develop between teacher and students over six weeks is suddenly compromised, and an adversary relationship develops. A manifest function of the typical classroom procedure—the teacher assigns work and students listen, read, obey, memorize, and give back on tests what the instructor has said—is to facilitate the learning process. A latent function of this procedure may be to train students to take orders and become good bureaucrats in the large organizations they will join in the future.

Now that you have some idea of how it works, try a functional analysis of some events or types of behavior—for example, prostitution, professional football, a funeral ceremony, or punishment for a crime. The discussions of family and religious institutions in Chapters Seven and Eight are clear examples of functional analysis.[6]

Conflict Theory   In contrast to the social order and stability stressed in functional analysis, another theory focuses on conflict and discord. **Conflict theory** suggests that competition and conflict are common in social interaction and that the study of these processes is the most appropriate way to understand society. In the mid-19th century, Karl Marx wrote that the capitalists' or industrialists' exploitation of the workers would inevitably lead to a conflict that could be resolved only by a workers' revolution, resulting in a classless society. Marx felt that economic factors were the basic cause of conflict. Today, conflict theorists tell us that the potential for conflict exists in all social situations. In every grouping of people an imbalance of power and authority exists. Some lead, others follow; some make decisions, others take orders. It's true in the factory, the classroom, the family, the football team, the small group, and the large organization. People continually compete for scarce resources such as wealth, status, power, or authority, and this leads to conflict. Conflict also arises over differences in values and interests, for example, between superiors and subordinates, management and labor, and in-groups and out-groups. In examining social interaction, conflict theorists ask their basic questions: Who benefits? Who is deprived? Conflict theorists focus on how power is used by special-interest groups for their own benefit.

Some conflicts are minor and of little consequence. Others are deeply felt and may result in the division of whole societies into hostile classes as, for example, Marx suggested. Conflict may range from outright violence to hostility, tension, competition, and rivalry; even humor is often a form of conflict. Conflict tends to be perceived as negative, but its consequences may also be positive. Conflict with an outside force will probably bind the members of a group more closely together. Conflict over issues may lead to beneficial social change. Lengthy discussions between consumer advocates and manufacturers have led to improved auto safety. Likewise, improved wages and working conditions for labor and changes in women's rights and in the civil rights of minorities have emerged out of both discussions and open battles between groups with differing interests and conflicting viewpoints. Thus, conflict theory attempts to examine and explain how social change takes place, an issue with which functional analysis has some difficulty.

Getting back to our sociology class, what would a conflict theorist see in this situation? The testing and grading system necessary to facilitate the transmission of information creates a system of domination (instructor) and subordination (students) and sets up conflicting interests. Consequently, the conflict theorist would not be at all surprised with the sudden change in mood as the first test is given. Competition for scarce resources (A's) leads to predictable types of behavior: cheating to get good grades, memorization of masses of soon-to-be-forgotten material, and rivalries with other students. The conflict theorist might also call attention to the tendency of the educational system to reinforce the current stratification system. Students from lower-class backgrounds, who often have less adequate educational preparation, may do less well in college classes, receive lower grades, or perhaps flunk out; because of this, they may obtain lower-level jobs and as a consequence stay near the bottom of the social-class ladder.

Again, try using conflict theory on some events or types of behavior like prostitution, professional football, or punishment for a crime. Ask who benefits and who is deprived. Conflict theory will come up several times in this book, particularly in Chapters Five, Six, and Nine in the discussions of stratification, power, prejudice, and discrimination.[7]

Symbolic Interaction   Symbolic interaction focuses on interactions among people and on the processes by which individuals come to develop viewpoints about themselves and to relate to their associates. In comparison with functional analysis and conflict theories, which examine social structure in groups, institutions, and societies, **symbolic interaction** theory narrows the focus to person-to-person interactions in everyday life. People interpret and define the symbols, gestures, and words of people around them, and they modify their own behavior accordingly. This activity is the fundamental concern of symbolic interactionists, who see social life as a *process*. They are interested in describing patterns of interaction; how parties to the interaction interpret what is going on; the use of signs, symbols, and other forms of communication; the meanings actions have for others; and the processes of socialization and development of the self.

Let's return once more to the sociology class, this time as symbolic interactionists. We would be interested in the type of interaction between instructor and students and would look at the signals, meaning the verbal and nonverbal bits of communication, that pass back and forth. Suppose a difficult point is being made. The instructor attempts to evaluate how well the class is getting the point and adjusts the lecture according to his or her interpretations of the students' responses. Back in the corner is a student totally disgusted with this instructor, and yet class participation is required. Notice the way the student balances competing values—hostility, deference, and the desire for an A—in his gestures, tone of voice, body language, asides to other students, and so on.

The instructor has a particular self-concept; how is it expressed in his or her behavior? One instructor might be informal and innovative; another might be vastly self-confident, secure in the knowledge that he or she is the best instructor in the Western world. Imagine that a student in the classroom who has always seen herself as an A student gets a C on a test. What responses does she make that help her maintain her original self-concept? If a change in mood appears after the first test, the symbolic interactionist would be interested in how the students express it and how the instructor reacts to it. The symbolic interactionist, knowing that interaction is a two-way process, would be interested not only in changes in the students' behavior, but also in how the instructor's behavior changes as a consequence of his or her interaction with the class.

The following examples of the symbolic interactionist approach come to mind: the discussion of nonreactive techniques later in this chapter, all of Chapter Two on socialization and self, parts of the discussion of groups in Chapter Four, discussions on self-fulfilling prophecy throughout, and much of Chapter Fourteen on deviation.[8]

These theories represent contrasting approaches used to explain or organize the material of sociology. Each theory emphasizes certain aspects and ignores others, and each has particular strengths and weaknesses. It is a good idea when reading sociology to be aware of the author's theoretical perspective, for it will suggest to you why certain factors are stressed and why others are not. Keep in mind that a theory represents the starting point for organizing ideas. We will continually return to these theories in the chapters that follow.

RESEARCH METHODS IN SOCIOLOGY

Like other social sciences, sociology attempts to understand human behavior from the scientific point of view. This means that the sociologist is guided by the principles and assumptions of science that were outlined earlier in this chapter. To get some idea of what this means, let's consider some of the ways in which sociological research has been done. First, we must distinguish between two categories of research, *reactive* and *nonreactive*.[9] **Reactive research** refers to situations in which the observer or researcher is a part of the research situation. Much of social science research is based on surveys, specifically questionnaires and interviews. The researcher creates a questionnaire or interview, administers it to a group of people—the subjects—and measures the attitudes or behaviors of the subjects responding to the questionnaire or interview. These reactive measures are widely used because they are relatively inexpensive, are easy to administer, can be done on a group basis, and are less complicated than other methods. Also, in some cases they require less time to develop and to administer than do other methods.

Situation 4 describes the use of a reactive technique. Trained interviewers knock on doors or call people and ask them to answer certain questions in what are called public opinion polls. These days when we ask people what they think about the death penalty, we find that most favor it. The problem is that this type of approach tends to influence attitudes and behavior. Here's a simple illustration: If you *know* you're being observed and your behavior is being recorded, what happens? Well, you behave differently; you are uneasy, on guard, and tentative. The situation is not normal, and you don't behave the way you usually would. The researcher may be liked or disliked, the subjects may cooperate or not, and they may be ignorant or knowledgeable of the subject matter of the questionnaire; any of a number of things may happen. Whatever the result, the issue with reactive research remains the same: A foreign element is introduced into a situation, and attitudes are created as well as measured.

Nonreactive research has no intruding observer and doesn't use questionnaires or interviews. Instead, nonreactive research focuses on physical traces and signs left behind by people: on records and archives, such as hospital records, census data, and government records, and through simple observation in which the researcher observes but does not intrude. Some simple comparisons between reactive and nonreactive studies may help at this point.

1. Suppose you want to know which radio station is most listened to in your city.

Reactive study: Go door to door and ask people what their favorite radio station is.

Nonreactive study (using physical traces): Go to a busy gas station and as each car comes in, look in and note the spot on the dial the car radio is turned to.

2. Suppose you want to determine who are the most liberal and most conservative members of Congress in your state.

Reactive study: Devise an in-depth interview or questionnaire that measures liberalism/conservatism and give it to each member of Congress.

Nonreactive study (using records): Choose several issues that came up during the last legislative session and that divided people along liberal and conservative lines, and see how the members of Congress actually voted on these issues.

3. Suppose you want to find out who uses more profanity in everyday conversation, males or females.

Reactive study: Design a questionnaire or interview that asks people about their language patterns (including how often they swear), and administer it to a group of males and to a group of females.

Nonreactive study (using simple observation): Gather several observers and have each of them listen in on several casual conversations of people in all-male or all-female groups. The observers could jot notes, although tape recording the conversations would be better. It is important that the observers *not* affect the group's behavior; observers could appear to be part of the groups or totally removed from them, perhaps sitting across the room apparently reading.

Situation 2 is an example of a nonreactive study. David Phillips's studies of "famous people" suicides used records. He looked at newspapers to get his sample of famous suicides. He obtained suicide rates in the nation from vital statistics records collected by the government. He marked the dates when the famous suicides occurred and then looked at the national data to see what happened then. He found that suicide rates in the country *went up sharply* in the period immediately after each famous suicide! In other studies, Phillips found that famous suicides and murder-suicides seemed to be related to other behaviors; that is, after famous people committed suicide and after highly publicized murder-suicides happened, certain other odd events took place. Can you imagine what they might be? (We'll return to this question later in this chapter.)

Another type of nonreactive study relates to **Situation 5.** Kai Erikson decided to study crime in the past to see if it bore any resemblance to crime in the present. He used historical records and analyzed deviant behavior in a 17th-century Puritan settlement in Massachusetts. In his book, *Wayward Puritans*, he tells us that the Puritans did indeed have crime waves—their "crime waves" were of a religious nature involving conflict with Quakers and the appearance of hysteria over supposed witchcraft. It is Erikson's view that although the crime waves sound quite different, the Puritans' way of viewing deviant behavior is not all that different from ours of today.

The value of nonreactive research is that it is unobtrusive. Either the traces or records of behavior are examined after the people have left them, or the act of observation is hidden in that the people don't know they are being observed. If people don't know they are being studied, it seems likely that we get a more accurate measurement of whatever it is we are studying. On the other hand, nonreactive studies tend to lack depth and flexibility. We may know about the *what* but be curious about the *why*, and nonreactive studies won't provide us with motives. The reactive study—the interview—is more flexible and can probe at some of the answers inaccessible to the nonreactive study. In summary, the contrasting reactive and nonreactive approaches each have strengths and weaknesses. Probably the most effective study is one that combines elements of each approach. Keep these two categories in mind as we look further at sociological research.

Three Research Techniques

In the previous paragraphs we have mentioned several sources of research information: questionnaires, interviews, census data, hospital records, government records, and so on. To get a better idea of how sociologists obtain knowledge, let's look at three typical research techniques in more detail: participant observation, survey research, and the experiment.

Participant observation, now often called field research, is a study in which the researcher is or appears to be a participant in the activity or group that is being studied. *Street Corner Society* by William F. Whyte is a classic participant-observation study. While a student at Harvard some years ago, Whyte decided he wanted to study a slum district of a large city. He walked around Boston until he came upon an interesting area he called "Cornerville." How should he study the area? He first tried a door-to-door questionnaire, but he was totally dissatisfied with this approach. After several other false starts, Whyte sought help from settlement-house social workers in Cornerville, and they introduced him to "Doc." Doc was a native of Cornerville and knew everybody in the area. He took Whyte around as "my friend," and Whyte became involved in the community in ways he never could have achieved as an outsider. Shortly after, he rented a room in Cornerville, further establishing himself in the community. He found that as he got to know key members of groups, he could get information he wanted from other members of the groups. He learned how to join in streetcorner conversations, and it wasn't long before Doc told him that he was as much a fixture on the street corner as the lamppost.

Everything did not go smoothly, however. On one occasion Whyte was talking about police payoffs to a gambling operator. When he became too inquisitive, the gambler became suspicious. On another occasion while talking with some Cornerville people, he decided to swear and use obscenities like they did. The conversation immediately stopped, and he was gently reminded that although they talked that way, they knew that he did not.

Whyte's relationship with Doc changed as his research continued. At first Doc was a friend and entrée into situations and groups. Later Doc became more of a collaborator as he helped Whyte interpret what he was seeing. In Whyte's case, many of the citizens in Cornerville knew he was something more than just another fellow on the street. It was reasonably well known that he was "working on a book." However, he was well enough integrated into the community that the knowledge that he was an observer as well as a participant didn't seem to change the way people behaved around him. By using participant observation, Whyte learned facts about Cornerville that he probably never could have learned otherwise. Through participant observation, he was able to analyze "the structure and functioning of the community through intensive examination of some of its parts—*in action.*" [10]

As described in **Situation 6,** David Rosenhan and some friends individually went to mental hospitals, gave accurate case histories, and added (inaccurately) that they had been hearing some odd voices. What happened? Rosenhan and his friends were admitted to the hospitals, usually diagnosed as schizophrenic, and held from one to more than seven weeks. Although I'm not sure that's what he had in mind, it allowed Rosenhan and his friends to study the hospitals and mental-health care as participant observers. In the paper he wrote, Rosenhan comments on faulty diagnosis, inadequate care, misuse of medication, and the tendency of mental-health labels to go unquestioned once they've been applied.

We touched earlier on **survey research,** a very common social science research technique, in discussing reactive studies. The survey involves the systematic collection of information from or about people through the use of self-administered questionnaires or interviews. The researcher is interested in the general characteristics of a sample of people or in some experience or event in which they have been involved. A survey can be mainly descriptive or it can look for relationships among variables in the hope of getting to explanation and cause. Recently, I administered a short questionnaire to my criminology class that asked four background questions (gender, age, marital status, and major in school) and three opinion or attitude questions dealing with subjects we had discussed during the semester (death penalty, gun control, and major social problems). The items I asked about could be called variables. (A **variable** is a condition that changes or has varying values, such as age, I.Q., marital status, opinion about the death penalty, and so on.) The results are as follows:

Gender	Age	Marital Status		Major	
Male 53	Range 20 to 50	Single	76%	Criminal justice	74%
Female 50	Average 24.8	Married	19%	Sociology/social	
		Divorced/separated	5%	science	21%
				Other	5%

	Yes	No
"I am in favor of the death penalty (that is, I think death is an appropriate punishment for those who commit certain major crimes)."	80%	20%

"The type of hand-gun control I favor is:"*	
Registration	68%
More severe punishment for use	41%
Confiscation	18%
None—I don't favor hand-gun control	3%

They ranked eight social problems from most serious to least serious as follows:

1. Violent crime
2. Poverty
3. AIDS
4. Pollution of the environment
5. Racism
6. Economic instability
7. Overpopulation
8. Political/religious terrorism

So far we have a simple descriptive survey that tells us that this class is about evenly divided between males and females, is a little bit older than the average college class, is mainly single, and contained mostly criminal justice majors (this class is required in their major). They support the death penalty, seem to favor several types of gun control, and, in their evaluation of eight social problems, find violent crime to be the most serious. We can leave it at this, but why not go a step further and look for other relationships that might exist? A simple thing would be to see if males differ from females, or if young students are different from older students. So, I compared the males with the females and the 25 oldest students with the 25 youngest ones:

Males and females were similar on most items, but although they both favored the death penalty, females favored it less (74 percent) than did males (85 percent). And on ranking social problems, females considered poverty and economic instability to be more serious problems than the males did. Males viewed violent crime as the most serious.

The youngest and oldest students were similar on most variables except how they ranked social problems. The younger students saw violent crime and AIDS as serious problems, whereas the older students ranked poverty at the top and ranked AIDS fairly low in seriousness.

*Adds to more than 100% because more than one answer could be selected.

Depending on the size of the sample and the number of variables in the survey, we can get more and more detailed, and perhaps doing so will help us move from description to cause.

Several times in this chapter we have mentioned *cause,* and this brings to mind the story (perhaps true) of the student who was comparing auto traffic over a bridge with water level of the bay under that bridge. After many careful measurements, the student discovered that there was a relationship, or correlation, between the two—specifically, the more cars on the bridge, the higher the water under the bridge. The student found this to be amazing and assumed the following cause-effect relationship: The heavy traffic on the bridge must be compacting the land somehow and forcing the water higher. A wise instructor intervened at this point and suggested that, contrary to the apparent cause-effect relationship, perhaps some other variables were involved. Further study indicated that traffic was heaviest during morning and evening rush hours and that high tides often occurred at about those same times. The lesson is that correlation doesn't necessarily mean cause, and relationship may come from sources other than those first assumed.

The **experiment** represents a third kind of research technique. In an experiment, the researcher manipulates one variable (the *causal* or **independent variable**) and watches for changes in the other variable (the *effect* or **dependent variable**). For example, a researcher might find two groups of people in good health, then have one group run several miles a day and have the other group do no physical exercise. Finally, the researcher would watch the two groups for incidence of heart disease. The independent variable, exercise, is manipulated for two groups, and then possible changes in the dependent variable, heart disease, are observed.

As you can see, the experiment searches for explanatory or causal relationships. It systematically alters or manipulates aspects of the world to see what changes follow. Suppose you are interested in the relationship between studying and academic performance; that is, you want to improve your grades. You could keep a record of how much you study for each test and the grades you receive. This would be a type of survey. Or you could systematically vary your studying—10 hours for one test, 20 hours for another, 40 hours for a third— and check the effect on your grades. This would be an experiment.

Situation 7 describes an experiment done by Robert Rosenthal and Lenore Jacobson (R&J). R&J were doing research in elementary-school classrooms in South San Francisco. At the beginning of the semester they casually told each teacher that five of her (his) students were very sharp—were "academic spurters" who could be expected to do well during the year. R&J named the students so that the teachers would have no doubt. In fact, R&J had misled the teachers—the so-called academic spurters had been *selected at random* from each class. R&J then observed the students and the teachers through the rest of the school year. In the first and second grades, the selected students showed great gains in I.Q., much greater than gains by nonselected students. Further, the teachers described these students as much more capable, more appealing, more likely to be successful, and so on. R&J report that the teachers treated the selected kids differently, not in obvious ways such as spending more time with them but in very subtle ways—tone of voice, facial expression, touch, and posture. This experiment illustrates a number of things, not the least of which is ethics. Do you think it was ethical for R&J to mislead the teachers the way they did? We'll discuss another aspect of this research in Chapter Two.

In the last few pages we have described three research techniques: participant observation, survey research, and the experiment. You might find it interesting and challenging to try out one or more of these techniques. For example, take one of the groups you belong to, such as an athletic team, a dorm group, a social club, or a political organization, and study it. A short questionnaire could give you a descriptive survey of attitudes on the death penalty, on political conservatism, on types of deviant behavior, or on presidential candidates. Or, more informally, as participant observer, watch the group in action. Who interacts with whom? Who are the leaders of the group and why? Is disruption caused by the same people, or is it spread throughout the group? Are some people left out by the rest of the group? See whether, by using controlled observation, you can discover something new about your group.

A Study in Response Bias

Several students and I performed an experiment that centered on two ideas. First, the women's liberation movement was receiving wide attention in our area, and we were interested in the viewpoints college women held on some of the issues of the movement. Second, we were interested in response bias. For example, will your responses to a questionnaire or interview be the same if you are questioned by a white person as compared with a black person? By a young person as compared with an older person? By a man as compared with a woman? Probably not. Your responses on the questionnaire vary not so much because of the questions asked (although that is important, too) but because of *who* is asking them. This, of course, greatly affects the accuracy of the study.

As we discussed earlier, theories (in this case, theories about response bias) provide a framework for understanding phenomena. From this general framework we derive more specific testable statements, or propositions, which are used to guide the study. These statements are called **hypotheses.** The hypothesis for our study stated the following: "Response to statements concerning women's roles will be affected by the gender of the interviewer." A series of items dealing with these attitudes was collected, and 90 married college women were interviewed. They were asked whether they agreed or disagreed with statements such as "A woman's place is in the home," "Motherhood and a career should not be mixed," and "Males have higher I.Q.'s and better abilities than do females." The subjects were selected randomly and were divided into two groups. Half of the subjects were contacted by a male interviewer; the other half were asked the same questions by a female interviewer. When we analyzed their answers, we found that the response to the male interviewer was much different from the response to the female interviewer. I'll return to this later, but for now consider this question: Would you anticipate a pro–women's-liberation response to the male interviewer ("the enemy") or to the female interviewer ("one of us"), and why?

Let's review some of the factors in this study. We were trying to test a cause-effect relationship. The gender of the interviewer was the independent variable, or cause; the women's response on the items was the dependent variable, or effect. By having a man interview one group and a woman interview the other group, we manipulated, or altered, the independent variable. By selecting the sample randomly, we hoped to control for factors we weren't interested in but that might affect the results.

The term **random sample** has been used several times and perhaps should be explained further. A random sample is a selection of subjects from a larger population; each subject in the population has an equal chance of being selected. The subjects of our study were women, but to interview all the women in the United States was impossible. So we narrowed it down to a group of 900 women living in a campus housing project. It can be seen that this choice represents substantial compromises. In return for specifying a group we could manage, we lost other geographical areas as well as age, education, and marital-status differences. Almost any attempts at reducing a large group will involve a compromise of sorts, although not necessarily as severe as ours. A sample of 900 was still too large for us. We wanted a small sample of 90, but at the same time we wanted it to be representative of the total group. How would we do this? If all people in the sample were alike, we could pick anybody—the first 90 we found—and it wouldn't matter. But people aren't alike; they vary tremendously. To be valid, our sample of 90 had to show the same variations and peculiarities as the original 900. A random sample in which each element has an equal chance of being selected should, we decided, give us what we wanted. This could be done in a practical way by placing 900 names in a barrel, scrambling them, and then picking out 90. We started with an address (Apartment 7) and went to every tenth apartment after that (17, 27, 37, 47, and so on). This gave us a one-in-ten sample, or 90 out of 900. Careful sampling, whereas it can only approximate the entire population, is a vital step toward an accurate and manageable experiment.

In describing our project to others, we included the detailed characteristics of the subjects and interviewers and the questions that were asked, as well as the errors or weaknesses of the study. The major weakness, by the way, lay in the area of controls. For example, it was assumed that the interviewer's gender was the cause of the differing responses, but the cause could have been a difference in the personalities of the two interviewers, or in the clothes they were wearing, or in something else that had not been controlled. We should have used more interviewers of both genders and of various types. It may have been that *another* cause produced the apparent effect we noted; or perhaps *several* causes, of which the interviewer's gender was only one, were operating. The study should have been repeated by others to verify the results and control for the additional factors we missed.[11]

Problems in Social Science Investigation

We have been describing sociological research efforts—how sociologists come to know what they know. We have found that by working within the general

PREDICTING THE FUTURE

In February 1979, we had a total eclipse of the sun. Astronomers handled it beautifully; they told us when it would happen, how long it would last, and what we should expect to see depending on where we were watching from. They compounded that achievement by telling us when the *next* eclipse of the same type would happen—February 16, 2017! Then they asked a social scientist to tell them about social conditions in the United States in 2017—family size, migration patterns, form of government, crime patterns, economic conditions, types of social movements, popular social problems, and so on. The social scientist told them he would check out his data and get back to them in 2018. . . .

---

guidelines of science, sociologists develop special techniques to study human behavior. All sciences do the same thing. Subject matter varies, of course, as do the problems sciences face. In the case of the social sciences, the subject is *people.* Variations in humans and their behavior are vast. Multiple causation is the rule, not the exception, and digging out all the related factors is always difficult and often impossible. One problem shared by social and natural scientists alike results from the *interactive* nature of research: The very act of observation can change the nature or performance of the subject matter. This is especially true in the social science situation: When people are studied, they usually know it, and this may affect their behavior in ways that are difficult to predict. (We commented on this earlier in discussing reactive and nonreactive studies.) For example, look at a study done some years ago at the Hawthorne Works of the Western Electric Company. Some researchers were interested in increasing production among women who wired electrical relays. They experimented with two groups. For one group, lighting was improved so they could see their work better. For the other group nothing was done. Output improved equally in *both* groups. Next, lighting was reduced to the point at which the workers had difficulty seeing what they were doing. But there was no decrease in efficiency or speed of production! What was going on? It became apparent that there was much more to working efficiency than just physical working conditions. In fact, the workers were responding to *being studied* by the researchers, and this had more effect on their output than changing supposedly important physical factors had. This finding, sometimes called the **Hawthorne effect,** was the impetus for a series of studies dealing with psychological and social aspects of the work situation.

The problem begins with multiple causation of human behavior, and now we find that one of these causes is one *we* introduce by trying to study it. It is difficult enough as it is, and we make it worse—that is, more complicated—by looking at it.

Social scientists are involved in the unique situation of *humans studying humans,* and this leads to problems. It is difficult to be objective under the best of circumstances, and in this field, scientists study people just like themselves.

If it were oxygen or laser beams, one would be less likely to identify with the subject. But in social science the subject is people, and scientists may identify with their feelings, their reactions, and their beliefs. I suppose we are asking, Can people possibly be objective and unbiased when, in effect, they are studying *themselves*? This question has bothered social scientists for decades, and their responses to it vary. Some believe that one can put aside biases and beliefs and become a dispassionate observer. Others say no—humans can't study other humans and be objective. If this last group is correct, there isn't much future for social science, or at least we will always have to bear in mind its limitations. Who's right? Well, certainly there are problems with objectivity when studying humans. Most social scientists probably see this as another variable in their studies that needs to be taken into account but one that by itself doesn't compromise the scientific effort. Valuable knowledge about human behavior has been obtained and will continue to be obtained in spite of the various problems we have just outlined.

Is It Only Common Sense?

Students of introductory sociology, and many others as well, frequently describe sociology as mere common sense. Sociologists deal with contemporary human behavior and patterns of interaction between people; because most of us are human we feel quite familiar with the topic. Atoms and rockets and relativity are one thing, but when it comes to understanding people, we're all experts.

In some aspects of human behavior, our experiences and our intuition may serve us well. None of us, however, has experienced all situations. Because we live in a particular society, area, social class, and community at a particular time in history, our experiences are necessarily limited. We in fact know much less about human behavior than we think we do. For example, remember **Situation 2?** Suicide rates in the nation went up after a famous-person suicide. What else happened after a famous suicide or after a publicized murder-suicide? Can you imagine? Well, auto accidents (especially single-car accidents) and auto fatalities increased! The more publicity for suicide, the more auto accidents there are; and the age of the people in the auto accidents is significantly correlated with the age of the suicide victim. Apparently, people imitate suicidal behavior, and a substantial proportion of certain types of auto accidents are suicide attempts. Would your intuition or common sense have told you that? If so, try this one: After highly publicized (front page of the paper, on network nightly news) murder-suicide cases, there is an increase in private and commercial *plane crashes!* Again, violence is imitated and "accidents" aren't necessarily accidents.[12]

For one more example, let's return to the study of women's attitudes described earlier. The women were more pro–women's liberation to the male interviewer. To the female interviewer asking the same questions, they were much less pro–women's liberation. Did you guess right? Perhaps you did and had a logical explanation as well, but a peculiarity of common sense is that it can provide a rationale for almost *any* response. It takes a more sophisticated analysis to predict human behavior accurately and to answer the complex questions of why people do what they do. To put what we know by means of common sense further in perspective, try this adaptation of Gerald Maxwell's social awareness test.[13] How well does your common sense serve you? (The answers appear at the end of this chapter.)

Social Awareness Test   Directions: Answer true or false to the following questions:

1. There are five times as many arrests of men as there are of women.

2. Because of discrimination and depressed living conditions, more blacks commit suicide proportionately than do whites.

3. With the exception of movie stars, people who divorce are slow to remarry.

4. Land is generally less expensive in the suburbs than in other parts of the city.

5. Women are more likely to commit suicide than are men.

6. There are more Hindus in the world than Protestants.

7. Cities are where the jobs are; consequently, there are generally more men than women in cities.

8. Children from divorced or unhappy homes are usually more careful in selecting a mate, and they make better marriages.

9. Panic is a common response for people confronted by disaster.

10. People with some college education are more likely to attend church than are people with only a high-school education.

SUMMARY

This book focuses on the *concepts* of sociology. Concepts provide a way of generalizing about phenomena. Sociologists use concepts such as role, motivation, anticipatory socialization, culture, in-group, cult, institution, and mob to help them organize and understand the events they are studying. In fact, one could say that the major task of sociology (and probably of other disciplines as well) is to build concepts—that is, to isolate certain unifying, abstract qualities that underlie and thus "explain" behavior. This book attempts to gather together some of the major concepts of sociology and to define and illustrate them. The discussion of concepts is divided into three major sections. These sections discuss the individual in society (Socialization and Culture), the organization of society (Social Organization), and society in flux (Social Change and Social Deviance).

This chapter has looked at the sources of knowledge—how do we know what we know? To illustrate a particular case, the following reading examines the career and recent work of social-psychologist Dane Archer.

TERMS FOR STUDY

applied science (9)

cause-effect relationship (7)

conflict theory (13)

control (8)

dependent variable (21)

dysfunction (13)

empiricism (6)

experiment (21)

function (13)

functional analysis (12)

Hawthorne effect (24)

hypothesis (22)

independent variable (21)

latent function (13)

manifest function (13)

multiple causation (7)

mysticism (5)

nonreactive research (17)

operational definition (8)

participant observation (18)

pure science (9)

random sample (23)

rationalism (6)

reactive research (16)

replication (9)

science (6)

sociology (11)

survey research (19)

symbolic interaction (15)

variable (19)

Reading 1.1

MURDER IN MIND

John Wilkes

As this chapter has suggested, social research is done in a variety of ways. It is also true that the questions and hypotheses that researchers study come from a variety of sources. The article that follows illustrates both these points as it describes the ideas and research in which social-psychologist Dane Archer has been involved. We wondered in **Situation 4** *whether the death penalty deterred people from committing murder. Archer's interest in that question is illustrated in this article. We also see how research questions are formulated and some of the problems encountered in gathering data and doing research.*

Dane Archer is a professor at the University of California, Santa Cruz, and John Wilkes is director of the Science Communication Program at the same university.

[Dane] Archer is a world authority on homicide, having earned that distinction by completing, with the help of colleague Rosemary Gartner, a 10-year international study of criminal violence. The study established Archer as a premier, cross-national psychologist, one whose work is done entirely outside the traditional laboratory of the experimental psychologist. Archer's study, *Violence and Crime in Cross-National Perspective,* was published as a book in 1984 by Yale University Press. The study, which has won four major awards in psychology and sociology, explores such elusive but critical social questions as: "Does the death penalty deter potential killers?" "Does violence increase in a nation that has just concluded a war?" "Do large cities have higher homicide rates than small cities?"

Drawing on statistics from 110 nations and 44 of their most cosmopolitan cities, Archer and coauthor Gartner provide the following answers: No, the death penalty does not deter homicidal criminals; yes, violence does increase in a nation that has just finished fighting a war, especially if it has won that war; and yes, large cities do have higher homicide rates than small cities in the same nation.

To explain most of their seemingly unrelated findings, Archer and Gartner proposed a single hypothesis: When a nation does violence to human beings, by conducting wars or executing criminals, it incites its citizens to more criminal violence than they would otherwise commit. Some people might reason, for example, that if the president was commanding the military to kill enemy soldiers, and if judges were ordering prison authorities to execute convicted murderers, why shouldn't the private citizen follow suit and use deadly force on personal enemies? In other words, in Archer and Gartner's hypothesis, the state can make violence the coin of its realm.

The Archer and Gartner study not only clarified controversial social issues, it challenged some obviously correct assumptions. In fact, the data they gathered sometimes showed that common-sense notions about violence were often dead wrong. Many people, for instance, believe that crowding causes violence. But Archer and Gartner found that Hong Kong is far more densely populated, yet far less violent, than Detroit. Archer hopes that research results such as theirs will provide a window on such previously unanswerable questions as why some cities are particularly violent. "Without international data such as ours," Archer says, "folklore reigns supreme."

War and urban homicide have haunted Archer since his high school days, which he spent idyllically as a scholar-athlete in a tiny southern New England town. Still a trim, muscular and enthusiastic basketball player at 40—an opponent calls his playing "fierce"—he remembers trudging home through the snow after practice trying to reconcile the evil of homicidal violence in the world with the human warmth and goodness he saw all around him. "Everybody not only knew everybody else—they cared about everybody," he says of his hometown.

During the next decade, as the U.S. homicide rate nearly doubled, Archer's high school musings turned into a commitment to understanding human violence. Between 1964 and 1972 he earned a Yale University undergraduate degree, won a Fulbright scholarship to New Zealand and took a Harvard University Ph.D. in social psychology. He then began teaching at the University of California, Santa Cruz.

At Yale, Archer had originally intended to major in either English or physics. But although he loved Chaucer's *Canterbury Tales,* he couldn't understand—and still can't—why he had to suffer through Spenser's *Faerie Queene.* And one morning early in his freshman year he realized he didn't belong in physics, either. He was watching a professor show how a bowling ball suspended on a chain from the lecture hall's high ceiling would swing away from his nose and back again, stopping just short of where it had started. Archer found himself hoping the principle of conservation of energy, which the professor was demonstrating, would be violated—just that once.

The following year Archer wrote a "passionate but spectacularly uninformed" paper on violence for a class taught by clinical psychologist Michael Kahn, who blended Freud, Margaret Mead and other social thinkers into a heady brew. Archer was hooked. By his junior year, he found himself enrolling in one social science class after another— psychology, sociology and anthropology.

As an undergraduate he was already highly suspicious of studies purporting to show that human violence stemmed directly and unavoidably from a genetically based tendency toward aggression. He came to believe, as he still holds, that human violence is a product of social forces rather than the direct result of a biological drive."How else," he says, "can you explain the fact that the United States has a murder rate 50 times as high as that of New Zealand, which is also a multicultural, urban, frontier-based society. New Zealanders have the same endocrine systems we do, but they don't kill each other."

In Archer's view, any account of human behavior that doesn't address the wide variation in homicide rates that exists across nations involves faulty logic. If it can't explain those variations, then it merely shows a meaningless correlation rather than a cause-and-effect relationship. "Even if it turns out that there is a specific hormonal or genetic type of violent person," he says, "we should stay optimistic and try to find out why these potentially violent hormonal types do not become violent people in some societies."

Under the influence of Harvard social scientists Robert Rosenthal, Freed Bales, Thomas Pettigrew and David McLelland, Archer began to conceive his international study of violence in 1972. In the fall of that year, he began his first term teaching at Santa Cruz. In his new undergraduate course on "Violence and War" he discovered his quiet, shy colleague-to-be Rosemary Gartner. "Rosemary hadn't said more than three things in class all term," Archer says. "And at the end of the course, she told me she hadn't been able to make up her mind on a paper topic. But instead of asking me for an extension, she astonished me by handing me two papers instead of the required one. Both of them were brilliant, and I signed her up right away to help me with the research."

Gartner stayed on for a year after graduation to help Archer with the project, then continued working on it while completing her doctoral work at the University of Wisconsin. Now a member of the sociology faculty at the University of Iowa, Gartner is helping Archer plan a follow-up study.

In the original study, the first question Archer wanted to test was whether the U.S. government's escalation of the Vietnam War was connected to a twofold increase of violence at home. "I thought at first that I'd just run off to the Inter-university Consortium for Political and Social Research at the University of Michigan and find volumes of data on the crime rates of various nations at war. Of course, the information wasn't there."

By then Archer realized that international data on violence could be invaluable to all kinds of researchers, not only psychologists but sociologists, criminologists and legal scholars as well. "I was determined to collect the data myself, no matter what it took. I recently sent the consortium the computer tape containing exactly the data I had hoped to find there when I started my study 14 years ago."

Collecting the data was fraught with problems—some large, some small and nearly all unexpected. One of the first occurred some months after Archer and Gartner applied to the National Institute of Mental Health for research funding for their homicide research. They received an agitated call from an institute official who remonstrated, "You don't have a clearance form to use human subjects in your research. Are you using human subjects? The committee wants to know." Archer replied, "Of course we're using human subjects, hundreds of thousands of them. But they're all dead." Archer and Gartner got their grant.

Another problem was the seemingly obvious sources of information that turned out to be blind alleys. Archer and Gartner knew, for example, that the United Nations maintained international statistics on a wide variety of social phenomena, including some crimes. But the published U.N. crime statistics were useless to Archer because they showed only the percentage changes in crime in member nations, while omitting the actual crime rates. "The percentage increases were ludicrously misleading in some cases," says Archer. "For example, the homicide rate rose by 100 percent in New Zealand in one year. That's an alarming figure," he explains, "until you learn that it meant two people were killed in the entire nation—one more than the year before."

When Archer tried to obtain the member nations' homicide rates from the United Nations, he got the runaround. "They made me very cranky with their evasions," Archer recalls. "They'd say, 'The information is for member states only.' And I'd reply, 'As far as I know, the United States is a member state.' Then they'd say, 'We can't afford the Xeroxing.' And I'd reply, 'I'll fly to New York and do the Xeroxing myself.' It went on and on. But now," says Archer with a triumphant grin, "we have better data than they have."

Archer and Gartner then tried to contact various national offices of crime statistics by mail. Officials in many nations claimed they didn't have the data Archer asked for, or they returned his letters stamped "No such address." In these cases, Archer and Gartner would simply invent four new official-sounding titles for the bureau they sought—such as "The Department of Prisons" and "The Bureau of Police Statistics"— for each nation that had disappointed them and send out a new barrage. By creative variation on official titles for bureaus of crime statistics, the researchers usually got the information they requested, eventually.

One respondent from a "despotic" West-bloc nation, which Archer declines to name, was touchingly idealistic. The official made clear in a clandestine letter that he had spent months trying to get approval for sending the requested data but had failed. Convinced of the importance of the research, he was sending the data anyway, at great personal risk. "The world needs to know these facts," he wrote Archer.

The responses of many statistical offices ran true to national stereotypes. The French data, for example, reflected a Gallic fascination with relationships and subtleties in crimes of passion. Unlike the many nations that simply break down their capital-crime statistics into homicide and lesser crimes, such as manslaughter, the French use nearly every category of relationship imaginable between killer and victim, including patricide, matricide, fratricide, sororicide, and infanticide.

The Germans, for their part, more than live up to their reputation for efficiency and thoroughness. "Ten years after my letter of inquiry I'm still getting huge blue statistical volumes in German every few months," Archer says. Nor did the British disappoint him. A correct official of a British bureau of crime statistics sent a copy of the excellent United Kingdom homicide records along with a disdainful stab at Archer's study. He wrote: "We have grave doubts about the value of the study you are conducting, given the circumstances of data collection in some nations." Archer is confident, though, that his book's chapter on methodology, which explains in detail the strengths and weaknesses of different kinds of comparisons, will satisfy even the punctilious British.

Perhaps the most controversial hypothesis addressed by Archer and Gartner is that the death penalty deters people from killing other people. Public support for the death penalty in the United States is now at an unprecedented high, having risen from roughly 40 percent in the 1960s to more than 70 percent today. "Politicians are now embracing the issue so eagerly," Archer says, "that I half expect some of them to start asking to be allowed to throw the switch on the electric chair themselves."

Despite the public's enthusiasm for the death penalty, Archer and Gartner found no evidence that it deters murder, even after they devoted a large share of their international study to this question. They even devised sophisticated tests that would compensate for such variables as a time lag between the enactment of a new capital-punishment law and the public's becoming aware of it. Yet they consistently found that, if anything, the homicide rates tended to decline slightly after the death penalty had been abolished.

In fact, says Archer, only one investigator—Isaac Ehrlich, a professor of economics at the State University of New York at Buffalo—has claimed to have found evidence that the death penalty deters killers. And that study, Archer adds, is suspect; other investigators reanalyzed these data and found no such deterrent effect. Meanwhile, Archer says, not one of the other 50-odd U.S. studies of the deterrence hypothesis has produced evidence that the death penalty deters murderers. "In a mature social science you don't draw conclusions from a single study," he says, "because that study could be fatally flawed for any one of a host of reasons. Instead, you draw your conclusions from the preponderance of evidence. And, in my view, the overwhelming pattern in deterrence research shows that the death penalty has no effect on the homicide rate."

Even though Archer and many social scientists are now convinced that the deterrence argument for the death penalty has virtually no empirical support, Archer believes that the controversy will continue because many people favor the death penalty for other reasons. They want to see the offender suffer; they don't believe convicted killers can be rehabilitated; they fear having paroled killers living among them; or they simply don't want to pay living expenses for such people.

"The debate over capital punishment held in the revolutionary French Constituent Assembly of 1791 is being echoed almost word for word by American politicians today," says Archer. He adds, "Revenge is a powerful undercurrent in all societies, including our own. I believe that the deterrence hypothesis is frequently nothing more than a veneer for revenge."

Although his research has led him to oppose the death penalty, Archer considers himself a hard-liner on crime. "I'm in favor of rehabilitation, but when it comes to choosing between the safety of the community and the well-being of the violence-prone criminal, someone who has torn up the social contract and thrown it away, I would place the community's good above the criminal's. But in deciding on appropriate measures to take against violent criminals, we mustn't forget that capital punishment is homicide. Just because there's an official lineup of people pressing the button doesn't change the fact. And putting someone to death seems an odd way to teach lessons about the value of human life."

The body of accepted research about the origins of human violence is growing, but Archer feels that those findings should not be translated directly into public laws. He notes, for example, that research has established beyond doubt that some people imitate the violence they see on television, yet he wonders whether the findings warrant laws to limit the amount of violence the media may show. "You get into First Amendment problems there. But those research results have clearly fueled citizens' action groups. Those groups use research results, including mine, to target and bring pressure to bear on advertisers who sponsor the most violent TV series. And the groups are having an effect. Some sponsors have checked out of those series. If enough do, the series will fold."

The most important lesson U.S. psychologists and other social scientists can learn from studies like his, Archer says, is that they should take the broadest possible view of human behavior when formulating research questions. "The complexity of cross-national research costs you a lot of sleepless nights," he admits, "but it's the only way to get at some of the important questions."

As an example of a serious research question that was answered by cross-national study, he cites the recent controversy over simple versus radical mastectomy as a therapy for breast cancer. European surgeons had long been routinely performing simple mastectomies to treat breast cancer, while U.S. surgeons were performing radical mastectomies in the same circumstances. When U.S. researchers realized this and compared the survival data on the European and U.S. patients, they discovered that the survival rates of the two groups were practically identical. "A natural cross-cultural experiment had been conducted," Archer says, "and it revealed that American surgical procedures were unnecessarily severe."

Archer believes that, like the American surgeons who ignored the work of their European colleagues, psychologists who limit themselves to studying the "American male college sophomore volunteer" run the risk of encapsulating themselves in a cultural cocoon. The real question, he believes, is why the United States has a homicide rate grossly higher than that of other industrial nations, and this is a question that cannot be answered by studying Americans alone.

Cross-national research holds great promise, Archer says, but it requires considerable imagination in formulating testable hypotheses and extreme care in collecting and analyzing data. "In nonexperimental research, you just have to work harder to come up with valid empirical tests that will either support or disconfirm your hypothesis. When you're looking at data such as the international crime statistics I use, you have to apply very rigorous after-the-fact controls. You have to look for the evidence that confirms it. In other words, you have to play the game fairly. If a hypothesis isn't capable of disproof, it's not science; it's ideology or even theology."

Archer is optimistic that he can find the common causal relationships he believes underlie violence in all societies. And he is optimistic, too, about discovering how peaceful societies manage to salve the frustrations that can lead to violent behavior. Given the enormous variation in the homicide rates of different societies, Archer believes strongly, even passionately, that only international studies can allow researchers to obtain meaningful answers to these questions.

For Archer, laboratory experiments that test human responses to electric shocks and buzzers can never come close to matching the conditions of the real world and thus can contribute little of value to the study of human violence. "If we're really interested in questions such as 'Why is this a violent society?' we have to get away from shocks and electric buzzers in laboratories," he says. "There were 19,000 homicides in the United States last year, and I don't think one was committed with an electric buzzer."

QUESTIONS 1.1

1. If Archer's experience is typical, where do researchers get most of their ideas about topics they decide to study?

2. What type of research is this? Describe and categorize Archer's work using concepts from this chapter. How is his different from most social science research?

3. List what you think are the three or four most interesting/important points from Archer's research.

4. What sort of problems did Archer face in his research? Do you imagine these are typical of most social science research? Why or why not?

5. What characteristics of science discussed in this chapter do you find in Archer's work?

NOTES

1. For more detail on the fascinating story of Stonehenge, see Gerald Hawkins, *Stonehenge Decoded* (New York: Dell, 1965).

2. There is a short discussion of phlogiston in *On Understanding Science*, by James Conant

(New York: New American Library-Mentor, 1951), ch. 3.

3. For a more detailed discussion of the Puritans and their problems, see Kai Erikson, *Wayward Puritans* (New York: Wiley, 1966).

4. This and other examples of control failures are discussed in *The Logic of Survey Analysis,* by Morris Rosenberg (New York: Basic Books, 1968), pp. 29–33.

5. Robert Merton, *Social Theory and Social Structure,* 2d ed. (New York: Free Press, 1957), pp. 64–65.

6. From among the numerous sources available on functional analysis, the following have been especially helpful: Neil Smelser, ed., *Sociology* (New York: Wiley, 1967), pp. 706–708; and Robert Merton, *Social Theory and Social Structure,* 2d ed. (New York: Free Press, 1957). See the section on functional analysis in the *International Encyclopedia of the Social Sciences* for a summary and an excellent bibliography.

7. For a background on conflict theory, see the works of Karl Marx; Lewis Coser, *The Functions of Social Conflict* (Glencoe, Ill.: Free Press, 1956); Ralf Dahrendorf, *Class and Class Conflict in Industrial Society* (Stanford, Calif.: Stanford University Press, 1959); see also the section on conflict in the *International Encyclopedia of the Social Sciences* for an overview and bibliography.

8. For more on symbolic interaction, see the works of Charles Cooley or George H. Mead (see index) or a social psychology text such as *Social Psychology,* 4th ed., by Alfred Lindesmith, Anselm Strauss, and Norman Denzin (Hinsdale, Ill.: Dryden Press, 1975).

9. This distinction is best made in *Unobtrusive Measures: Nonreactive Research in the Social Sciences,* by Eugene Webb, Donald Campbell, Richard Schwartz, and Lee Sechrest (Chicago: Rand McNally, 1966).

10. William Foote Whyte, *Street Corner Society,* 2d ed. (Chicago: University of Chicago Press, 1955). Also see a good summary of Whyte's work in *The Origins of Scientific Sociology,* by John Madge (New York: Free Press, 1962), ch. 7.

11. For more detail on this study, see Judson R. Landis, Daryl Sullivan, and Joseph Sheley, "Feminist Attitudes as Related to Sex of the Interviewer," *Pacific Sociological Review,* vol. 16, no. 3 (July 1973), pp. 305–314.

12. A number of papers by David Phillips deal with various aspects of imitation of violent behavior. For example, see *American Sociological Review* 39 (1974), pp. 340–354; *Science* 196 (1977), pp. 464–465; *American Journal of Sociology* 84 (1979), pp. 150–173; *Social Forces* 58 (1980), pp. 1001–1024; *American Sociological Review* 48 (1983), pp. 560–568.

13. Gerald Maxwell describes a social awareness test that he has devised (*The American Sociologist,* November 1966, pp. 253–254). I have used his title as well as one or two of his questions.

Answers to the social awareness test:

1. True. (See *Uniform Crime Reports, 1992*)
2. False. (See p. 188.)
3. False. Divorced persons are generally more likely to marry than are single persons of the same age.
4. True. (See p. 337.)
5. False. (See p. 435.)
6. True. (See p. 254.)
7. False. Most cities have more women than men.
8. False. Couples whose parents had unhappy marriages are more prone to divorce.
9. False. (See p. 403.)
10. True. (See p. 260.)

PART ONE

Socialization and Culture

In this section, we discuss how the infant develops into a social being. The socialization process is of major importance. Through this process, values are transmitted from one generation to the next, and infants grow and learn to adapt to their environments. The self, the person's conception of what and who he or she is, slowly emerges from interaction with others. Through this interaction, individuals are introduced to norms as well as to roles—those behaviors that are expected of them in specific situations and positions. This personal and social development takes place within the context of a specific culture. The culture provides a set of behaviors, traditions, customs, habits, and skills that also become a part of the individual. These processes—socialization, development of the self, internalization of norms, roles, and other aspects of the culture—are particularly evident in young children, although they continue to be important throughout life.

CHAPTER TWO

Socialization and Self

TWO RESEARCH QUESTIONS

1. *Is a child's mental age (I.Q.) affected by the number of brothers and sisters he or she has? In families with two or more children, are there predictable mental-age differences among the children? Whose I.Q. is higher—firstborn, last born, in between? Or is there any consistent pattern?*

2. *Students are assigned the task of teaching rats to learn a maze. Some students are given rats that are said to be stupid, poor at learning mazes. Other students are given rats said to be very sharp, fast learners. In fact, all the rats are alike and have equal capabilities. How will the two groups of rats perform in learning mazes. Why?*

FEW PEOPLE SPEND much time thinking about the transformation they went through in the process of becoming mature human beings. Not only are we not concerned with the steps required to carry us from an early blob of protoplasm to a complicated, interacting individual, but it is even difficult for us to believe that we were ever at the stage of egg and sperm. Beyond a certain idle curiosity as we watch a baby grow up, we take these changes in the developing human for granted. They are like the phases of the moon or the coming of summer—they seem just to happen.

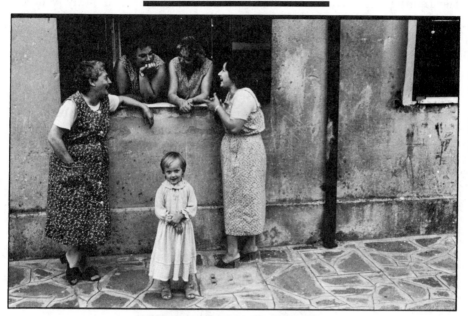

However, if one decides to analyze the development of the individual in some detail, several approaches are available. A biological approach would emphasize physiological maturation. As a man's hair changes from brown to gray to gone, so does the person grow and mature in other ways. This approach might suggest that maturity is merely a matter of cell changes and that, given enough time, these changes produce a mature social being.

But there is much more to human development than just biological and physiological changes. Interaction with other humans is of critical importance. Isolation from (or inadequate) human interaction can affect our normal growth in a number of ways. Examples of extreme isolation are unusual, but occasionally such cases are described. Some years ago, Kingsley Davis described the cases of Anna and Isabelle, who, because they were illegitimate children, were kept in nearly total isolation the first years of their lives.[1] Anna's mother was mentally retarded and left Anna alone in an attic room most of the time. Isabelle's mother was a deaf mute who stayed in a dark room with Isabelle, shut off from the rest of the family. When Anna was discovered, she was about six years old and could not walk, talk, or do anything that showed intelligence. She was in bad physical condition as well. After about two years in a county home, she had learned to walk, understand simple commands, feed herself, and achieve some neatness, but she still did not speak. She spent the next three years of her life (she died at the age of ten and a half) in a private home for retarded children, and there she made some further progress.

Her hearing and vision were normal, and she walked and ran fairly well. She could bounce and catch a ball, string beads, identify a few colors, and build with blocks. And she finally began to develop speech. She attempted conversations with others, although she spoke in phrases rather than in sentences. In summary, when Anna was found at the age of six, she had the mental capacity of a newborn infant. Four and a half years later she had progressed to a mental level of two and a half or three years.

Isabelle was six and a half years old when she was found. Like Anna, Isabelle was in bad shape both physically and mentally. For speech she made a strange croaking sound. She reacted to strangers, especially to men, with much fear and hostility. In many ways she behaved like a deaf child, and her mental capacity was no more than that of a six-month-old baby. Specialists working with her at first believed her to be feebleminded and uneducable. An intensive training program was started. At first it seemed hopeless, but gradually Isabelle began to respond. Then suddenly she began to learn rapidly. Two months after starting to speak, she was putting sentences together. Sixteen months later she had a vocabulary of 1,500–2,000 words and was asking complicated questions. She covered in two years the stages of learning that ordinarily require six. Her I.Q. tripled in a year and a half. Davis concluded his description of Isabelle by noting that she was over 14 years old and had passed the sixth grade. She was bright, cheerful, and energetic, and her teachers reported that she participated in school activities as normally as other children.

What do these cases tell us? It seems apparent that inadequate social interaction, near isolation in these cases, inhibits the development of the individual. Although they were six years old, neither Anna nor Isabelle had developed beyond the infant level, and both were believed to be feebleminded. Yet when effective interaction with others began, both girls progressed. In Isabelle's case the development was remarkable. This is probably because she had concentrated, expert training from the beginning and because she learned language quickly. Perhaps this is the key. Or perhaps Anna was more isolated—Isabelle's mother stayed with her, whereas Anna was often by herself. Or perhaps Isabelle's mental capacity was superior to Anna's to begin with. This we don't know. But in each case, the effect of inadequate social interaction on human development is clearly demonstrated. Sociologists are convinced that the development of the individual is, at least in part, a *social process*.

SOCIALIZATION: A SOCIAL PROCESS

The process by which an organism becomes a social being is called socialization. **Socialization** refers to the learning of expectations, habits, skills, values, beliefs, and other requirements necessary for effective participation in social groups. Biological maturity is necessary, hereditary factors may set limits, and the process takes place in a physical environment, but the crucial aspect of socialization is that it is a social process. The individual develops through interaction in the social environment; the social environment determines the result. This is not to imply that if baby Michael Jordan and baby Michael Jackson had been switched in the nursery shortly after birth, Jackson would have been a spectacular professional basketball player slam dunking with his tongue sticking out and Jordan would have been a famous singer with a high-pitched voice and one glove. We would have to look at other factors as well, and biology and heredity certainly play an important part.

The exact extent of the role of biological factors is difficult to determine. In fact, there is a heated and interesting debate going on now on this very issue—biology versus social environment. Edward Wilson, an entomologist from Harvard, proposed in a book published in 1975 that a new science called sociobiology be established. **Sociobiology** is defined as "the systematic study of the biological basis of all forms of social behavior, including sexual and parental behavior, in all kinds of organisms, including man." Sociobiologists are convinced that the behavior of lower organisms is biologically determined, and they believe that the same may be true of much of human behavior. In their writings they cite apparent genetic constraints on human social behavior in such areas as choice of sexual partners (the incest taboo and homosexuality), gender roles, aggression, kinship rules, and infant development.[2] The ideas of the sociobiologists are controversial, some say inflammatory; they have delighted some social scientists and threatened and angered others. Probably most social scientists recognize that biology places certain limits or conditions on human development. Beyond this, however, social scientists and especially sociologists believe that the crucial elements for understanding human behavior are the social environment and social interaction. By and large, according to sociologists, these are the primary determinants of who we are.

By **social interaction** we mean the process of being aware of others when we act, of modifying our behavior according to others' responses. Social interaction occurs in a variety of ways, and its patterns are complex. When a student asks a teacher a question, he speaks differently than he does when he talks to the student next to him in class. The teacher evaluates the question as fairly sophisticated, and this affects the teacher's response to the student. If the interaction continues, each response is modified and determined by the response of the other immediately preceding it. The pattern of interaction may be affected by manner, body language, deference, relative status, degree of acquaintance, eye contact, and numerous other factors in addition to the spoken words. Social interaction occurs between two people—husband and wife, strangers on the street—and in groups and even in large organizations. Try watching some examples of social interaction, and see whether you can identify the more subtle aspects of communication. How do the parties show agreement and disagreement, interest and boredom, and happiness and unhappiness through the use of body language, spatial distance, and facial expressions?

Socialization is continuous: It takes place throughout life, starting as soon as the infant leaves the womb and continuing until death. It occurs through interaction with other humans. We could call these other humans *agents of socialization.* Who these agents will be depends on one's place in the life span. For the baby, the socialization agents are the parents, who begin very early to communicate accepted and expected modes of behavior to the infant. Brothers and sisters also become socialization agents. An aspect of the socialization process can be observed in children as they pretend to be mothers, fathers, airplane pilots, or truck drivers. They are already anticipating and playing future roles. Later, peers become socialization agents, and, regardless of whether they are school friends or members of a delinquent gang or teammates on the football team, their influence becomes very important. Still later, one's colleagues on the job, one's spouse, and even one's own children become agents of socialization. Also, indirect influences such as vicarious participation or involvement in what we are reading or watching on TV continually play a part in the socialization process. In fact, in our interaction with others, we all become agents of socialization.

The first research question posed at the beginning of this chapter asked whether mental age or I.Q. might be related to family size and birth order. It turns out that sociologists have been looking at this for some time and have come to some interesting conclusions. They have found that mental age tends to decline as family size increases. They have also found that mental age declines as birth order increases. In other words, later-born children tend to have lower I.Q.'s than their older brothers and sisters. The researchers suggest that perhaps the confluence of two factors—family intellectual environment and a teaching function—may in part explain the differences. They suggest that the family's intellectual environment—its average mental age—declines with each new baby, especially if babies are closely spaced. The more young children around, the less intellectually rich the environment, at least during the time that they are young. The "teaching function" effect refers to the fact that last-born children lack the opportunity to teach children coming after them. Teaching a younger brother or sister stimulates the intellectual development of the older child, and because last-born children have no one to teach, they suffer from a "last-born handicap."[3]

As you can see, socialization is extremely important. Through this lifelong process we gradually take specific points of view. Our views of religion, politics, sports, style of family life, how to commit particular crimes, and how to raise children are all learned through the socialization process. I throw a volleyball to my son. He catches it in his hands and throws it back. Very easy; he doesn't even think about it. But suppose I throw the ball to a French or English boy—what happens? If the throw is above his shoulders, he will probably "head" it back to me. If it's lower, he will control the ball with his legs and feet and kick it back to me. Each child has been socialized—has learned—to respond in a specific way. Children respond differently from each other because different factors are emphasized where each child lives. But again, the socialization process "tracks" them into a certain response.

Research with parents and their newborn babies tells us that from the moment of birth, female and male infants are treated differently. They are described differently (girl babies are seen as softer, smaller, and more beautiful and delicate than boy babies), dressed differently, given different types of toys, and expected to behave in different ways, all because of their gender. Gender-role socialization begins at birth and continues through childhood. One effect of this socialization is illustrated in a study by Philip Goldberg.[4] Goldberg

SOCIALIZATION—WHEN DOES IT START?

A Los Angeles mother found her pregnancy boring, so she began watching the soap opera "Ryan's Hope." Shortly after her daughter's birth the mother noticed that the baby seemed to recognize the "Ryan's Hope" theme song and would stop fussing or crying when the program began!

Peter Hepper, an Irish medical researcher, has been studying this phenomenon, which he calls "fetal soap addiction." He compared the reactions of the babies of a small group of mothers who watched soap operas on TV while pregnant with the reactions of babies whose mothers didn't watch the soaps. He wrote a letter to the editor of the British medical journal *Lancet* in June 1988 describing his findings. Hepper suggests that "the results indicate that the reaction of a newborn baby to 'watching' televison may reflect long-term exposure to the theme tune of the program during pregnancy. This attention exhibited by the newborn may be a result of prenatal learning."

chose written articles from six fields—law, city planning, elementary education, dietetics, linguistics, and art history—and asked a group of college women to evaluate each article. The catch was that Goldberg manipulated the names of the authors of the articles. For half the women, the author of an article was a man ("John McKay"); for the other half of the women, the *same* article was written by a woman ("Joan McKay"). In other words, the women were reading the same six articles, but in half the cases the authors were listed as men, and in the other half the authors were women. What happened? In nearly all cases, the women rated the article supposedly written by the man *higher* than the same article supposedly written by a woman! Several fields were selected by Goldberg because some were traditionally viewed as women's areas (elementary education, dietetics); but even here, articles supposedly written by men were ranked higher than were those supposedly written by women. The explanation is inescapable: Women have been socialized to see themselves as inferior to men. A major goal of the feminist movement is to counteract this "inferior woman" socialization, in women as well as in men.

Socialization may be better understood as occurring at two levels: primary and secondary. According to Berger and Luckmann, **primary socialization** refers to the first socialization an individual experiences in childhood, through which he or she becomes a member of society. **Secondary socialization** refers to any subsequent process that inducts an already socialized individual into new sectors of his or her society.[5] Primary socialization usually takes place in the family. Here, the individual has no choice about the important or significant socializing agents; the child almost automatically and inevitably accepts and internalizes the family's view of the world. As primary socialization proceeds, the child's referents move from specific to general. A progressive abstraction occurs whereby at one stage the child understands that her parent specifically wants her to do certain things (for example, not to throw her food on the floor), and at a later stage she understands that people in general expect her to behave in certain ways (to use correct table manners). Through primary socialization, the individual's first world is constructed. Consequently, according to Berger and Luckmann, this first level of socialization is the most important, for the basic structure of all secondary socialization must resemble that of primary socialization.

Secondary socialization, which takes over where primary socialization leaves off, involves the individual's moving into and internalizing knowledge of new areas or sectors of life. Secondary socialization takes place when the individual decides to learn to read or to write, to take up skydiving, to become a police officer, or to raise a family. This level of socialization does not need an emotionally charged atmosphere to succeed, nor does it presuppose a high degree of identification, and it does not possess the quality of inevitability. In secondary socialization, the individual can be more objective than in primary socialization.

A person entering a new profession illustrates adult secondary socialization; an example might be the steps taken in the socialization of a new police officer, outlined by Arthur Niederhoffer. The rookie police officer proudly accepts the symbols of his new job—uniform, revolver, and badge. At the police academy he learns about law, police procedures, human relations, rules and regulations, first aid, and how to fire weapons. He graduates, becomes a police officer, and faces a hostile and unappreciative public. The shock of reality occurs. The more experienced officers tell him that to become a real police officer he will have to forget everything he learned at the academy. The officer learns a new language, he learns to treat people in various ways according to the areas of the city, and he learns to react to people's language, attitudes, and dress as well as to their criminal behavior. He learns when to enforce the law and when to look the other way. He learns how to make an effective arrest, and he learns about graft. According to Niederhoffer, his attitudes change the longer he is on the force; if the officer has a college education but is slow to get promoted, he becomes more and more cynical. And generally as officers near retirement, they become less cynical.[6]

RESOCIALIZATION AND ANTICIPATORY SOCIALIZATION

Socialization at the adult level (secondary socialization) is usually a gradual process; that is, the changes in the individual are more minor than are those that occurred during primary socialization. Occasionally, however, major modifications or reconstructions of people are attempted. This could be called **resocialization.** It usually represents a concentrated effort by a group or organization. Consider prisons as an example. Individuals commit crimes; they are caught, convicted, and sent to reformatories or prisons. The idea is that the discomfort of their confinement will help them see the error of their ways, counseling and training programs there will help them change their behavior, and criminals will be *resocialized* into good citizens.

What actually happens? Resocialization certainly does take place, but its effects are far more diversified than one might imagine. Prisons are very effective crime schools for some prisoners. Regardless of the inmates' expertise upon entering, they can improve their crafts while inside. Through association with other inmates, they can learn new skills and techniques of crime. They can set up networks of important criminal friends and contacts for when they are released. Inmates learn a new language in prison. For example, "gleaning," "low riders," "bonaroos," "shots," "politicians," "merchants," "hoods," and "toughs" are prison slang for types of convicts.[7] The inmates learn new sexual practices, sometimes against their will, and they must learn how to get what they want in a society that is not at all like the one they are used to. Some inmates even *like* prison; they have found a new home. These inmates find it easier to get along inside than outside, and they take steps to make sure they can return quickly once they are released. These people, too, have been resocialized. Also, people convicted of minor offenses may be resocialized into a criminal role while in prison.

When marines or soldiers go through basic training, they are being resocialized. Recruits are taught to get up at times when they would ordinarily go to bed, to cut their hair in peculiar ways, and to spend enormous amounts of time on the maintenance and appearance of their shoes, clothing, and equipment. They are told that a patch of dirt, rock, and sand outside the barracks is in fact a plot of grass that must be raked and cleaned every morning. They are taught to obey authority unquestioningly, to believe that the individual is unimportant, and to act as a unit with their fellow recruits at all times.

There are other examples of resocialization. Parents whose children have joined religious cults are usually astounded at the abrupt changes in people they thought they knew well. And the entire world was shocked by the mass suicide committed by the followers of the Reverend Jim Jones, in Jonestown, Guyana. Brainwashing of war prisoners through isolation and sensory deprivation is another type of resocialization. One technique used by the Chinese in Korea to lower morale was to allow only collection notices, divorce subpoenas, "Dear John" letters, and other demoralizing mail to reach the prisoners. They held back positive letters and let the negative ones through.[8] Organizations such as Alcoholics Anonymous, dedicated to helping people recover from dependence on alcohol, are also involved in attempts at resocialization.

When a person adopts the values, behavior, or viewpoints of a group he or she would like to belong to, this is called **anticipatory socialization.** This is a common occurrence and explains the behavior of the lower-class person who develops middle-class values, or the behavior of the "gung-ho" military recruit who wants to become an officer, or the new behavior and viewpoints of law or medical students who anticipate becoming lawyers or doctors. Anticipatory socialization has the interesting function of easing the transition from one stage in life to another. When individuals practice for a new role ahead of time, they often assume that the new role is much less difficult than it actually is. On the other hand, if the new stage in life is never reached, the trauma may be even greater than a sudden, unexpected change would have been.

RESOCIALIZATION AT VMI

Virginia Military Institute (VMI) is a state-run military college. In Virginia it is called the "West Point of the South."

At the very bottom of the VMI pyramid are those cadets known as rats. Rats are incoming freshmen. The first thing they're told is that they're the lowest form of life imaginable. Then they're given a week-long experience that bears this out. This week—VMI's version of freshman orientation—is known as cadre. Cadre begins the evening after matriculation, after mom and dad have left for home. The rats have all had their heads sheared and have been given baggy fatigues and shirts and baseball caps to wear. They look stupid and feel stupider. The only upperclassmen present are the officers and noncoms who run cadre. They are immaculately uniformed, polished to a glistening shine, and in total command. The rats quickly learn to fear them.

"You get a bunch of guys out of high school," explained one first classman, "they're all going every which way. You get 'em all in here together, you get 'em all looking alike, and they may feel miserable, but it serves a purpose. It gets everybody moving in the same direction—doing the same thing at the same time. And you can't be nice and do it, because some of them are going to tell you where you can get off. So you have to be hard. You've got to put the fear of God into 'em."

From cadre until sometime after Christmas, "ratline" prevails. During ratline, rats are not permitted to drink liquor, to walk to the bathroom or the showers except in full uniform, or to say the name of their class—the class of '82, for instance. They are given a rat bible with assorted facts to memorize, such as the names of the cadets who were killed

at New Market, and they are forced to strain in barracks—to hold chest and face perfectly rigid, with chin tucked into throat. Hazing is no longer permitted; sweat parties—late-night workouts that turn the fourth-class stoop slippery with sweat—are the rule now. Only regulation Army exercises are allowed, but that's little consolation to the rat who has just been ordered to do 250 pushups.

The honor system, one hears from every quarter, is the foundation on which VMI is built. It governs all academic work, all official statements, all theft, no matter how petty. It's administered by a cadet honor court, elected by the corps, which conducts all its proceedings (investigations, pretrial hearings, and trials) in absolute secrecy. The accused has the right to civilian counsel. If nine of the court's eleven voting members find him guilty, he is immediately dismissed. Once he has gone—in the very early hours of the morning, as it usually happens—his departure is announced by a low drum roll that grows steadily in intensity until it is broken by a boom on the bass drum. As the drum rolls and the intermittent booms continue, the members of the honor court march through the arch and up to the topmost stoop. From there, they work their way around and down, their president pronouncing as they go the name of the disgraced cadet, the charge on which he was convicted, and the admonition never to utter his name again. By this time he is far from Lexington, on his way alone toward whatever home he might have.

From Esquire, *January 1980, pp. 38–44. Copyright © 1980 by Frank Rose. Reprinted by arrangement with Virginia Barber Literary Agency. All rights reserved.*

SELF AND PERSONALITY

Psychology and sociology are most closely linked when they attempt to define the individual and to explain developmental processes. A separate discipline called social psychology, which deals with theories of socialization and individual development, has evolved. *Self* and *personality* are the major concepts in this analysis. The words are similar in meaning, but **personality** is a somewhat more general term and may be defined as the sum total of the physical, mental, emotional, social, and behavioral characteristics of an individual. **Self** refers to one's awareness of and ideas and attitudes about one's own personal and social identity. Psychologists are perhaps more inclined to use the term *personality*; sociologists usually prefer to use *self*.

Generally, psychologists have placed more emphasis on hereditary and biological factors in personality development, whereas sociologists focus on social interaction and the social environment. Psychologists place more emphasis on early childhood than do sociologists. Although numerous psychological theories have been developed, the most famous concepts are those of Sigmund Freud. Freud was a Viennese physician and founder of psychoanalysis. According to Freud, the personality is made up of three major components: the *id*, the *ego*, and the *superego*. The **id** is the most primitive aspect of the personality and represents the basic instinctual drives with which a person is born. Sexual and aggressive desires in the id constantly seek expression. The **ego** is the acting self; it is the mediator between the id and the outside world. Whereas the id operates by a pleasure principle—maximizing pleasure for the organism—the ego operates by a reality principle—balancing the instinctual needs of the organism against the conditions in the immediate environment. The **superego** is an internalized set of rules and regulations that represents the values and ideas of society initially interpreted for individuals by their parents. The superego, commonly referred to as the *conscience*, is the aspect of personality that controls behavior. We can think of id, ego, and superego as three sets of processes whose sometimes conflicting purposes combine to function as personality. Freud believed that the individual passed through a series of psychosexual stages and that personality was fixed very early in life, around the age of six. Sociologists believe that although the early years are extremely important, socialization and development of self are lifelong processes. Because Freud was a psychoanalyst and was working with people who were mentally ill, he emphasized the restricting nature of parents and society and the conflicting aspects of group life. Sociologists have focused more on the positive roles played by parents and groups in socialization and development of self.

With this brief view of Freud as an introduction, let's turn to sociological theories of the self. Sociologists believe that self-development is rooted in social behavior and not in biological, hereditary, or instinctual factors. The self is developed during the socialization process. Through interaction and association with others, you as an individual develop an image of *what you are*. This involves a perception of your role requirements and of your position and behavioral expectations in the various social groupings with which you identify.

As sociologist Charles H. Cooley put it, the word *self* means simply that which is designated in common speech by the pronouns of the first person singular: *I, me, mine,* and *myself*.[9] Aspects of one's self might include the following: I am male, I am a college student, I am of medium height, I am athletic, I am usually happy, I am intelligent. The self might include the following as well: I can never do anything right, I am not worth much, nobody listens to me, I am inadequate, I am unloved, I am not as capable as my older brother.

Individuals develop these images or viewpoints about themselves from the way others respond to them. More specifically, they develop them from the way *they think* others respond to them. Development of the self is a two-way interactive process that takes place throughout the period of socialization, the entire life span. Development of the self is an extremely subjective process involving interpretations of others' evaluations. Current evaluations are based on past evaluations, so that earlier mistakes become additive. A father may continually tell his daughter that she is extremely intelligent. This becomes part of her self, one of the ways she views herself, and she behaves and has certain expectations because of it. If she flunks bonehead English as a college freshman, contradictory information comes in. What happens next? She makes excuses and rationalizes her failure to support the view of self that she held earlier. She will probably seek reinforcement from other sources (father, girl friend, I.Q.-test scores) to support her self-view. If evidence is lacking or if reinforcement becomes more difficult to find (other bad grades, low I.Q.-test scores), this part of her self-conception will probably begin to change.

Aspects of self constantly undergo change. Often these changes occur slowly. Occasionally changes may occur more rapidly, suddenly, and dramatically, as we mentioned earlier, in cases of resocialization of inmates of prisons and concentration camps and of recruits in military basic training. All of these changes are a result of our evaluations of others' evaluations of us.

Cooley described the development of the self in his concept, the **looking-glass self.** The looking-glass self contains three elements: the imagination of our appearance to the other person, the imagination of his or her judgment of that appearance, and some sort of self-feeling, such as pride or mortification. Suppose, for example, a person eating in a crowded restaurant accidentally knocks his plate off the table. It makes a huge crash, and the food spills all over him. According to Cooley, he first "steps outside himself" and observes himself from the viewpoint of others in the room (". . . a well-dressed fellow with spaghetti all over him . . ."). Next, still examining himself as object, he imagines that others are evaluating his behavior (". . . must be a rather clumsy and awkward person . . ."). Finally, the individual as subject develops feelings and reactions to these imaginary evaluations of himself as object. He gets embarrassed, his face reddens, and he tries unsuccessfully to pretend it didn't happen. The process is made more interesting when we note that probably the same things would have happened if the person had been the only one in the restaurant. We are carrying our judges around with us in our heads, and they are constantly evaluating our behavior.

George Herbert Mead[10] described the development of the self this way: Individuals will conceive of themselves as they believe significant others conceive of them. They will then tend to act in accordance with expectations they impute to these significant others concerning the way "people like them" should act. As children develop the ability to examine and control their behavior in accordance with others' views and attitudes about their behavior, they are learning to "take the role of the other." At this first stage of development, called *primary socialization* by Berger and Luckmann, children do not cooperate with others; they relate only to specific individuals such as parents and perhaps an older brother or sister or a teacher—people whom sociologists call **significant others.** Later, as individuals develop, they are less likely to continue to react to *individual others,* and they learn instead to react more to a less personalized *grouping of others.* This grouping of others, which Mead called the **generalized other,** represents the sum of the viewpoints of the social group or community of people to which one belongs. The second stage of development, *secondary socialization,* begins when children are mature enough to cooperate with others in joint activities and to be able to react to the idea of people in general, rather than to specific others.

Although our internalized attitudes of others are crucial, they do not determine all behavior. Mead made a distinction between two aspects of the self, the "I" and the "me." The "I" represents the subjective, acting self, which may initiate spontaneous and original behavior. The "**me**" sees self as object and represents a conception of others' attitudes and viewpoints toward self. The "I" is freer, more innovative; the "me" is more conventional. "I" behaves; "me" judges and evaluates. "I" is in part governed by "me," but not completely. The abstract scrawling on the wall or in the dirt by the young child might represent "I" behavior; the same child's in-school drawing of the houses or trees that seem to please the teacher so much would be closer to "me" behavior. The driver involved in an auto accident who leaps out of the car and punches the other driver, swears at the passenger, and kicks a passing dog is probably

SHOOTING AN ELEPHANT

George Orwell tells the story of the English police officer on duty in a Burmese town who is told one day that an elephant has run amok, is ravaging the bazaar, and has killed a native—"caught him with its trunk, put its foot on his back and ground him into the earth." The officer gets a rifle and seeks out the elephant. When he finds the elephant, he realizes immediately that he ought not to shoot him—the elephant is eating peacefully and is not dangerous. "But at that moment I glanced round at the crowd that had followed me. It was an immense crowd, two thousand at the least and growing every minute. I looked at the sea of yellow faces all happy and excited over this bit of fun, all certain that the elephant was going to be shot. They did not like me, but with the magical rifle in my hands I was momentarily worth watching. And suddenly I realized that I should have to shoot the elephant after all. The people expected it of me and I had got to do it; I could feel their two thousand wills pressing me forward, irresistibly. Here was I, the white man with his gun, standing in front of the unarmed native crowd—seemingly the leading actor of the piece; but in reality I was only an absurd puppet pushed to and fro by the will of those yellow faces behind."

The officer did not want to shoot the elephant, but he did, many times, and the elephant finally died. Later there was much discussion about whether he had done the right thing. "The older men said I was right, the younger men said it was a damn shame to shoot an elephant for killing a coolie. And afterwards I was very glad that the coolie had been killed; it put me legally in the right and it gave me a sufficient pretext for shooting the elephant. I often wondered whether any of the others grasped that I had done it solely to avoid looking a fool."

Excerpts from "Shooting an Elephant" in Shooting an Elephant and Other Essays *by George Orwell are reprinted by permission of Harcourt Brace Jovanovich, Inc.*

showing us the "I" aspect of self; another passenger who goes to the nearest phone to call the highway patrol illustrates the "me." Or imagine the football player about to catch a punt as a group of monstrous players from the other team race down the field to smash him. The player's "I" says, "I don't want to catch that thing. It's coming down hard and is going to hurt my hands; look at all those guys going to tackle me—I think I'll let the ball bounce off my helmet." The "me," however, says, "Oh no you don't—there are a lot of people watching. You're supposed to *catch* that ball and *run* with it. The team and the coach are depending on you. *Do* it!"

THE IMPORTANCE OF LANGUAGE

It was also Mead who first emphasized that socialization and development of the self could not occur without *language*. Indeed, many argue that for the most part language distinguishes humans from other animals. The essential factor in language and symbolic communication is that a symbol arouses in one's self the same meaning it arouses in another. Primitive communication begins with a conversation of gestures, such as the mating dance of birds, the snarl of dogs, the roar of lions, or the cry of an infant. At this level, meanings of gestures are not shared. To be sure, the gestures may bring forth a response from some observer, such as from another dog or lion or from the infant's parent. But the important fact, according to Mead, is that the gesture does not arouse in the actor the same response it arouses in the observer. For example, a very young baby feels discomfort and cries. The parent interprets this as hunger and feeds the baby. If the parent is correct, this is communication of a sort, but for the baby, the act of crying had no communicative intent; it was merely a biological response. This represents primitive communication because the meanings were not shared by both parties.

As children mature they learn to use symbols and words. The words are poorly formed at first, and probably only a parent could understand them, but they are the beginnings of language—something beyond primitive communication. As their use of language improves, children learn that what they say and do elicits responses from others. Language allows them to replace behavior with ideas; they can now say they are hungry rather than acting it out by crying or pointing at food. Language allows them to think, makes possible the internalization of attitudes of others, and allows individuals to control their responses to others. We can imagine a child's internal conversation: "I would really like to have ice cream and cake, but my parents seem to like me better if I eat that other stuff they put out. . . ." Mead suggests that the use of symbolic communication continues to develop as a child grows, and through language, individuals are able to think, to develop shared social meanings, to take themselves as objects, and to evaluate their own behavior as they think others do. In summary, then, language makes possible the development of mind and self.

Language is important in another way. The language and symbols we use influence us to "see" the world and to "think" in particular ways. Imagine trying to balance your checkbook or to do your algebra or geometry homework using Roman numerals instead of the number system we are used to; much modern math probably could not have developed using the Roman system. The Inuit has many ways of expressing the phenomenon we describe by the single term *snow*. Navaho language is more literal, concrete, and specific than is English. Spanish has two verbs for *to be* whose different uses reflect a type of thinking unique to people who speak the language. French makes distinctions between a familiar and a formal *you*. Even color is "seen" differently in different societies depending on the words or symbols available; some societies recognize only two or three colors. Numerous examples make this same interesting point: The language and symbols we use act as a "filter" through which we experience the world. It is a filter *we* construct, which nevertheless leads us to see what we see in particular ways, to focus on certain elements, and to miss others entirely.[11]

DEFINITION OF THE SITUATION

An important principle of socialization and self-concept is that our understanding of reality is subjective and socially structured. We respond to what we *think* is so, rather than what really is so. When we peer into Cooley's looking glass the view may be quite distorted, but we believe what we see. Two ideas, *definition of the situation* and *self-fulfilling prophecy,* help illustrate this point.

Sociologist W. I. Thomas introduced **definition of the situation** by saying, "If men define situations as real, they are real in their consequences." The point he was making is that reality is socially structured and that people respond as much or more to the meaning a situation has for them than to the objective features of a situation. If a student defines passing a certain exam as the most important thing in the world, then for her it is, and she goes to great extremes to ensure that she does pass it. People define certain races as inferior and certain cultural beliefs or practices (cannibalism, polygamy) as peculiar, regardless of the objective reality of the situation. Or recall the old club initiation stunt: The victim is blindfolded and fed cold cooked spaghetti. Halfway through the meal he is told he is eating worms. What happens? He gags and is sick to his stomach. Why? He responds to what he thinks is true. That response has been socially defined and includes what his feeders tell him, how important initiation is to him, and what his background has taught him regarding the relative merits of angleworms and spaghetti. Even worse and horrible to consider, suppose we feed our friend angleworms and tell him it's spaghetti. That, I'm afraid, would work too.

The **self-fulfilling prophecy** is an extension of the concept of definition of the situation. A self-fulfilling prophecy occurs when a *false* definition of a situation evokes a new behavior that makes the originally false conception come *true.* A student is worried that he will flunk an exam. This makes him so nervous and anxious that he is unable to study effectively, and he flunks the exam. Or the police believe Gorples to be very criminal types. Consequently, more police patrols and surveillance are concentrated in those sections of the city where the Gorples live. Increased police surveillance leads to greater visibility and reporting of crime, so the crime rate of Gorples increases. As Merton explains, "Confident error generates its own spurious confirmation."[12] If we look carefully, we can see this phenomenon frequently in everyday situations.

Psychologist Robert Rosenthal performed several interesting experiments that illustrate what we are discussing here.[13] In one experiment, to which we referred in the second research question at the beginning of this chapter, six students were given five rats each. They were told that their rats were very bright ones, bred especially for running a maze. Six other students were given five rats each and were told that their rats were genetically inferior and probably would be poor at running through mazes. Then both groups of students attempted to train their rats to run the maze. What the students didn't know was that there were no "bright" rats and no "dull" rats; they were all the same with equal capacities for learning. What happened? Well, right from the beginning of the training the "bright" rats performed better. The "dull" rats made little progress and sometimes would not even budge from the starting position in the maze. And there were differences in more than just performance. The students found the "bright" rats to be brighter, more pleasant, and more likable than the "dull" rats. The self-fulfilling prophecy explains what happened: The students with "bright" rats helped, coaxed, and encouraged their animals to do what they expected them to, and the "dull" rats got little encouragement. Or perhaps because the students expected "brights" to do well and "dulls" to do poorly, that is what they saw, regardless of what was actually happening.

In Chapter One we discussed another of Rosenthal's experiments. As introduced in **Situation 7**, you remember that Rosenthal and Lenore Jacobson (R&J) tested the intelligence of school children. R&J then told teachers that certain of their students were very sharp ("academic spurters") and could be expected to do quite well in school. As with the rats in the earlier study, the children were all alike. The "spurters" were *not* of higher intelligence, and in fact R&J had chosen them at random. The children were tested again later in the year, and the teachers evaluated the students. You can guess what happened. The students in the first and second grades who were labeled "academic spurters" were described as happier than other students, more curious, more interesting, more appealing, and better adjusted. They also showed greater gains in intelligence than did the other children. The teachers acted in terms of what they thought to be true (definition of the situation) and behaved so that what was expected to occur, did occur (self-fulfilling prophecy).[14] This illustrates our point once again: Our reality is subjective and socially structured.

In the preceding paragraph we introduced the process of **labeling,** which is an important one in sociology. Labeling refers to the tendency of people to stamp, typecast, or categorize others. Often these categories are negative, and understanding the concept of labeling is important for precisely the reasons we have just been discussing. If one is labeled "bright," "dull," "academic spurter," "stupid," "ex-con," "deviant," "lesbian," "mentally ill," or "sex offender," it is bound to have an effect on that individual. The label may be accurate or it may be false. But as we have just discovered, accuracy is of little consequence because reality is subjective and people believe what they are told. Labeling leads to different treatment by others (who respond to the label rather than to the individual), to effects on the individual's self-concept, and perhaps to a self-fulfilling prophecy. We will return to labeling in more detail in Chapter Fourteen.

DEBATES AND ISSUES IN SOCIALIZATION

I may have implied that there is general agreement about the topics and issues we have discussed so far. I wish it were so, but the process of becoming a socialized person is a complicated one, and here, as in other areas of human behavior, there are great differences of opinion. One of the debates concerns the relative importance of biology and heredity on the one hand and of the social environment on the other. Sociologists see socialization as a *social* process. One's biology may set limits, but its overall effect in the development of the human being is relatively small. The key factor is the social context—the interaction between oneself and one's family, friends, and peers—and the social influence of one's religion, social class, race, gender, age, occupation, education, and so on. The diagram in Figure 2.1 illustrates this view. The social context is created by input from family, groups, friends, and numerous other sources (in fact, many more than the diagram's lines and arrows show here). The effect of the social context on behavior is great; the effect of biology (B) is small. Now the argument starts. Is the effect of B really so small, or should it be written much larger in comparison with the social context? Many feel that the key is biology, that the B should be written very large because human behavior, like animal behavior, is basically genetically determined. They find biological explanations for all sorts of behavior, from parental and sexual behavior to mental illness, aggression, shyness, and even alcoholism, to mention only a few. Sociologists grant that in certain types of cases—for example, a person born with a birth defect, a child with Down's syndrome, a very tall or very short person, a very old person—the B must be written larger; in such cases one's biology may have a greater influence on behavior. These cases are exceptional however, and even here, the effect of the social context is great.

Another issue concerns the importance of the early years of life. Sociologists believe that these years from birth to age six or seven are important but that the socialization process is lifelong; we continually change through interaction in the social environment. Psychologists tend to place much more emphasis on the early years. Many learning theorists believe that intelligence and learning capacity are determined in the first year or two of life. Isolation from or impaired interaction with people during these years, as in Anna's case discussed at the beginning of this chapter, causes irreparable damage. Examples of isolated children, and even studies with monkeys, are cited to show the effects of deprivation. But the evidence is ambiguous. Some children who are isolated never recover, and others do, as occurred with Isabelle. In a part of Holland there is a local custom of isolating a child in a room outside the house for the first ten months; the child is tightly bound and is given no toys and offered minimal interaction. At the age of one the child appears to be retarded,

Social Context

FIGURE 2.1 *Effects of social context and biology on behavior.*

but by the age of five he or she is fully recovered. Jerome Kagan describes a village he visited in Guatemala in which the children are kept isolated in dark huts. They have no toys, and although parents may hold their children, they don't talk or interact with them.[15] At the age of one and a half, the children are retarded by our standards. Yet 11-year-olds in this same village appear normal—they are alert, active, and intelligent. This leads Kagan to believe that too much emphasis has been placed on the first two years of life. He suggests that because children learn and develop at different rates, perhaps schools should rethink their policy of rank-ordering and labeling students in the early grades. We should keep an open mind on the importance of the early years and remember that socialization is a lifelong process.

THEORY AND RESEARCH: A REVIEW

In Chapter One we stated that scientists use theory and research in their search for understanding. In Chapter Two we have seen some examples of their attempts. This chapter focuses on theories about the concepts of socialization, self, and personality. The ideas of Berger and Luckmann, Freud, Cooley, and Mead represent theoretical positions or viewpoints about how these basic concepts can best be understood. In this chapter at least, the *symbolic interactionist approach* seems to be the most useful theoretical view. Research discoveries may then help support or reject a particular position. Davis's *case histories* of isolation (Anna and Isabelle) tell us something about inadequate social interaction and its effect on the socialization process. Goldberg's *survey research* refines our understanding of gender differences in the socialization process. Niederhoffer's *participant observation study* of police officers provides research support for theoretical ideas about secondary socialization. Rosenthal's interesting *experiments* with students and their rats, and with teachers and their students, illustrate and support the ideas of definition of the situation and self-fulfilling prophecy.

Of course, there can be problems in the search for understanding. Some theories, such as Freud's id, superego, and ego and Mead's generalized other, are difficult to test. Sometimes research efforts yield conflicting results or are open to different interpretations. For example, why did Isabelle and Anna turn out so differently, and why are the repeats of Rosenthal's studies not consistent? The viewpoint of scientists, however, is that theory and research work together in a way that is both self-correcting and cumulative. Conflicting results prove the value of continued studies, and those theories that today seem difficult to test will submit to new methodological approaches in the future.

SUMMARY

Our study of sociology starts with the assumption that humans are social animals. The obvious question is, How do they become that way? In this chapter we have discussed some of the processes that sociologists consider essential if we are to understand how one changes from a blob of matter into a complex, interacting social being.

The process through which this transformation occurs is called socialization. During socialization, the self is developed. The self is the individual's set of images about who or what he or she is. The development of self is a social process; it arises out of our interpretations of others' reactions to us. Several theories were examined that describe in detail how the self is developed. Cooley's explanation involves the looking-glass self; Mead describes self-growth through the generalized other and through assuming the role of the other. In Mead's view, language is a complex method of symbolic communication that is essential in socialization and in development of the self.

Although the individual is a combination of biological, hereditary, and social factors, sociologists focus their studies on interaction in the social environment—the stage or setting in which socialization and self-development take place. Determining the effects of the social environment on the individual, which include the interpretation and internalization of others' reactions to us, is also a very subjective process. There are substantial variations in how individuals perceive, interpret, and react to their social environment. Suppose 20 people observe an accident at a busy intersection, and then each person writes a description of what happened. How many different descriptions might we get? Probably at least 10 or 15, maybe even 20. Or observe how different students react to a good or bad grade on an exam; again, there is great variation. The effect of the social environment, then, is a very subjective and interpretive phenomenon; no two people react in exactly the same way. Differences multiply because future perceptions and interpretations are based on past perceptions and interpretations. In like manner, socialization and development of the self are building processes that take place throughout the life span.

The four readings that follow will help illustrate some of the concepts we have discussed in this chapter. In an excerpt from her book *The Story of My Life,* Helen Keller, who was deaf and blind from early childhood, recalls meeting her new teacher and beginning her education. This reading illustrates how important language and symbolic communication are in the processes of socialization and self-development. The second reading is an excerpt from Malcolm X's autobiography in which he describes how some events that occurred during his early teens seemed to have a great and lasting impact on him. In the third reading, a sad story about the death of a boy, Jean Mizer illustrates socialization, agents of socialization, and the self-fulfilling prophecy. The fourth reading reflects on the nature versus nurture, biology versus social environment debate. The author suggests that if a gene explains aggression (as has been suggested), perhaps we should also look for a gene that explains racism, hatred of gays, and the greed of white-collar criminals.

TERMS FOR STUDY

anticipatory socialization (43)	generalized other (47)
definition of the situation (50)	I (47)
ego (45)	id (45)

labeling (51)

looking-glass self (46)

me (47)

personality (45)

primary socialization (41)

resocialization (42)

secondary socialization (41)

self (45)

self-fulfilling prophecy (50)

significant other (47)

social interaction (39)

socialization (38)

sociobiology (39)

superego (45)

Reading 2.1

THE STORY OF MY LIFE

Helen Keller

Socialization is defined as the process by which the organism becomes a social being—the learning of habits, skills, and other requirements for effective participation in social groups. Socialization is a continuing process that occurs through interaction with other humans—typically parents and other family members, teachers, and friends.

Helen Keller was deaf and blind from early childhood. In this excerpt from one of her books, she describes the period when her new teacher first arrived and a new form of learning began to take place.

The most important day I remember in all my life is the one on which my teacher, Anne Mansfield Sullivan, came to me. I am filled with wonder when I consider the immeasurable contrasts between the two lives which it connects. It was the third of March, 1887, three months before I was seven years old.

Have you ever been to sea in a dense fog, when it seemed as if a tangible white darkness shut you in, and the great ship, tense and anxious, groped her way toward the shore with plummet and sounding-line, and you waited with beating heart for something to happen? I was like that ship before my education began, only I was without compass or sounding-line, and had no way of knowing how near the harbour was. "Light! give me light!" was the wordless cry of my soul, and the light of love shone on me in that very hour.

I felt approaching footsteps. I stretched out my hand, as I supposed to my mother. Someone took it, and I was caught up and held close in the arms of her who had come to reveal all things to me, and, more than all things else, to love me.

The morning after my teacher came she led me into her room and gave me a doll. The little blind children at the Perkins Institution had sent it and Laura Bridgman had dressed it; but I did not know this until afterward. When I had played with it a little while, Miss Sullivan slowly spelled into my hand the word "d-o-l-l." I was at once interested in this finger play and tried to imitate it. When I finally succeeded in making the letters correctly I was flushed with childish pleasure and pride. Running downstairs to my mother I held up my hand and made the letters for doll. I did not know that I was spelling a word or even that words existed; I was simply making my fingers go in monkey-like imitation. In the days that followed I learned to spell in this uncomprehending way a great many words, among them pin, hat, cup, and a few verbs like sit, stand, and walk. But my teacher had been with me several weeks before I understood that everything has a name.

One day, while I was playing with my new doll, Miss Sullivan put my big rag doll into my lap also, spelled "d-o-l-l" and tried to make me understand that "d-o-l-l" applied to both. Earlier in the day we had had a tussle over the words "m-u-g" and "w-a-t-e-r."

Miss Sullivan had tried to impress it upon me that "m-u-g" is mug and that "w-a-t-e-r" is water, but I persisted in confounding the two. In despair she had dropped the subject for the time, only to renew it at the first opportunity. I became impatient at her repeated attempts and, seizing the new doll, I dashed it upon the floor. I was keenly delighted when I felt the fragments of the broken doll at my feet. Neither sorrow nor regret followed my passionate outburst. I had not loved the doll. In the still, dark world in which I lived there was no strong sentiment of tenderness. I felt my teacher sweep the fragments to one side of the hearth, and I had a sense of satisfaction that the cause of my discomfort was removed. She brought me my hat, and I knew I was going out into the warm sunshine. This thought, if a wordless sensation may be called a thought, made me hop and skip with pleasure.

We walked down the path to the well-house, attracted by the fragrance of the honeysuckle with which it was covered. Someone was drawing water and my teacher placed my hand under the spout. As the cool stream gushed over one hand she spelled into the other the word water, first slowly, then rapidly. I stood still, my whole attention fixed upon the motions of her fingers. Suddenly I felt a misty consciousness as of something forgotten—a thrill of returning thought; and somehow the mystery of language was revealed to me. I knew then that "w-a-t-e-r" meant the wonderful cool something that was flowing over my hand. That living word awakened my soul, gave it light, hope, joy, set it free! There were barriers still, it is true, but barriers that could in time be swept away.

I left the well-house eager to learn. Everything had a name, and each name gave birth to a new thought. As we returned to the house every object which I touched seemed to quiver with life. That was because I saw everything with the strange, new sight that had come to me. On entering the door I remembered the doll I had broken. I felt my way to the hearth and picked up the pieces. I tried vainly to put them together. Then my eyes filled with tears; for I realized what I had done, and for the first time I felt repentance and sorrow.

I learned a great many new words that day. I do not remember what they all were; but I do know that mother, father, sister, teacher were among them—words that were to make the world blossom for me, "like Aaron's rod, with flowers." It would have been difficult to find a happier child than I was as I lay in my crib at the close of that eventful day and lived over the joys it had brought me, and for the first time longed for a new day to come.

I recall many incidents of the summer of 1887 that followed my soul's sudden awakening. I did nothing but explore with my hands and learn the name of every object that I touched; and the more I handled things and learned their names and uses, the more joyous and confident grew my sense of kinship with the rest of the world.

When the time of daisies and buttercups came Miss Sullivan took me by the hand across the fields, where men were preparing the earth for the seed, to the banks of the Tennessee River, and there, sitting on the warm grass, I had my first lessons in the beneficence of nature. I learned how the sun and the rain make to grow out of the ground every tree that is pleasant to the sight and good for food, how birds build their nests and live and thrive from land to land, how the squirrel, the deer, the lion, and every other creature finds food and shelter. As my knowledge of things grew I felt more and more the delight of the world I was in. Long before I learned to do a sum in arithmetic or describe the shape of the earth, Miss Sullivan had taught me to find beauty in the fragrant woods, in every blade of grass, and in the curves and dimples of my baby sister's hand. She linked my earliest thoughts with nature, and made me feel that "birds and flowers and I were happy peers."

I had now the key to all language, and I was eager to learn to use it. Children who hear acquire language without any particular effort; the words that fall from others' lips they catch on the wing, as it were, delightedly, while the little deaf child must trap them by a slow and often painful process. But whatever the process, the result is wonderful. Gradually from naming an object we advance step by step until we have traversed the vast distance between our first stammered syllable and the sweep of thought in a line of Shakespeare.

At first, when my teacher told me about a new thing I asked very few questions. My ideas were vague, and my vocabulary was inadequate; but as my knowledge of things grew, and I learned more and more words, my field of inquiry broadened, and I would return again and again to the same subject, eager for further information. Sometimes a new word revived an image that some earlier experience had engraved on my brain.

I remember the morning that I first asked the meaning of the word, "love." This was before I knew many words. I had found a few early violets in the garden and brought them to my teacher. She tried to kiss me; but at that time I did not like to have any one kiss me except my mother. Miss Sullivan put her arm gently round me and spelled into my hand, "I love Helen."

"What is love?" I asked.

She drew me closer to her and said, "It is here," pointing to my heart, whose beats I was conscious of for the first time. Her words puzzled me very much because I did not then understand anything unless I touched it.

I smelt the violets in her hand and asked, half in words, half in signs, a question which meant, "Is love the sweetness of flowers?"

"No," said my teacher.

Again I thought. The warm sun was shining on us.

"Is this not love?" I asked, pointing in the direction from which the heat came. "Is this not love?"

It seemed to me that there could be nothing more beautiful than the sun, whose warmth makes all things grow. But Miss Sullivan shook her head, and I was greatly puzzled and disappointed. I thought it strange that my teacher could not show me love.

A day or two afterward I was stringing beads of different sizes in symmetrical groups—two large beads, three small ones, and so on. I had made many mistakes, and Miss Sullivan had pointed them out again and again with gentle patience. Finally I noticed a very obvious error in the sequence and for an instant I concentrated my attention on the lesson and tried to think how I should have arranged the beads. Miss Sullivan touched my forehead and spelled with decided emphasis, "Think."

In a flash I knew that the word was the name of the process that was going on in my head. This was my first conscious perception of an abstract idea.

For a long time I was still—I was not thinking of the beads in my lap, but trying to find a meaning for "love" in the light of this new idea. The sun had been under a cloud all day, and there had been brief showers; but suddenly the sun broke forth in all its southern splendour.

Again I asked my teacher, "Is this not love?"

"Love is something like the clouds that were in the sky before the sun came out," she replied. Then in simpler words than these, which at that time I could not have understood, she explained: "You cannot touch the clouds, you know; but you feel the rain and know how glad the flowers and the thirsty earth are to have it after a hot day. You cannot touch love either; but you feel the sweetness that it pours into everything. Without love you would not be happy or want to play."

The beautiful truth burst upon my mind—I felt that there were invisible lines stretched between my spirit and the spirits of others.

QUESTIONS 2.1

1. Without language and communication, socialization cannot take place. Discuss.

2. What skills were developed by Helen Keller during the period discussed in this reading? Could she have learned these skills without the help of others? Is the same true for a person who doesn't have the handicaps that Helen Keller had?

Reading 2.2

MASCOT

Malcolm X

Socialization involves many things. Part of it is learning that you are male or female, short or tall, black or white; a major part is learning how people react to you because you are male or female, short or tall, black or white. These reactions and interpretations become an important part of the self. Many people probably do not recall the stage in the socialization process when they became aware of their gender or race. To remember, one must be very perceptive, and the new learning must be so unique or unexpected that it creates a lasting impression. In this excerpt from his autobiography, Malcolm X recalls learning what it means to be black in America.

They told me I was going to go to a reform school. I was still thirteen years old.

But first I was going to the detention home. It was in Mason, Michigan, about twelve miles from Lansing. The detention home was where all the "bad" boys and girls from Ingham County were held, on their way to reform school—waiting for their hearings.

The white state man was a Mr. Maynard Allen. He was nicer to me than most of the state welfare people had been. He even had consoling words for the Gohannas and Mrs. Adcock and Big Boy; all of them were crying. But I wasn't. With the few clothes I owned stuffed into a box, we rode in his car to Mason. He talked as he drove along, saying that my school marks showed that if I would just straighten up, I could make something of myself. He said that reform school had the wrong reputation; he talked about what the word "reform" meant—to change and become better. He said the school was really a place where boys like me could have time to see their mistakes and start a new life and become somebody everyone would be proud of. And he told me that the lady in charge of the detention home, a Mrs. Swerlin, and her husband were very good people.

They were good people. Mrs. Swerlin was bigger than her husband, I remember, a big, buxom, robust, laughing woman, and Mr. Swerlin was thin, with black hair, and a black mustache and a red face, quiet and polite, even to me.

They liked me right away, too. Mrs. Swerlin showed me to my room, my own room—the first in my life. It was in one of those huge dormitory-like buildings where kids in detention were kept in those days—and still are in most places. I discovered next, with surprise, that I was allowed to eat with the Swerlins. It was the first time I'd eaten with white people—at least with grown white people—since the Seventh Day Adventist country meetings. It wasn't my own exclusive privilege, of course. Except for the very troublesome boys and girls at the detention home, who were kept locked up—those who had run away and been caught and brought back, or something like that—all of us ate with the Swerlins sitting at the head of the long tables.

They had a white cook-helper, I recall—Lucille Lathrop. (It amazes me how these names come back, from a time I haven't thought about for more than twenty years.) Lucille treated me well, too. Her husband's name was Duane Lathrop. He worked somewhere else, but he stayed there at the detention home on the weekends with Lucille.

I noticed again how white people smelled different from us, and how their food tasted different, not seasoned like Negro cooking. I began to sweep and mop and dust around in the Swerlins' house, as I had done with Big Boy at the Gohannas'.

They all liked my attitude, and it was out of their liking for me that I soon became accepted by them—as a mascot, I know now. They would talk about anything and everything with me standing right there hearing them, the same way people would talk freely in front of a pet canary. They would even talk about me, or about "niggers," as though I wasn't there, as if I wouldn't understand what the word meant. A hundred times a day, they used the word "nigger." I suppose that in their own minds, they meant no harm; in fact they probably meant well. It was the same with the cook, Lucille, and her husband, Duane. I remember one day when Mr. Swerlin, as nice as he was, came in from Lansing, where he had been through the Negro section, and said to Mrs. Swerlin right in front of me, "I just can't see how those niggers can be so happy and be so poor." He talked about how they lived in shacks, but had those big, shining cars out front.

And Mrs. Swerlin said, me standing right there, "Niggers are just that way. . . ." That scene always stayed with me.

It was the same with the other white people, most of them local politicians, when they would come visiting the Swerlins. One of their favorite parlor topics was "niggers." One of them was the judge who was in charge of me in Lansing. He was a close friend of the Swerlins. He would ask about me when he came, and they would call me in, and he would look me up and down, his expression approving, like he was examining a fine colt, or a pedigreed pup. I knew they must have told him how I acted and how I worked.

What I am trying to say is that it just never dawned upon them that I could understand, that I wasn't a pet, but a human being. They didn't give me credit for having the same sensitivity, intellect, and understanding that they would have been ready and willing to recognize in a white boy in my position. But it has historically been the case with white people, in their regard for black people, that even though we might be *with* them, we weren't considered *of* them. Even though they appeared to have opened the door, it was still closed. Thus they never did really see *me*.

This is the sort of kindly condescension which I try to clarify today, to these integration-hungry Negroes, about their "liberal" white friends, these so-called "good white people"—most of them anyway. I don't care how nice one is to you; the thing you must always remember is that almost never does he really see you as he sees himself, as he sees his own kind. He may stand with you through thin, but not thick; when the chips are down, you'll find that as fixed in him as his bone structure is his sometimes subconscious conviction that he's better than anybody black.

But I was no more than vaguely aware of anything like that in my detention-home years. I did my little chores around the house, and everything was fine. And each weekend, they didn't mind my catching a ride over to Lansing for the afternoon or evening. If I wasn't old enough, I sure was big enough by then, and nobody ever questioned my hanging out, even at night, in the streets of the Negro section.

I was growing up to be even bigger than Wilfred and Philbert, who had begun to meet girls at the school dances, and other places, and introduced me to a few. But the ones who seemed to like me, I didn't go for—and vice versa. I couldn't dance a lick, anyway, and I couldn't see squandering my few dimes on girls. So mostly I pleasured myself these Saturday nights by gawking around the Negro bars and restaurants. The jukeboxes were wailing Erskine Hawkins' "Tuxedo Junction," Slim and Slam's "Flatfoot Floogie," things like that. Sometimes, big bands from New York, out touring the one-night stands in the sticks, would play for big dances in Lansing. Everybody with legs would come out to see any performer who bore the magic name "New York." Which is how I first heard Lucky Thompson and Milt Jackson, both of whom I later got to know well in Harlem.

Many youngsters from the detention home, when their dates came up, went off to the reform school. But when mine came up—two or three times—it was always ignored. I saw new youngsters arrive and leave. I was glad and grateful. I knew it was Mrs. Swerlin's doing. I didn't want to leave.

She finally told me one day that I was going to be entered in Mason Junior High School. It was the only school in town. No ward of the detention home had ever gone to school there, at least while still a ward. So I entered their seventh grade. The only other Negroes there were some of the Lyons children, younger than I was, in the lower grades. The Lyons and I, as it happened, were the town's only Negroes. They were, as Negroes, very much respected. Mr. Lyons was a smart, hardworking man, and Mrs. Lyons was a very good woman. She and my mother, I had heard my mother say, were two of the four West Indians in that whole section of Michigan.

Some of the white kids at school, I found, were even friendlier than some of those in Lansing had been. Though some, including the teachers, called me "nigger," it was easy to see that they didn't mean any more harm by it than the Swerlins. As the "nigger" of my class, I was in fact extremely popular—I suppose partly because I was kind of a novelty. I was in demand, I had top priority. But I also benefited from the special prestige of having the seal of approval from that Very Important Woman about the town of Mason, Mrs. Swerlin. Nobody in Mason would have dreamed of getting on the wrong side of her. It became hard for me to get through a school day without someone after me to join this or head up that—the debating society, the junior high basketball team, or some other extracurricular activity. I never turned them down.

And I hadn't been in the school long when Mrs. Swerlin, knowing I could use spending money of my own, got me a job after school washing the dishes in a local restaurant. My boss there was the father of a white classmate whom I spent a lot of time with. His family lived over the restaurant. It was fine working there. Every Friday night when I got paid, I'd feel at least ten feet tall. I forget how much I made, but it seemed like a lot. It was the first time I'd ever had any money to speak of, all my own, in my whole life. As soon as I could afford it, I bought a green suit and some shoes, and at school I'd buy treats for the others in my class—at least as much as any of them did for me.

English and history were the subjects I liked most. My English teacher, I recall—a Mr. Ostrowski—was always giving advice about how to become something in life. The one thing I didn't like about history class was that the teacher, Mr. Williams, was a great one for "nigger" jokes. One day during my first week at school, I walked into the room and he started singing to the class, as a joke, "Way down yonder in the cotton field, some folks say that a nigger won't steal." Very funny. I liked history, but I never thereafter had much liking for Mr. Williams. Later, I remember, we came to the textbook section on Negro history. It was exactly one paragraph long. Mr. Williams laughed through it practically in a single breath, reading aloud how the Negroes had been slaves and then were freed, and how they were usually lazy and dumb and shiftless. He added, I remember, an anthropological footnote on his own, telling us between laughs how Negroes' feet were "so big that when they walk, they don't leave tracks, they leave a hole in the ground."

. . . Then, in the second semester of the seventh grade, I was elected class president. It surprised me even more than other people. But I can see now why the class might have done it. My grades were among the highest in the school. I was unique in my class, like a pink poodle. And I was proud; I'm not going to say I wasn't. In fact, by then, I didn't really have much feeling about being a Negro, because I was trying so hard, in every way I could, to be white. Which is why I am spending much of my life today telling the American black man that he's wasting his time straining to "integrate." I know from personal experience. I tried hard enough. . . .

That summer of 1940, in Lansing, I caught the Greyhound bus for Boston with my cardboard suitcase, and wearing my green suit. If someone had hung a sign, "hick," around my neck, I couldn't have looked much more obvious. They didn't have the turnpikes then; the bus stopped at what seemed every corner and cowpatch. From my seat in—you guessed it—the back of the bus, I gawked out of the window at white man's America rolling past for what seemed a month, but must have been only a day and a half.

When we finally arrived, Ella met me at the terminal and took me home. The house was on Waumbeck Street in the Sugar Hill section of Roxbury, the Harlem of Boston. I met Ella's second husband, Frank, who was now a soldier; and her brother Earl, the singer who called himself Jimmy Carleton; and Mary, who was very different from her older sister. It's funny how I seemed to think of Mary as Ella's sister, instead of her being, just as Ella is, my own half-sister. It's probably because Ella and I always were much closer as basic types; we're dominant people, and Mary has always been mild and quiet, almost shy.

Ella was busily involved in dozens of things. She belonged to I don't know how many different clubs; she was a leading light of local so-called "black society." I saw and met a hundred black people there whose big-city talk and ways left my mouth hanging open.

I couldn't have feigned indifference if I had tried to. People talked casually about Chicago, Detroit, New York. I didn't know the world contained as many Negroes as I saw thronging downtown Roxbury at night, especially on Saturdays. Neon lights, night-clubs, poolhalls, bars, the cars they drove! Restaurants made the streets smell—rich, greasy, down-home black cooking! Jukeboxes blared Erskine Hawkins, Duke Ellington, Cootie Williams, dozens of others. If somebody had told me then that some day I'd know them all personally, I'd have found it hard to believe. The biggest bands, like these, played at the Roseland State Ballroom, on Boston's Massachusetts Avenue— one night for Negroes, the next night for whites.

I saw for the first time occasional black-white couples strolling around arm in arm. And on Sundays, when Ella, Mary, or somebody took me to church, I saw churches for black people such as I had never seen. They were many times finer than the white church I had attended back in Mason, Michigan. There, the white people just sat and worshiped with words; but the Boston Negroes, like all other Negroes I had ever seen at church, threw their souls and bodies wholly into worship.

Two or three times, I wrote letters to Wilfred intended for everybody back in Lansing. I said I'd try to describe it when I got back.

But I found I couldn't.

My restlessness with Mason—and for the first time in my life a restlessness with being around white people—began as soon as I got back home and entered eighth grade.

I continued to think constantly about all that I had seen in Boston, and about the way I had felt there. I know now that it was the sense of being a real part of a mass of my own kind for the first time.

The white people—classmates, the Swerlins, the people at the restaurant where I worked—noticed the change. They said, "You're acting so strange. You don't seem like yourself, Malcolm. What's the matter?"

I kept close to the top of the class, though. The top-most scholastic standing, I re-member, kept shifting between me, a girl named Audrey Slaugh, and a boy named Jimmy Cotton.

It went on that way, as I became increasingly restless and disturbed through the first semester. And then one day, just about when those of us who had passed were about to move up to 8-A, from which we would enter high school the next year, something happened which was to become the first major turning point of my life.

Somehow, I happened to be alone in the classroom with Mr. Ostrowski, my English teacher. He was a tall, rather reddish white man and he had a thick mustache. I had gotten some of my best marks under him, and he had always made me feel that he liked me. He was, as I have mentioned, a natural-born "advisor," about what you ought to read, to do, or think—about any and everything. We used to make unkind jokes about him: why was he teaching in Mason instead of somewhere else, getting for himself some of the "success in life" that he kept telling us how to get?

I know that he probably meant well in what he happened to advise me that day. I doubt that he meant any harm. It was just in his nature as an American white man. I was one of his top students, one of the school's top students—but all he could see for me was the kind of future "in your place" that almost all white people see for black people.

He told me, "Malcolm, you ought to be thinking about a career. Have you been giving it thought?"

The truth is, I hadn't. I never had figured out why I told him, "Well, yes, sir, I've been thinking I'd like to be a lawyer." Lansing certainly had no Negro lawyers—or doctors either—in those days, to hold up an image I might have aspired to. All I really knew for certain was that a lawyer didn't wash dishes, as I was doing.

Mr. Ostrowski looked surprised, I remember, and leaned back in his chair and clasped his hands behind his head. He kind of half-smiled and said, "Malcolm, one of life's first needs is for us to be realistic. Don't misunderstand me, now. We all here like you, you know that. But you've got to be realistic about being a nigger. A lawyer—that's no realistic goal for a nigger. You need to think about something you *can* be. You're good with your hands—making things. Everybody admires your carpentry shop work. Why don't you plan on carpentry? People like you as a person—you'd get all kinds of work."

The more I thought afterwards about what he said, the more uneasy it made me. It just kept treading around in my mind.

What made it really begin to disturb me was Mr. Ostrowski's advice to others in my class—all of them white. Most of them told him they were planning to become farmers. But those who wanted to strike out on their own, to try something new, he had encouraged. Some, mostly girls, wanted to be teachers. A few wanted other professions, such as one boy who wanted to become a county agent; another, a veterinarian; and one girl wanted to be a nurse. They all reported that Mr. Ostrowski had encouraged what they had wanted. Yet nearly none of them had earned marks equal to mine.

It was a surprising thing that I had never thought of it that way before, but I realized that whatever I wasn't, I *was* smarter than nearly all of those white kids. But apparently I was still not intelligent enough, in their eyes, to become whatever *I* wanted to be.

It was then that I began to change—inside.

I drew away from white people. I came to class, and I answered when called upon. It became a physical strain simply to sit in Mr. Ostrowski's class.

Where "nigger" had slipped off my back before, whenever I heard it now, I stopped and looked at whoever said it. And they looked surprised that I did.

I quit hearing so much "nigger" and "What's wrong?"—which was the way I wanted it. Nobody, including the teachers, could decide what had come over me. I knew I was being discussed.

In a few more weeks, it was that way, too, at the restaurant where I worked washing dishes, and at the Swerlins'. . . .

QUESTIONS 2.2

1. Malcolm X believes he was treated as a "mascot." What does he mean by this? Why was he treated this way? Discuss.

2. Outline the important aspects of socialization that occurred during the period described by Malcolm X. Who were the socialization agents in this period?

3. Review Chapter Two, and using examples from this reading, illustrate development of the self using Cooley's and Mead's theories.

Reading 2.3

CIPHER IN THE SNOW

Jean E. Mizer

Socialization and development of self are interactive processes that are critically important in the growth and future of every individual. Seldom, however, are these processes so graphically and dramatically displayed as they are in this reading. Jean Mizer was a school teacher in Idaho when she wrote this memoir of a personal experience.

It started with tragedy on a biting cold February morning. I was driving behind the Milford Corners bus as I did most snowy mornings on my way to school. It veered and stopped short at the hotel, which it had no business doing, and I was annoyed as I had to come to an unexpected stop. A boy lurched out of the bus, reeled, stumbled, and collapsed on the snowbank at the curb. The bus driver and I reached him at the same moment. His thin, hollow face was white even against the snow.

"He's dead," the driver whispered.

I didn't register for a minute. I glanced quickly at the scared young faces staring down at us from the school bus. "A doctor! Quick! I'll phone from the hotel. . . ."

"No use. I tell you he's dead." The driver looked down at the boy's still form. "He never even said he felt bad," he muttered, "just tapped me on the shoulder and said, real quiet, 'I'm sorry. I have to get off at the hotel.' That's all. Polite and apologizing like."

At school, the giggling, shuffling morning noise quieted as the news went down the halls. I passed a huddle of girls. "Who was it? Who dropped dead on the way to school?" I heard one of them half-whisper.

Reprinted from the November 1964 issue of *NEA Journal* by permission of the National Education Association and the author.

"Don't know his name; some kid from Milford Corners," was the reply.

It was like that in the faculty room and the principal's office. "I'd appreciate your going out to tell the parents," the principal told me. "They haven't a phone and, anyway, somebody from school should go there in person. I'll cover your classes."

"Why me?" I asked. "Wouldn't it be better if you did it?"

"I didn't know the boy," the principal admitted levelly. "And in last year's sophomore personalities column I note that you were listed as his favorite teacher."

I drove through the snow and cold down the bad canyon road to the Evans place and thought about the boy, Cliff Evans. His favorite teacher! I thought. He hasn't spoken two words to me in two years! I could see him in my mind's eye all right, sitting back there in the last seat in my afternoon literature class. He came in the room by himself and left by himself. "Cliff Evans," I muttered to myself, "a boy who never talked." I thought a minute. "A boy who never smiled. I never saw him smile once."

The big ranch kitchen was clean and warm. I blurted out my news somehow. Mrs. Evans reached blindly toward a chair. "He never said anything about bein' ailing."

His stepfather snorted. "He ain't said nothin' about anything since I moved in here."

Mrs. Evans pushed a pan to the back of the stove and began to untie her apron. "Now hold on," her husband snapped. "I got to have breakfast before I go to town. Nothin' we can do now anyway. If Cliff hadn't been so dumb, he'd have told us he didn't feel good."

After school I sat in the office and stared bleakly at the records spread out before me. I was to close the file and write the obituary for the school paper. The almost bare sheets mocked the effort. Cliff Evans, white, never legally adopted by stepfather, five young half-brothers and sisters. These meager strands of information and the list of D grades were all the records had to offer.

Cliff Evans had silently come in the school door in the mornings and gone out the school door in the evenings, and that was all. He had never belonged to a club. He had never played on a team. He had never held an office. As far as I could tell he had never done one happy, noisy kid thing. He had never been anybody at all.

How do you go about making a boy into a zero? The grade school records showed me. The first and second grade teachers' annotations read "sweet, shy child," "timid but eager." Then the third grade note had opened the attack. Some teacher had written in a good, firm hand, "Cliff won't talk. Uncooperative. Slow learner." The other academic sheep had followed with "dull"; "slow-witted"; "low I.Q." They became correct. The boy's I.Q. score in the ninth grade was listed at 83. But his I.Q. in the third grade had been 106. The score didn't go under 100 until the seventh grade. Even shy, timid, sweet children have resilience. It takes time to break them.

I stomped to the typewriter and wrote a savage report pointing out what education had done to Cliff Evans. I slapped a copy on the principal's desk and another in the sad, dog-eared file. I banged the typewriter and slammed the file and crashed the door shut, but I didn't feel much better. A little boy kept walking after me, a little boy with a peaked, pale face; a skinny body in faded jeans; and big eyes that had looked and searched for a long time and then had become veiled.

I could guess how many times he'd been chosen last to play sides in a game, how many whispered child conversations had excluded him, how many times he hadn't been asked. I could see and hear the faces and voices that said over and over, "You're dumb. You're dumb. You're nothing, Cliff Evans."

A child is a believing creature. Cliff undoubtedly believed them. Suddenly it seemed clear to me: When finally there was nothing left at all for Cliff Evans, he collapsed on a snowbank and went away. The doctor might list "heart failure" as the cause of death, but that wouldn't change my mind.

We couldn't find ten students in the school who had known Cliff well enough to attend the funeral as his friends. So the student body officers and a committee from the junior class went as a group to the church, being politely sad. I attended the services with them, and sat through it with a lump of cold lead in my chest and a big resolve growing through me.

I've never forgotten Cliff Evans nor that resolve. He has been my challenge year after year, class after class. I look up and down the rows carefully each September at the unfamiliar faces. I look for veiled eyes or bodies scrouged into a seat in an alien world. "Look, kids," I say silently, "I may not do anything else for you this year, but not one of you is going to come out of here a nobody. I'll work or fight to the bitter end doing battle with society and the school board, but I won't have one of you coming out of here thinking himself into a zero."

Most of the time—not always, but most of the time—I've succeeded.

QUESTIONS 2.3

1. How does the self-fulfilling prophecy apply in this situation?

2. Who were the agents of socialization in this story? Why did they define Cliff Evans as they did?

Reading 2.4

BORN TO RAISE HELL?

Dennis Overbye

Which is more important—social environment or biology? The nature versus nurture debate continues. Clearly, biological factors are an important part of who we are. For example, we see the biological stamp on such factors as skin and eye color and perhaps on susceptibility to certain diseases. However, sociologists believe that behavior is not biologically determined but is a product of one's social environment, past experiences and relationships, and the socialization process. Dennis Overbye gets involved in the debate and examines some of the implications of assuming that biology explains behavior.

Some of us, it seems, were just born to be bad. Scientists say they are on the verge of pinning down genetic and biochemical abnormalities that predispose their bearers to violence. An article in the journal *Science* last summer carried the headline EVIDENCE FOUND FOR A POSSIBLE "AGGRESSION" GENE. Waiting in the wings are child-testing programs, drug manufacturers, insurance companies, civil rights advocates, defense attorneys and anxious citizens for whom the violent criminal has replaced the beady-eyed communist as the boogeyman. Crime thus joins homosexuality, smoking, divorce, schizophrenia, alcoholism, shyness, political liberalism, intelligence, religiosity, cancer and blue eyes among the many aspects of human life for which it is claimed that biology is destiny. Physicists have been pilloried for years for this kind of reductionism, but in biology it makes everybody happy: the scientists and pharmaceutical companies expand their domain; politicians have "progress" to point to; the smokers, divorcés and serial killers get to blame their problems on biology, and we get the satisfaction of knowing they are sick—not like us at all.

Admittedly, not even the most rabid sociobiologists contend that babies pop out of the womb with a thirst for bank robbing. Rather, they say, a constellation of influences leads to a life of crime, among them poverty, maleness and a trait known as "impulsivity," presumably caused by bad brain chemistry, caused in turn by bad genes. What, you may ask, is impulsivity? The standard answer tends to involve people who can't control their emotions or who get into bar fights. A study conducted in Finland found that men so characterized tend to be deficient in the brain hormone serotonin—one of several chemical messengers that transmit signals between nerve cells. In another study researchers found that the men in a Dutch family with a history of male violence seemed to lack the ability to break down certain neurotransmitters, including serotonin, that build up in the brain during flight-or-fight situations. Couple this with persistent statistical surveys purporting to show that criminals tend to run in families and you have the logic behind the Violence Initiative dreamed up a couple of years ago by the Department of Health and Human Services, which included research to discover biological markers that could be used to distinguish violence-prone children as early as age five. Any doubts about the potential for abuse in such a program were erased when Frederick Goodwin, then director of the Alcohol, Drug Abuse and Mental Health Administration, lapsed into a comparison of inner-city youth to murderous oversexed monkeys during a speech about the initiative. "Maybe," he said, "it isn't just the careless use of the word when people call certain areas of certain cities jungles . . ." Amid the ensuing outcry, the National Institutes of Health canceled its financial support for a planned conference on the biology of violence. Now, however, bolstered by a cautious approval from the National Academy of Sciences, the conference is back on, and Goodwin is head of the National Institute of Mental Health.

Science marches on. Or does it? The whole affair is uncomfortably reminiscent—as the scientists admit—of the 1960s, when researchers theorized that carriers of an extra Y, or male, chromosome were predisposed to criminality, or of earlier attempts to read character from the bumps on people's skulls. As any doctor who ever testified for a tobacco company in one of those trials knows, the statistical association of two things, like, smoking and cancer, or guns and murders, does not necessarily imply cause and effect. We know too little about the biochemical cocktail that is the brain, and all too much about how stigmatized children live down to our expectations.

Being a great fan of science, I'm all in favor of more research, testing, poking, drugging, jailing, genetic therapy, amniocentesis, just as soon as someone can give me a scientific definition of impulsivity, one that provides, say, a cultural- and color-blind distinction between a spirited child and an impulsive one. If bar fights are the criteria, it's nice to think that a simple blood test might have spared us the antics of people like Billy Martin or George Steinbrenner.

My real complaint is that the violence initiative doesn't go nearly far enough. Some laboratory should be looking for the racism gene, or the homophobia gene. Goodwin was right; the inner city is a jungle. But so are the corporation, the newsroom and the White House staff. The language of trial lawyers or bond traders in full testosterone fury is as blood-curdling as any mugger's. When it comes to social carnage, the convenience-store stickup can't compare with a leveraged buyout, trickling-down unemployment, depression, anger, alcoholism, divorce, domestic abuse and addiction. I'd like to see white men with suspenders and cellular phones tested for the greed gene. The genomes of presidential candidates should be a matter of public record.

I'm a middle-aged white man as afraid as anyone else of being not quite alone on a dark New York City street, and I've stared into the stone-cold eyes of a mugger while he told me, "It's just you and me." When you hear those words, it's too late to fight the war on crime. The rush to define criminals as sick obscures an uncomfortable truth about our society, which is that crime and violence often pay handsomely. Just ask the conquistadores, the Menendez brothers, Oliver North or the comfortable and respected descendants of bootleggers and slaveholders. Ask the purveyors of the most violent television program in recent memory: the Gulf War.

QUESTIONS 2.4

1. List a number of your own characteristics. Now try to assess their origin— are they (B)iological (I was born this way), or (E)nvironmental (this was a product of my life experiences).

2. If there is an aggression gene, there must be a greed gene. Argue this statement pro and con.

3. The author suggests that the search to discover biological markers has the "potential for abuse." What might he have in mind?

NOTES

1. This section is summarized from "Final Note on a Case of Extreme Isolation," by Kingsley Davis, *American Journal of Sociology* 52 (1947), pp. 432–437. Also see Bruno Bettelheim, "Feral Children and Autistic Children," *American Journal of Sociology* 64 (March 1959), pp. 455–467.

2. Edward C. Wilson's books include *Sociobiology: The New Synthesis* (Cambridge, Mass.:

Harvard University Press, 1975) and *On Human Nature* (Cambridge, Mass.: Harvard University Press, 1978). His newer book devotes much more attention to human behavior than his earlier book did. Also see *Society* 15 (September/October 1978). The whole issue is on sociobiology. The quotation cited in the text comes from p. 10 of the *Society* issue.

3. Quite a bit of research has been done on this topic. For some of the most recent, see the two lead articles on confluence theory in the *American Sociological Review* 56 (April 1991), pp. 141–165.

4. This study is described in "Are Women Prejudiced Against Women?" by Philip Goldberg, *Transaction* 5 (April 1968), pp. 28–30.

5. This discussion follows that contained in *The Social Construction of Reality,* by Peter Berger and Thomas Luckmann (Garden City, N.Y.: Doubleday, 1966), especially pp. 129–147.

6. See Arthur Niederhoffer, *Behind the Shield: The Police in Urban Society* (Garden City, N.Y.: Doubleday, 1967).

7. Numerous books and articles on prison life describe the language or argot of inmates. For example, see Gresham Sykes, *Society of Captives* (New York: Atheneum, 1958). The example I have used here comes from *The Felon,* by John Irwin (Englewood Cliffs, N.J.: Prentice-Hall, 1970), pp. 66–85. Also Peter Letkemann, *Crime as Work* (Englewood Cliffs, N.J.: Prentice-Hall, 1973).

8. *The Brain Benders,* by Charles Brownfield (New York: Exposition Press, 1972), has an interesting description and analysis of brainwashing. Brownfield deals with the actual occurrence in wartime and with scientific experiments. The example cited here comes from p. 55.

9. Charles Horton Cooley's books are *Human Nature and the Social Order* (New York: Charles Scribner's, 1902) and *Social Organization* (New York: Charles Scribner's, 1909).

10. George Herbert Mead's writings include *Mind, Self, and Society* (Chicago: University of Chicago Press, 1934) and "The Genesis of the Self and Social Control," from *International Journal of Ethics* 35 (April 1925), pp. 251–273.

11. This discussion on the importance of language and the Sapir-Whorf school of thought draws in part from *Social Psychology,* rev. ed., by Alfred Lindesmith and Anselm Strauss (New York: Holt, 1956), ch. 8.

12. This discussion of definition of the situation and self-fulfilling prophecy follows that of Robert Merton in *Social Theory and Social Structure,* 2d ed. (New York: Free Press, 1957), pp. 421–436. The quote is from p. 128.

13. The study is first reported in "The Effect of Experimenter Bias on the Performance of the Albino Rat," by Robert Rosenthal and Kermit Fode, *Behavioral Science* 8 (July 1963), pp. 183–189. The discussion here is summarized from another report, "Teacher Expectations for the Disadvantaged," by Robert Rosenthal and Lenore Jacobson, *Scientific American* 218 (April 1968), pp. 19–23. Also see Rosenthal and Jacobson's book *Pygmalion in the Classroom: Teacher Expectation and Pupils' Intellectual Development* (New York: Holt, Rinehart & Winston, 1968).

14. We mentioned in Chapter One the importance of verification in science, and the Rosenthal study provides an interesting sidelight in this regard. When the study appeared in 1968 it drew wide interest. Then, a series of critical reviews appeared that took Rosenthal to task for the research methods he used in his study. Next, numerous attempts were made to repeat Rosenthal's study on other samples. Some researchers have been able to repeat his findings and conclusions, but many have not been able to do so, and the debate goes on about the original study. If you would like to get involved in the discussion, see Glen Mendels and James Flanders, "Teachers' Expectations and Pupil Performance," *American Educational Research Journal* 10 (Summer 1973), pp. 203–212, which describes a replication and cites a number of other attempts and critiques. It should be mentioned that the differences between "academic spurters" and others described in the text were distinct for the first and second graders but were mixed or nonexistent in grades three through six.

15. See "A Conversation With Jerome Kagan," *Saturday Review,* March 10, 1973.

CHAPTER THREE

Norms, Roles, Culture, Society

TWO RESEARCH QUESTIONS

1. For every 100 girls born worldwide, between 105 and 106 boys are born. But an official Chinese fertility survey in 1988 showed gender ratios of up to 120 boys per 100 girls among children born in China between 1980 and 1987. Why did the Chinese suddenly give birth to more boys than girls? How can this be explained?

2. Some people are involved in many different jobs, activities, or positions in society; other people are more focused, are committed to one job or position. Would their approach to life, as indicated by the number of activities in which they are involved, have any effect on how long they live? Would the more focused live longer, would those with more varied lives live longer, or would their life span be unrelated to style?

IN THE PREVIOUS chapter we focused on how individuals develop into social beings. We discussed the socialization process, the development of the self and personality, and the importance of language. Our analysis was detailed, and it concentrated on the individual's growth from a helpless infant to a person capable of participating in complex social situations. The focus, again, was on the individual. In this chapter, we move away from the individual to the *stage* or *setting* in which the processes described in the previous chapter take

Elliott Erwitt/Magnum Photos Inc.

place. We could call this stage the *social environment*. Within this environment, individuals live in specific societies and cultures, they occupy positions, they play parts, they are governed by rules, and they behave in predictable ways. Concepts like *culture, norm, society,* and *role* help us better understand and explain what happens on this stage.

THE MEANING OF CULTURE

The importance of culture in sociology is comparable to the importance of evolution in biology, of gravity in physics, and of disease in medicine. It is a central concept that allows us to organize and explain what we see. Much of human behavior can be understood and indeed predicted if we know a people's culture, their design for living.

For social scientists, the concept of culture is very important. By **culture** we mean that complex set of learned and shared beliefs, customs, skills, habits, traditions, and knowledge common to the members of a society. Culture is viewed as the social heritage of a society. According to anthropologist Clyde Kluckhohn, culture represents the distinctive way of life of a group of people, their complete design for living.[1]

There are many cultures in the world, some similar to each other and some very different from most others. A culture evolves through time, it is continually modified, and its complexity reflects its sources in human creativity. A culture is shared by the members of a society and is learned through the socialization process. The culture in which we live determines for us what we will want to eat, whom we will like or hate, what we will fear (snakes and mice, but probably not evil spirits), how we will express our emotions, how we will dress (blue jeans and bikinis, but not turbans or saris), what types of manners we will develop, and how we will celebrate New Year's Eve. I mentioned earlier that when I throw a volleyball to my son, he is likely to catch it with his hands and throw it back, whereas a French or English teenager is likely to head or kick the ball back to me. We can now see that these are *cultural* differences. In American culture, ball sports emphasize the use of hands—catching and throwing or hitting in basketball, baseball, volleyball, and handball, for example. In France and England and in many other cultures, the major sport is soccer, which emphasizes controlling the ball with one's feet and head.

When the details of a culture are examined, a distinction can be made between material and nonmaterial aspects. **Material culture** refers to the concrete things that a society creates and uses: screwdriver, house, classroom, desk, car, plane, telephone. **Nonmaterial culture** refers to the abstract creations of a society: customs, laws, ideas, values, beliefs. The nonmaterial culture would include beliefs about religion and courtship, ideas about democracy and communism, definitions of good manners, and rules for driving a car. One aspect of the nonmaterial culture that we don't usually think much about are the rules for conduct—what we are supposed to do in specific situations. We call these *norms.*

NORMS

A sociology instructor is lecturing to a class of college freshmen and sophomores. At one point in her lecture, she asks a student to help her illustrate a point by standing up. The student is belligerent and upset at being interrupted in his reading of the school newspaper. The student replies that he would rather not help and that the teacher should ask someone else. The teacher asks the student once more to please help her with the experiment and to stand up. The student replies in a louder voice that he doesn't think much of this stupid class or the teacher and that he is not about to participate in some silly experiment. The now-angry teacher responds, "Either stand up or get out of this classroom!" The student replies, "I'm not standing up—and if you want me out of here you'll have to throw me out!" The two then glare at each other for what seems like a year.

If we were watching the scene and could take our eyes off the principal characters, we would notice that the rest of the class is behaving strangely. Nearly all appear to be very uncomfortable, and many are obviously embarrassed. They are twisting and squirming and are avoiding meeting the eyes of the instructor. At the beginning of the confrontation some may have laughed nervously; now all are quiet. They act as if they are trying to deny that what is happening is actually happening. . . . Why?

All societies have rules that specify what people should do in specific situations. Sociologists call these shared behavioral standards **norms.** Norms describe the accepted or required behavior for a person in a particular situation. For example, when a person gets into a car, that person unconsciously begins to behave according to a whole set of norms relating to driving procedure. In America the driver knows that he or she should drive on the right side of the road, pass on the left, signal before turning, and so on. The norms vary from place to place, but they always exist. These shared standards for behavior allow us to predict what other people will do. Without even thinking about it or knowing the individual, I know how the driver of a car approaching me will behave. The system of norms allows us to predict what other people will do in specific situations and to pattern our own behavior accordingly.

Where do norms come from? Like much of what we discuss in this book, norms emerge through the process of social interaction. As people interact, they come to understand that certain ways of behaving are proper and acceptable and that other ways are improper and unacceptable. These general opinions and beliefs are called **values.** Values represent the broad perspectives people carry regarding usually intangible concepts such as patriotism, progress, equality and democracy, and the importance of telling the truth. Although norms are based on values, the difference between the two concepts is important: Values are more broadly based and represent general perspectives, whereas norms represent specific rules for particular situations. Both norms and values can be long-standing, born of custom and tradition, or they can be of shorter duration.

Norms take many forms, from written laws to informal agreements among group members as they try to complete a task or to manage emotion and interaction among members. Superstitions, customs, and myths may act as norms: Many buildings don't have a 13th floor; people take care not to walk under ladders or on cracks in the sidewalk; alcohol consumption is much higher around some holidays; and June is the month in which to get married (which in turn probably influences the sale of rice). China's population grew by about 15 million people in 1991. This is obviously quite a lot, but it's not nearly as many as expected and 1.25 million *less* than the increase in 1990. Amazingly, the birth rate in 1991 was even lower than the government's planned quota. Why the decline? No major external incident—for example, disease, war, or drastic change in the standard of living—seems to be responsible. It appears that even China's severe family planning laws were not the cause. Actually, the answer seems to be superstition! 1991 was the Year of Goat, and many rural families deliberately avoided starting families in a year considered to be unlucky. 1992, however, is the highly auspicious Year of Monkey. A bright, favorable future can be forecast for babies born in such a year. Many experts predict that the official records, when they become available, will show a significant increase in births, perhaps more than 24 million babies for 1992.

It is disturbing when the normative system breaks down and one can't predict behavior. This may explain the strange behavior of the classroom students described in the hypothetical example that began this section. We have norms that govern the student-teacher relationship. These norms are so obvious and ingrained that we do not even think about them. Characteristically, norms do not become apparent until they are violated. According to norms concerning student-teacher relationships, the student will show respect for the teacher, will laugh at his or her jokes, and will generally obey the authority that the teacher represents; that is, students will respect any reasonable demands made by the teacher without resorting to mutiny. When, as in our previous example, behavior is contrary to the norms, the situation becomes uncomfortable for others. Instructors who have tried this "stand up" experiment on classes (with the aid of a willing villain) report that the experience is almost as hard on the two of them (instructor and villain), who know it's a fake, as it is for the rest of the class, who don't know that it has been set up. Thus even planned and legitimate norm violation may be hard on individuals when done in the context of a group that is unaware of the new definitions.

Other experiments in norm violation have been devised.[2] For example, get somebody you don't know very well in a game of ticktacktoe. Invite the other person to make the first move. After he or she makes a mark, erase it, move it to another square, and make your own mark. Act as if nothing unusual has happened. Or select a person (not a family member or very close friend) with whom you will engage in an ordinary conversation. During the course of the conversation, without indicating that anything unusual is happening, bring your face closer to the person's until your noses are almost touching. The first experiment deals with norms concerning game playing, and the second with norms concerning spatial invasion, or the appropriate distance between people. In each case, the subject will react noticeably, and possibly unpredictably, because common norms that ordinarily one doesn't even think about are suddenly being violated.

Norms continually change and may even break down completely. The most likely time to see norms breaking down is during periods of rapid social change, such as during wars or disasters. When norms cease to be effective at controlling people's behavior (when they don't "work" any more), we say a state of "normlessness" is present. One tends to lose all sense of stability, security, and orderliness when this happens. Imagine the consternation if we did not know what to expect of an oncoming driver—if we had *no* idea what he or she was going to do.

Types of Norms

Norm strength varies greatly. One way of determining the relative strength of norms is by the sanctions that the norms carry. A **sanction** is the punishment one receives for violation of a norm, or the reward one receives for correct norm performance. Sanctions take a variety of forms: a look on someone's face, an A on an exam, a sharp word from a spouse or boss, a kind word from a parent, a traffic ticket from a police officer, a life sentence in a penitentiary. The norms that are most important to a society, that tend to be obeyed without question, and that have harsh sanctions if they are violated are called **mores.** In our society, norms dealing with taking another person's life, with eating human flesh, and with sexual activity with one's parents are examples of mores. Mores are often traditional norms that are a part of the customs of a society—"things have always been that way." When a society feels it necessary, perhaps because increasing violations mean that the informal sanctions aren't working, mores may be translated into written law. Whether written or un-written, the emotional force of mores is strong; they still represent the "musts" of behavior.

We also have norms dealing with what we *should* do, rather than what we *must* do. These norms are less obligatory than mores, and the sanctions for violation are milder in degree. People may look at us rather strangely if we violate these norms, but they probably will not lock us up or banish us. These norms are called **folkways.**[3] Practices such as shaking hands when meeting someone or norms regulating the type of clothes one wears would be consid-ered folkways. A student who comes to class in a bathing suit and golf shoes would probably be allowed to stay in spite of the obvious violation of the folkways. At the same time, there would likely be ample private discussion of the student's character and intelligence.

A further distinction should be made between ideal norms and statistical norms. **Ideal norms** refer to what people agree *should* be done. **Statistical norms** refer to what they *actually* do. Ideal norms indicate that cheating on tests, extramarital sexual intercourse, and falsifying one's income-tax records are wrong. However, statistical norms—what people are actually doing—would indicate that a substantial number of people are performing these ac-tivities. Often, amazingly enough, these are the same people who previously told us these things were wrong. It shouldn't surprise us that there may be quite a discrepancy between talk and action—or, "Do as I say, not as I do. . . ."

Norm Conflict

As we said earlier, norms develop out of a set of beliefs and values people have about the way things should happen. Norms probably develop first at the group level when specific friendship groups, clubs, work groups, clans, organizations, or communities of people establish rules. Often these rules are informal and unwritten. Sometimes, if the people involved have sufficient power, their rules may become written down in the form of laws.

Norms vary in how widely they are supported. For example, norms dealing with taking a human life, stealing property, or eating human flesh are proba-bly looked at in the same way by most people in United States society. On the other hand, because norms emerge from groups and because each group has its own interests and viewpoints, we shouldn't be at all surprised to see norm conflict—the norms of one group conflicting with the norms of another. People who have been taught the norm of not killing other people may find themselves as soldiers being trained to kill. The norms of the consumer often conflict with the norms of the manufacturer. Car owners want a vehicle that runs cheaply, looks beautiful, is safe, and won't fall apart in the first six months. The manufacturer is interested more in reducing costs per unit and cutting corners where possible. One would hope that the producers of the food you eat are as concerned about its purity and wholesomeness as you are when you eat it. Perhaps they're not, however; consider that laws and agencies

OBEDIENCE AND DISOBEDIENCE TO AUTHORITY

Social psychologist Stanley Milgram conducted a fascinating series of experiments on what happens when people are faced with conflicting norms. A subject is told by the experimenter to teach a learner a task, to test the learner, and to punish the learner if the learner makes mistakes. The subject punishes the learner by administering electric shocks. Shocks are given by a special machine controlled by the subject. Shocks are increased in intensity (marked from "Slight Shock" to "Danger: Severe Shock") with each wrong answer. The learner continually makes mistakes, continually gets shocked, and pleads with the subject to stop the shocking: "I can't stand the pain," "I have a heart condition," "Get me out of here. ... I refuse to go on!" then shrieks of agony, and finally silence. The subject is reluctant to keep shocking but the experimenter tells the subject to continue. In reality, the experimenter and the learner are working together; the shocks, and the cries and screams are fake, but the subject doesn't know this. The subject has a problem, a norm conflict: Should the subject follow the orders of the official-looking white-coated experimenter, or heed the cries of the poor suffering learner? What would you do?

It turns out that the subjects were much more obedient to the experimenter than Milgram had imagined they would be. He had guessed that few would go beyond "Strong Shock," and yet in practice, many subjects were willing to give the most extreme shocks available. Milgram then added conditions to see if the degree of obedience would change. In one condition (Remote Feedback) the learner was in another room and couldn't be seen or heard by the subject, but as the shocks got higher the learner pounded on the wall. In a second condition (Voice Feedback) the learner couldn't be seen but could be heard. In a third condition (Proximity) the learner was in the same room, about two feet away from the subject. In a fourth condition (Touch-Proximity) the learner had to put his hand on a shock-plate, and if he refused, the subject had to force his hand on the plate. As you might guess, the closer the subject was to the learner, the more trouble the subject had going through with the shockings. But even in the last two conditions, 30 to 40 percent of the subjects remained obedient. Milgram also tried moving the authority figure, the experimenter, out of the room to see if that would have an effect. Obedience dropped off tremendously with the experimenter gone, and many of those subjects who remained obedient, cheated; they gave lower shocks than they were supposed to.

What does it all mean? Milgram suggests several reasons for why the subjects were as obedient as they were. For some subjects, perhaps the shocking provides a release for aggressive feelings. From a sociological viewpoint, the subjects became entangled in a complicated social situation from which they had trouble escaping. There were conflicting norms: Who should be heeded, the authority figure or the person in pain? Milgram somewhat pessimistically concludes: "A substantial proportion of people do what they are told to do, irrespective of the content of the act and without limitations of conscience, as long as they perceive that the command comes from a legitimate authority."

Excerpt from "Some Conditions of Obedience and Disobedience to Authority," Human Relations, *February 1965, pp. 57–76.*

(Food and Drug Administration, for example) are necessary to *force* food producers to maintain standards. So, because different groups have conflicting viewpoints and motivations, norm conflict is bound to occur.

Examples of conflict over norms are not hard to find. There seems to be almost constant conflict over norms dealing with sexual conduct. Prostitution is not unknown in our society, although it is against the law nearly everywhere. Groups upset by the existence of prostitution get together to stamp it out. Other groups get together to defend prostitution. Police agencies often don't know whether to attempt to enforce laws prohibiting prostitution or to ignore its existence. For a time, narcotics laws defined possession and use of marijuana as a serious crime. In some parts of society, a different set of norms existed that didn't view marijuana as dangerous and encouraged its use. The energy crisis and the related efforts to conserve fuel in 1973 and 1974 led to a law lowering the speed limit to 55 miles per hour. The law was difficult to enforce, and finally it was modified.

Society's laws usually reflect its mores, but sometimes they don't. As we have said, it is natural for various segments of society to disagree on norms. Occasionally, however, a special-interest group gains enough power to manipulate laws. Also, sometimes the change in norms is rapid, and the change in laws does not keep pace. When either of these situations occurs, laws may not reflect the mores of society. Usually, when laws and mores are in conflict, mores will win in the long run. The prohibition laws in the 1920s and 1930s apparently did not reflect the mores of the country, and the laws were flagrantly violated until they were changed. What often happens when there is conflict between norms and laws is that low priority is given to enforcement of existing laws. This won't resolve norm conflict, but it represents an attempt to achieve a compromise between conflicting viewpoints.

CULTURAL DIVERSITY

When we compare different cultures we find similarities and differences. All cultures seem to have some form of family arrangement, and they all have some system for educating their young. I suppose that cultures are more similar than they are different. But the differences attract our attention and are often fascinating. Why are there differences? A number of factors are probably at work. Cultural patterns may reflect responses to a harsh or benevolent physical environment (very wet or dry, agriculturally rich or poor, desert or mountain). Cultural patterns may be affected by accidents or variations in history—revolutions, wars, overpopulation, starvation, inventions, stage of development of neighbors, and so on. Finally, and perhaps most important, cultural patterns are a consequence of the almost random variation that emerges out of human creativity. As we look at a few examples of cultural diversity, try to figure out what produced these patterns.

Every four years, writers from England and Europe flock to the United States to observe a very peculiar ritual the like of which is unknown throughout the rest of the civilized world. These strange ceremonies are written up and read about by disbelieving audiences everywhere. The tribal celebrations being enacted are called locally the "American political conventions." Strange indeed.

Imagine that you are dropped onto a small South Sea island, and your only possession is a golf club—a two iron. To avoid being eaten alive you have to explain the function and purpose of the golf club to the natives. Where do you start?

Anthropologists specialize in the study of cultural patterns in different societies, especially in primitive or preliterate societies, and in their studies they have made many discoveries that seem strange to Americans. They describe a society in which very fat women are highly regarded. Women in this society spend weeks in fattening sheds where they eat starchy, fatty foods and have their bodies greased to make them more attractive. On festival day they are paraded before the king, who chooses the fattest and heaviest as his mate. Very peculiar people. But imagine for a moment that you have to explain to a member of that society the popularity in America of the various dieting and health spas where men and women spend great sums of money to lose weight.

Some Indian tribes living along the Amazon have an interesting reaction to childbirth. The woman breaks off from work in the fields and returns home for only two or three hours to give birth to the child. Meanwhile, her husband has been lying at home in a hammock, tossing about and groaning as if in great pain. Even after the birth when the woman has returned to the fields, the husband remains in bed with the baby to recuperate from his ordeal. It appears that pain is determined by something more than the nature of the wound.

The first research question at the beginning of this chapter described a curious circumstance concerning the gender ratio in China—many more boy babies or many fewer girl babies were born during the 1980s than should have been the case. Demographers (those who study population patterns) are still working on this one, but the answer seems to be related to two factors. In 1979 the Chinese government, in attempts to control rapid population growth, introduced a strict policy limiting Chinese families to one child. The second factor, clearly a cultural one, was the desire of Chinese parents to have a boy. In that culture, a male baby is much preferred to a female baby. If the family was allowed to have only one child, it was very important that it be a boy. Birth statistics and the peculiar gender ratio implied that between 1980 and 1987 about 2.5 million female births were never reported. The experts first concluded that because boy babies were desired, girl babies were killed (female infanticide) or their births somehow hidden and not recorded. Perhaps that's what happened, but Swedish statistician Sten Johansson thinks there is another answer. He studied adoption records and found that the vast majority of children adopted during that time were girls. He concludes that more than 1 million girls were, in some way, left by their natural parents and adopted, some secretly and some openly, by foster parents. By the way, is Chinese culture alone in its preference for boy babies?

Ethnocentrism and Cultural Relativism

All cultures differ to some extent, and yet because we are so used to our own, we often forget that basic fact. The tourist is reminded of this even in countries similar to our own when, for example, he or she encounters the English motor-

A REPORT FROM THE GULF

Though dressed modestly in civilian clothes, her arms and head covered, a U.S. Navy lieutenant was stopped in a Riyadh [Saudi Arabia] market by a member of the Saudi Religious Police and berated for not having her face veiled, and hit with a stick. When she protested that she was an American service-woman, the policeman scolded her in English for having violated Saudi morality, and might have hit her again if she hadn't ducked into a store and phoned for help.

Reported by The New Republic, *January 28, 1991.*

ist driving on the left side of the road, the German driver approaching at a very high speed with lights flashing, Volkswagens and Porsches used as police cars, Spanish restaurants not opening for dinner until after 9:30 in the evening, French children drinking wine, or English police officers not carrying guns. The failure to anticipate and plan for cultural differences is common. Professor David Ricks has found that companies doing business in foreign countries make the same mistake. Chevrolet had trouble selling Novas in Latin American markets: *"No va"* in Spanish means "does not go." Pepsodent's promise of white teeth did not go over well in a part of Southeast Asia where black, discolored teeth are a sign of prestige. "Body by Fisher" translated to "Corpse by Fisher" in a Belgian ad. "Come alive with Pepsi" in a German translation became "Come alive out of the grave." Some firms forgot when labeling their products that green is the color of disease in Africa and white the color of death in Japan. And a firm had difficulty selling a refrigerator to the mostly Moslem Middle East when it used an ad picturing the refrigerator full of food, including a giant ham on the middle shelf. Some other examples of odd notices in non-English-speaking countries include: a Hong Kong dentist—"Teeth extracted by the latest Methodists"; a Greek tailor shop—"Order your summers suit. Because is big rush we will execute customers in strict rotation"; a Mexican hotel—"The manager has personally passed all the water served here."

Ethnocentrism describes a type of prejudice that says simply, my culture's ways are right and other cultures' ways, if they are not like mine, are wrong. Racism (a particular race is superior to others) and sexism (one sex is superior to another) are related to ethnocentrism but are more specific types of prejudice. Informal practices and formal policies (who gets hired, who gets paid most) may emerge to support racism, sexism, and ethnocentrism, as well as other types of prejudice. The ethnocentric person says that the familiar is good and the unfamiliar or foreign is bad. An ethnocentric person in the United States might maintain, among other things, that non-Christians are barbarians; that Inuit tribes practicing sexual hospitality are totally lacking in moral fiber; that anybody who eats dogmeat, horsemeat, or human flesh is not civilized; that democracy is the only way of government; and generally that we are doing other cultures a favor when we go in and Americanize them.

Ethnocentrism is, to some degree, difficult to avoid. Informally, while interacting with family and friends, and formally through a kind of indoctrination within the educational system, we are frequently taught that our ways are best and, at least by implication, the ways of others are less good. The mass media encourage ethnocentrism by treating the foreign and unfamiliar in terms of easily recognized stereotypes. It is probably true that ethnocentrism is impossible to escape, for it is encouraged in one way or another by most societal institutions, such as the family, the church, the schools, and the government. This is at least partly due to the positive functions of ethnocentrism. Ethnocentrism probably leads to greater group solidarity, loyalty, and patriotism, and a certain degree of ethnocentrism may be essential for the survival of a culture. The effects of ethnocentrism are complicated: As we reinforce our belief in the goodness of our own ways, we make unfair and often derogatory judgments about the beliefs of others.

Even social scientists (who know better) sometimes run into difficulties when studying other cultures. Because most social scientists are white middle-class representatives of the dominant culture, they may have a tendency to describe minority cultures that are different—Amish, Inuit, delinquent gangs—from the viewpoint of their own value system rather than from the viewpoint of the people they are studying. As you come across descriptions of research in this book and elsewhere, see if the research seems to have been influenced by ethnocentrism. See if you can detect ethnocentrism in the choice of topic, in the way the research is done, or in the interpretation of results.

Related to ethnocentrism but opposite in meaning is the concept of **cultural relativism.** Cultural relativism suggests that each culture be judged from its own viewpoint without imposing outside standards of judgment. Behaviors, values, and beliefs are relative to the culture in which they appear. The cultural relativist believes that what is right in one society may be wrong in another and that what is considered civilized in one society may be seen as barbaric in another; however, the relativist believes basically that judgments should not be made about the "goodness" or "badness" of traits in cultures other than in one's own. Although cultures share many values, beliefs, and behaviors, differences between cultures do exist. When one evaluates particular traits, one should do it from the viewpoint of the society in which they appear.

Subculture and Counterculture

Most societies, especially large, complex societies like the United States, have groups that, by their traits, beliefs, or interests, are somewhat separated and distinct from the rest of society. Such groups may share many of the characteristics of the dominant culture, but they have some of their own specific customs as well. If these groups have definite boundaries and if their differences from the rest of society have some permanence, they are called **subcultures.** Sociologists generally use the term *subculture* to refer to groups that stand out in that some of their values and customs are different from or even at odds with those of the rest of society. The sociologist asks such questions as: How do the subculture's values differ from those of the dominant culture? Is there conflict? Does the dominant culture attempt to change the subculture? What does the subculture do to maintain its separate identity?

Some religious groups, the Amish for example, seem to qualify as subcultures. A major problem for the Amish, as for many subcultures, is to maintain their identity, even their existence, in the face of a dominant culture frequently hostile to their beliefs. The opposition of the Amish to electricity, which means no lights at night on their horse-drawn carriages, and their opposition to any formal education beyond the eighth grade has brought well-publicized confrontations with the authorities in several midwestern states. Other examples of subcultures might include terrorist organizations, college students, Hare Krishnas, Chicanos, the Hutterites, professional baseball or football players, and sociologists.

The line of distinction between culture and subculture is not always clear. Some believe that over a period of time societies go through a melting-pot experience in which different nationality, racial, and interest groups become so mixed and merged together that their subcultural differences cease to exist. The result of the melting-pot experience would be a society representing a mix of the remnants of former subcultures.

There is an element of conflict in some subcultural behavior, and this has led to the use of the term **counterculture.** The central element in a counterculture is opposition to, or conflict with, certain norms and values of the dominant culture. Examples of countercultures might include youth gangs, motorcycle gangs, revolutionary groups, and terrorist organizations. *Subculture* is the more general term, *counterculture* the more specific. Countercultures are subcultures, but only those subcultures that include the element of opposition to the dominant culture could be called countercultures.

Culture and Personality

Personality and temperament vary from one culture to another. In her book *Sex and Temperament,* Margaret Mead describes what she found when she visited three primitive societies in New Guinea. First, Mead describes the mountain-dwelling Arapesh, among whom both the men and women behave in a way that Americans would describe as maternal or feminine. Both parents devote their lives to raising the children. The men are gentle, and there is complete cooperation between both sexes at all times. Next, Mead describes a cannibalistic tribe, the Mundugumor, living along a river. Here both the men and women behave in a way that Americans would describe as masculine. Men and women work in the fields together and are aggressive individualists. Finally, Mead describes the lake-dwelling Tchambuli. In this society, the men behave in a manner that Americans would define as feminine or maternal, and the women are masculine by our standards. The women spend the days fishing and weaving, and they have all the power. The men spend their time dancing in ceremonies, dressing and making themselves up, and engaging in

---

HOW NATIONS PERCEIVE EACH OTHER

Do cultures have particular personalities? A survey conducted by Research International and the Henley Centre for Forecasting (reported in the *London Times,* November 16, 1992) found that personality stereotypes associated with nationalities certainly exist. The purpose of the research, interestingly, was to decide whether companies should play up or disguise the nationality of their products. (For example, a British firm made goods under the Japanese-sounding brand name Saisho to benefit from the Japanese reputation for high-quality products.) In the table below, "HIGH" and "low" describe how particular traits are associated with nations.

	England	Germany	Japan	USA	France	Italy
Successful	low	HIGH	HIGH	HIGH	low	low
Hardworking	low	HIGH	HIGH	low	low	low
Clever			HIGH			
Aggressive				HIGH	low	
Humorous	HIGH	low	low			HIGH
Arrogant			low	HIGH	HIGH	
Boring	HIGH			low		low
Untrustworthy						HIGH

artistic endeavors. The men gossip, quarrel, and become very jealous of each other over the affections of a woman. The women's attitude toward the men is one of kindly tolerance and appreciation; they watch the shows that the men put on.

We tend to believe that some patterns of behavior and temperament are automatically, necessarily related to gender. We may believe that to be male is to behave naturally in a certain way, and that female behavior is innate as well. The results of Mead's research suggest that the relationship between gender and the corresponding behavior and temperament is not necessarily biological, but is determined by the culture in which one lives—as are preferences for body shape and size, taste in food or art, belief in a supreme being, the nature of recreational activity, and many other characteristics.

David Riesman in *The Lonely Crowd* describes the effect of culture on personality in a somewhat different manner. His subject is American character, and he describes how character and personality change as other aspects within the culture change. According to Riesman, three types of social character have been dominant in American society. In earlier years the dominant type of character was *tradition-directed*. In a tradition-directed culture, behavior is carefully controlled. Routine orients and occupies the lives of everyone. Ritual, religion, and custom are dominant. New solutions are not sought, and change is very slow. Later, the *inner-directed* type of character appeared. Inner-directed people are taught early in life to have an inward focus, with emphasis on the self and its needs and gratifications. Other people are not of crucial importance. Individuals may be internally driven toward such ideals as power and wealth; they are encouraged to set their own goals and to be on their own. Their lives are concerned with self-mastery and accomplishment. Finally, and more recently, the *other-directed* type of character appeared. The chief interest for other-directed people is to be liked by other people. According to Riesman, other-directed people have built-in radar systems that search out the reactions and feelings of others so that they may adapt themselves to them. These people are more concerned with conformity, are shallower, friendlier, more unsure of themselves, and more demanding of approval from others. The peer group is all-important, as is the front that one puts up. Riesman believes that these character types result from other changes within the culture, such as population growth or change, changes in economic, industrial, and agricultural techniques, and urbanization. This major point applies to all of us: The culture in which we live plays an important part in determining who we are and how we behave, including patterns of personality and character.

Stanley Milgram designed an experiment to test cultural differences in conformity.[4] He instructed subjects to listen to two tones and then to tell him which tone was longer. Each experiment involved six subjects. Five of the subjects, however, were confederates of the experimenter. The five confederates answered first, and their decisions on the tones were heard by the real subject, who answered last. The confederates were instructed to give wrong answers on about half the trials. What did the subject (number six) do when he heard the five others respond one way and he heard the tones another way? The subject was faced with the clear choice of being independent and giving the right answer regardless of what the others said, or conforming and going along with the group. Milgram tried his experiment on 100 French students in Paris and on 100 Norwegian students in Oslo. He found that the Norwegians were more conforming, the French more independent. The differences weren't great, but they were consistent. Through a number of trials under varied conditions, the Norwegians voted with the group more often than did the French. By way of explanation, Milgram describes Norwegian society as highly cohesive. The Norwegians tend to identify with the group and have a strong sense of social responsibility. The French, on the other hand, show much less consensus in their social and political life and more of a tradition of dissent and critical argument. Milgram's research demonstrates cultural differences in personality characteristics between what many would consider to be very similar nationality groups.

American males have one of the highest heart-disease rates in the world. By contrast, Japanese males have one of the lowest rates. There is obviously a cultural factor at work here, but which cultural factor is it? For years scientists thought the explanation to be diet. The Japanese diet of fish and rice is much lower in cholesterol than is the American diet of meat and dairy products. However, a nine-year study completed in 1975 of 4,000 Japanese Americans living in this country found another cultural factor to be the culprit.[5] The key turned out to be *lifestyle*. The Japanese living in the United States who

THE AMERICANIZATION OF GEORGE

America is coming between me and my 12-year-old son. Actually, it started earlier, when George was hardly 10. Since then, I have helplessly watched this incursion, often with dismay and alarm.

Over two years ago we escaped the horrors of Lebanon, and chose quiet Princeton over the vibrancy of Harvard. . . . But little did I know that he was to face, at such an early age, the more subtle "terror" of American peer pressure, tantalizing communications media, and the unsettling dissonance of conflicting norms and expectations.

The family system in Lebanon is, on the whole, intimate, warm, and affectionate. A child there grows up in a nurturing atmosphere of extended kinship networks sustained by filial piety and mutual obligations.

The Lebanese, much like adjacent Mediterranean cultures, are very tactile. Touching, kissing, hugging, and the outward display of emotion—regardless of gender—are generously and spontaneously expressed. At least children of George's age indulge in these emotive expressions with little self-consciousness or feelings of shame or guilt.

I feel resentful that George should be disarmed of such harmless but reassuring expressions. I first noticed this transformation (or deformation) upon returning from a brief trip a few months after we had settled in Princeton.

Normally, even after the regular daily return home from work, George would interrupt his play and rush across the driveway to greet me; often, he would literally hurl himself into my open arms.

On that day, however, just as he was about to heed his normal impulse as he rushed across the driveway, he suddenly "froze" in mid-passage, looked in the direction of his watchful playmates, and with obvious hesitation and embarrassment, calmly walked over to greet me with a cold handshake and a casual "Hi, Dad." Bit by bit, even this gesture has been abandoned.

Such "frozen" moments have recurred and spilled over to other daily encounters with members of the family, and in particular acquaintances from Lebanon. I could see him fret as relatives and friends he has not seen for two years try, in vain, to solicit a hug or a kiss on the forehead. The reluctant denial has been transformed into a boast, that he is now an "American boy."

His "Americanization" was most forcefully conveyed by a recent incident on the tennis court. We were struggling in a doubles game against two other, more seasoned partners who normally beat us. After a long and heated game we won the set, partly because of two exquisite shots by George. He was ecstatic. As he rushed across to share his exuberance with me Lebanese style, he "froze" once again and treated me to a tamed version of the American "high five."

Samir Khalaf, the Christian Science Monitor, *September 22, 1987.* Reprinted by permission of the author.

---

maintained their traditional lifestyle—downplaying individual competition and accepting their place in family and society—had low heart-attack rates *regardless* of what they ate, how much they smoked, their blood pressure, or their weight. Those Japanese who adopted the American lifestyle (or personality) and became impatient, aggressive, hard-driving, competitive go-getters were five times as likely to have heart attacks as those who maintained Japanese ways. The Japanese culture is apparently better able to protect the individual against the effects of pressure and stress. Unhappily, then, one's culture dictates not only how one will live, but also how and why one will die.

SOCIETY AND SOCIAL STRUCTURE

We learned in Chapter One that sociologists study the origin, development, organization, and functioning of human society. In its most general meaning, the term *society* refers to human association and the existence of social relationships. This is too general for our use, however, because it suggests that any set of people interacting could be called a society. A more precise definition states that a **society** is a continuing number of people living in a specific area who are relatively organized, self-sufficient, and independent and share a common culture. *Continuing* means that there is some permanence to the society. The number of people can vary greatly: American society numbers more than 250 million people, whereas other societies number fewer than 1,000 people. By *organized* we mean that there is some systematization and structure to the patterns of social interaction. A society occupies an area; it has boundaries that separate it from other societies. These boundaries—an ocean, a river, a mountain, or even a line drawn on a map—are not only *geographical* but *social* as well; the vast majority of the social interactions of a society take place within a particular area. Finally, each society, because of its particular cultural beliefs, habits, and traditions, has certain unique characteristics that distinguish it from other societies.

As you can see, several of the terms in our definition involve matters of degree. There is often much interaction among societies, and probably no society is totally independent of others. Although interpretation will always be necessary, this definition should give us a reasonably good idea of what is meant by society. However, other uses of the term don't fit our definition. Special-interest groups such as professional organizations (Society for the Study of Social Problems), political groups (John Birch Society), social clubs (Second Street Skateboard Society), and other groups are *not* societies in the sociological sense. And whatever those people have whose pictures appear in the "society" pages of the newspaper, isn't a society. In fact, none of these groups alone constitutes a society; all of them are *part* of a society.

How do sociologists study human society? Let's suggest for a moment that any given society has a form, a composition, a structure. Imagine society as a building with many parts—wood, steel, bricks, and concrete—all somehow attached together in a coherent structure. If you look at the end product you'll see a five-story building, but thousands of elements actually make up the whole. Yet rather than a group of random parts, the building is a network, a structure. Now keep that image of society as a building: The many parts are people, and they are held together by the ways they relate to each other, their relationships to each other. The people of society, just like the steel and concrete of the building, are held together in very specific ways.

Just as we can take the five-story building apart to examine its elements, so can we take apart and examine society's elements. As we look at the ways people relate to each other, we find that these ways fit consistent patterns. The same important elements seem to appear again and again. **Social structure** refers to the network of ways people relate to each other in society. In their study of human society, sociologists concentrate on the important elements of social structure: status, role, group, and institution. Status and role will be discussed in this chapter; group and institution will be the focus of later chapters.

STATUS AND ROLE

When students and an instructor walk into class the first day of the term, they know without thinking what to expect of each other and how each will behave. They know these things even though they have not seen each other before. The students know that the instructor will stand in front of the class behind a lectern, probably call roll, assign reading, and dismiss them early the first day. The instructor knows that, unless the class is required, students will be shopping around. They will be trying to decide whether to take this class, and their decision will be based on course content, the viewpoint and personality of the instructor, the amount and type of work required, and how the instructor is known to grade.

We know these things about each other partly because of the system of norms discussed previously. The concepts of *status* and *role* are closely related to norms, and they play a major part in the situation described above. People typically use *status* to refer to one's rank in society—"she's a neurosurgeon, which means she has high status and a Mercedes Benz. . . ." This usage is not incorrect, and I may even use the term this way in Chapter Five when I'm not careful, but for the time being let's forget this usage. When sociologists are using the concepts status and role, we use the term *status* in a slightly different way. By **status** we mean a position in society or in a group. There are innumerable positions one may occupy: teacher, student, police officer, president, football player, father, wife, convict. Furthermore, each of us may occupy several positions at once: teacher, handball player, father, husband, and so on. By **role** we mean the behavior of one who occupies a particular status. As Robert Bierstedt puts it, a role is what an individual *does* in the status he or she occupies; statuses are occupied, roles are played.[6]

A set of norms surrounds each status and role. These norms, called **role requirements,** describe the behavior expected of people holding a particular position in society. Recalling our earlier example, the behavior of the student who refused to stand up was disturbing because it was unpredictable. He was occupying the status or position of student, but the role he played—his behavior—was contrary to the expected behavior of a person in that status. His behavior was outside the limits set by the norms, or role requirements.

Within the boundaries set by the role requirements, there is often extensive variation in how a role is played. On a football team, status would refer to the positions, role to the behavior of the incumbent of the position. One status would be quarterback. Role requirements of quarterbacks are generally to call the plays, to direct the team, and to try to move the ball down the field. But now look at the actual performance of several quarterbacks. One passes frequently, another seldom passes but often runs with the ball, and a third does neither but usually blocks. Compare four or five of your instructors in their role behavior. Although all occupy the status of college professor, no doubt their behavior varies markedly. One paces the floor; another stays behind the lectern while lecturing. One demands class discussion; the next dislikes having lectures interrupted. One has beautifully organized and prepared lectures; another has a disorganized, stream-of-consciousness presentation put together on the way to class. Or compare the behavior of the last three presidents of the United States. Again we see marked role differences within a given status. These differences in behavior obviously occur because people holding the same status define the role differently. This should sound familiar because it's essentially the idea we had in mind in discussing definition of the situation in Chapter Two. As a result of a particular pattern of socialization, each individual defines status and role in a particular way. Each individual brings to the situation a specific personality and a set of skills, interests, and abilities. Therefore, although each status carries with it certain role requirements, there is still variation and flexibility in actual behavior.

Role—behavior that is suited to a particular status—varies not only because of the style of a particular quarterback, college professor, or president; roles must also be seen in an *interaction* setting. While behaving, people are always socially interacting with others, and consequently their behavior adjusts to and is modified by the responses of others. This continues the viewpoint introduced in the previous chapter in which we pointed out that socialization and self-development occur through the process of social interaction. Roles we play are shaped by others' reactions to us. For example, after comments and complaints from an unhappy class, the unprepared, stream-of-consciousness professor mentioned earlier may modify his or her performance. It is a common occurrence to go into a situation prepared for one sort of role only to find that, in the process of the interaction, another sort of behavior is necessary.

The second research question at the beginning of this chapter compared people who had many different jobs, activities, or positions in society with those who were more focused on one. We can now see that we were talking about roles—about people who are involved in many roles compared with those who concentrate more on one. Would the number of roles one is involved in have any effect on life span? A recent study says yes. A number of women were randomly selected and interviewed in 1956 and again in 1986. Those who actively "played" many roles, such as worker, church member, friend, neighbor, relative, and club or organization member, were compared with those who focused more on one role. The study found that those involved in many roles lived significantly longer. Of the various roles studied, active participation in clubs or organizations seemed to have the greatest effect on longevity. The scientists involved in the study concluded that role enhancement—being involved in multiple roles—seems to enhance health. This supports earlier studies that have shown that integration into society in terms of "playing" multiple roles reduces the likelihood of psychological distress among both men and women.[7]

Achieved and Ascribed Status

How do we happen to occupy the statuses that we do? Some, probably most, statuses are earned or achieved in some way, and hence these are called **achieved statuses.** The astronaut, police officer, college professor, and truck driver represent achieved statuses. Some statuses are automatically conferred on us with no effort or choice on our part. These are **ascribed statuses.** One's gender, race, and nationality are ascribed (although occasionally some changes can be made). Sometimes it is difficult to tell whether a status is ascribed or achieved. Consider the student who feels forced to go to college because of the wishes of his or her parents: Is the status of student ascribed or achieved? Or how about the statuses a child inherits from his or her parents, such as political and religious affiliations: Are they ascribed or achieved?

Statuses are stratified, or ranked, at a number of levels. Some statuses are of high rank and bring much prestige to the occupant. The doctor, board director of a large corporation, college president, author, artist, scientist, and movie star represent statuses that have high value and prestige in our society. An evaluation of position is usually determined by the requirements one must have to fill that status: extensive education, wealth, beauty, skill, or some other extraordinary characteristic. Sometimes this ranking is based on a societal tradition that automatically ranks certain characteristics above others, such as a particular gender, race, religion, or aristocratic affiliation. Sometimes having high ascribed status makes it easier to obtain desired achieved statuses. For example, a child born in a middle-class family (ascribed status) will have a better chance of becoming a doctor or scientist (achieved status) than will a lower-class child. Sometimes having high ascribed status is limiting, however. People who are members of royalty may find that their freedom is severely restricted. Living up to their ascribed statuses may mean declining to pursue more attractive achieved statuses.

Role Strain and Role Conflict

Problems may occur when a person must play several roles simultaneously or when one role requires a person to perform in several different ways. These situations are called role conflict and role strain, and they may lead to personal stress and discomfort. **Role strain** refers to the situation in which there are differing and conflicting expectations regarding one's status or position. A student may experience role strain when he compares the expectations of his parents (to study, get A's, prepare for a vocation) with the expectations of his fraternity brothers (to be social, to be active in fraternity affairs, to be athletic). Police officers, who are trained to arrest people who have committed crimes, probably feel role strain when they are ordered by superiors who do not want to make a bad situation worse to stand by and watch looting take place during a riot. A typical situation on college campuses leads to role strain for young professors: Their students expect them to be good teachers, but the school tells them that keeping their jobs and getting promoted will depend on how many articles and books they publish. Doing one takes valuable time from the other—what to do?

Role conflict occurs when a person occupies several statuses or positions that have contradictory role requirements. Here there is no confusion or disagreement about the requirements of a single role, as in the first case. The requirements of the roles are clearly understood; the problem is that the requirements of two or more roles are contradictory. Imagine the dilemma of a police officer invited by friends to a party where marijuana is being smoked. Police officers are trained to respond to violations of the law with the authority of their position, both on and off duty. But the officer is a normal citizen who is expected by friends—and who wants—to behave like everyone else at the party and have fun. The requirements of the two roles are clearly contradictory. Doctors seldom treat members of their own families because of the role conflict that may occur. Similarly, the football coach whose son is trying out for the team experiences conflict between the contradictory requirements of two different roles: coach and father. And finally, if we have ever marveled at the phenomenon of the champion tennis player beating everyone in the world except her future husband, we should have recognized and sympathized with her attempt to resolve her role conflict.

THEORY AND RESEARCH: A REVIEW

This chapter primarily introduces and illustrates a series of concepts: society, social structure, norm, status, role, and culture. Of the three theoretical perspectives introduced in Chapter One, the discussions in this chapter are most influenced by *functional analysis*. Specifically, the discussion of society and social structure and the sections on norms, status, and role, and much of the rest of the chapter, revolve around the ideas of stability and unity while focusing on the structure and the integration of the parts of society. All of these elements are characteristic of functional analysis. Studies of the differing interests that lead to norm conflict (car owners versus manufacturers) and of the characteristics of countercultures turn more in the direction of *conflict theory*.

Of the research efforts described in this chapter, the comparative study of Japanese and American incidence of heart disease was a *nonreactive survey using records*. The example of the recalcitrant student who wouldn't stand up and Milgram's two studies (one on administering shocks, the other on differences in conformity between French and Norwegian students) are *experiments*. Milgram's studies are interesting in that, by varying selected conditions in his experiments (for example, distance between subject and learner, and presence or absence of experimenter), he introduced certain *controls* that sharpened his findings. Mead's research on primitive societies is an example of *participant observation*.

SUMMARY

In the first chapter in this section on socialization and culture, we focused on how individuals develop into social beings. In this chapter, we have turned our attention from individuals and the processes of socialization and self-development to the stage or setting where these processes take place. This setting is called the *social environment*. Within this environment, individuals exist in specific societies and cultures; they occupy numerous positions, are governed by rules, and behave in a variety of ways that are sometimes appropriate, sometimes inappropriate, but seldom unusual. The membership of individuals in a given society and culture, whose patterns and customs developed long before them and will probably long outlive them, affects and explains much of their behavior.

Culture is made up of the learned, shared patterns of behavior and knowledge common to a society, which includes both material and nonmaterial aspects. *Norms* are the rules of society. Because of these rules, we behave appropriately and behavior becomes predictable. Norm breakdown occasionally occurs, and when it does, it may produce crises both for society and for individuals. Norms vary in strength: "Shoulds" are called *folkways;* "musts" are called *mores*. There are *group norms* and *societal norms*, and sometimes these sets of norms conflict with each other. When this happens, those people influenced by the conflicting sets of norms experience problems. *Subcultures* refers to groups that share many of the traits of the dominant culture but have some unique customs and traits as well. Our cultural and subcultural affiliations are of crucial importance in determining who we are and what we do. The concept of *ethnocentrism* helps us to understand the familiar tendency to assume that the world everywhere is the same as it is here and that if by some chance it's not, it should be. On the other hand, *cultural relativism* is an attitude that judges each culture from its own viewpoint.

Society is defined as a continuing number of people who live in a specific area; who are relatively organized, self-sufficient, and independent; and who share a common culture. A *status* is a position in society, and *role* describes the behavior of one who occupies a status. Norms define the boundaries for role requirements, but within these boundaries, the performance of roles will vary. Statuses, which may be either achieved or ascribed, are ranked in value or prestige. When a person occupies several statuses with contradictory role requirements, role conflict may occur. Role strain may result from one who tries to play a role that includes conflicting expectations.

Three readings follow that deal with norms and culture. In the first, William Kephart describes the Gypsies, a group that he views as the most incredible of those discussed in his book *Extraordinary Groups*. In the second reading, Edward Hall shows that the way in which people view time and space determines how they communicate and that these norms vary more than we might think from one culture to another. And finally, in his classic article, "One Hundred Percent American," Ralph Linton explores whether new elements of the American culture are introduced from within through invention or whether they are adapted from other cultures through the process of diffusion.

TERMS FOR STUDY

achieved status (90)	role (89)
ascribed status (90)	role conflict (91)
counterculture (83)	role requirements (89)
cultural relativism (82)	role strain (91)
culture (73)	sanction (76)
ethnocentrism (81)	social structure (88)
folkways (77)	society (87)
ideal norms (77)	statistical norms (77)
material culture (74)	status (89)
mores (76)	subculture (82)
nonmaterial culture (74)	values (75)
norms (75)	

Reading 3.1

THE GYPSIES

William M. Kephart

The subtitle of William Kephart's book Extraordinary Groups *is "An Examination of Unconventional Life-Styles." His study includes the Amish, the Mormons, modern communes, the Shakers, and others. He finds the Gypsies "incredible" and even after much study finds it "difficult to grasp their culture patterns." In this excerpt from his chapter on the Gypsies, we see some of these "unconventional" culture patterns.*

Of all the groups discussed in [*Extraordinary Groups*], the Gypsies are the most incredible. For several years now, I have studied them, interviewed them, and—on occasion—mingled with them, but I still find it difficult to grasp their culture patterns. Other writers have experienced similar difficulties, for the Gypsies have a life-style that comes close to defying comprehension. On the dust jacket of Peter Maas' controversial and widely read *King of the Gypsies,* for example, the following blurb appears:

> *There are perhaps a million or more Gypsies in the United States—nobody knows exactly how many, not even the government. They no longer live in horse-drawn caravans on dusty roads; they live in cities, drive cars, have telephones and credit cards. Yet they do not go to school, neither read nor write, don't pay taxes, and keep themselves going by means of time-honored ruses and arrangements. Gypsies themselves recognize the contrast they make, and they are proud of it.*

Given the nature of modern journalism, can this statement be true? The answer is not a simple one, and each of the above points requires some explanation.

It is true that no one knows how many Gypsies there are in the United States, although the million figure commonly reported in the press may be too high. More reliable estimates place the figure closer to 500,000. If the latter figure is correct, it would mean that only two other countries in the world have larger Gypsy populations: Yugoslavia (750,000), and Romania (680,000).

Although 500,000 seems a reasonable estimate, it is unlikely that the real figure will ever be known. The fact of the matter is that Gypsies move about so much, have so many different names and aliases, and are generally so secretive that it is often difficult to pinpoint the numbers for a given city let alone for the nation at large.

Gypsies live in cities and drive cars? Indeed they do. They are not likely to be found on farms or in the suburbs. They will not be found on the water. They are urban dwellers—towns and cities—and they reside in nearly all of the fifty states. At the same time, Gypsies are, and always have been, great travelers. They may be the greatest travelers

Reprinted with permission of The Free Press, a Division of Macmillan, Inc., and with permission of Tavistock Publications from *Gypsies: The Hidden Americans* by Anne Sutherland. Copyright © 1975 by Anne Sutherland.

the world has ever known. (In England, the terms "Gypsy" and "Traveler" are often used interchangeably.) As we shall see, traveling serves as an integral part of the Gypsy life-style.

As for the cars, Gypsies not only drive them but sometimes make their living repairing them. The days of the horse-drawn wagons and caravans have long since gone, but Gypsies—as is their wont—have adapted remarkably well to motorized transportation. Indeed, despite the fact that they are a low-income group, Gypsies often drive Cadillacs.

Gypsies do not go to school? Not very often—and not for very long. They feel that formal education is not germane to their way of life and that the American school system would tend to "de-Gypsyize" their youngsters. Both claims are at least partially true, and the subject will be discussed later.

Gypsies neither read nor write? True. A large proportion of them are functionally illiterate. They cannot even read or write their own language, Romany, for it is a spoken rather than a written tongue. The literacy situation is improving, but so far progress has been slow. In spite of their self-imposed linguistic handicap, however, Gypsies have made a remarkable adaptation to their environment.

Gypsies do not pay taxes? Some observers would reply: "Not if they can help it." And it is true that many Gypsies do not pay property taxes because they have no taxable property. They often prefer to rent rather than to buy a dwelling-place. Also, many Gypsies work irregularly and have low-paying jobs, so that their income taxes would be negligible. A fair number are on welfare. On the other hand, at least some Gypsies are moving into white-collar occupations, and their tax payments are probably commensurate with those of other white-collar workers.

Gypsies keep themselves going by means of time-honored ruses and arrangements? A complicated question, surely, but then the Gypsies are a complicated people. As is true of all ethnic groups, there are honest Gypsies and there are dishonest Gypsies. Unfortunately, however, many Gypsies continue to believe that all *gadje* (non-Gypsies) are fair game. And more than occasionally this belief does culminate in ruses and petty swindles.

At the same time, Gypsy attitudes toward the *gadje* have been shaped in part by the *gadje* themselves. As will be shown, Gypsies have not been met with open arms by the various host countries. On the contrary, they have experienced near-universal prejudice and discrimination. Social distance studies in several countries, including the United States, simply confirm the obvious; namely, that Gypsies rank at the absolute bottom of the status scale.

Through it all, the Gypsies have survived. Gypsies always survive. If they haven't exactly flourished, they have in many ways given a very good account of themselves. It is not easy to be a Gypsy. As one writer put it: "Only the fit need apply.". . .

Marimé

Central to any understanding of the Rom [Gypsies] is their concept of *marimé*. It is *marimé* that is the key to their avoidance of the *gadje,* and it is *marimé* that serves as a powerful instrument of social control.

Marimé means defilement or pollution, and as used by the Gypsies it is both an object and a concept. And since there is really no comparable term used by non-

Gypsies, it is sometimes difficult for the latter to comprehend the meaning. "*Marimé*," writes Miller, "extends to all areas of Rom life, underwriting a hygienic attitude toward the world. . . . Lines are drawn between Gypsy and non-Gypsy, the clean and the unclean, health and disease, the good and the bad, all of which are made obvious and visible through the offices of ritual avoidance."

The most striking aspects of *marimé* have to do with the demarcation of the human body. The upper parts, particularly the head and the mouth, are looked upon as pure and clean. The lower portions, especially the genital and anal regions, are considered *marimé*. As the Rom see it, the upper and lower halves of the body must not "mix" in any way, and objects that come into contact with one half must not come in contact with the other.

There are countless examples of this hygienic-ritualistic separation. Ronald Lee, who is himself a Gypsy, writes that

> *you can't wash clothes, dishes, and babies in the same pan, and every Gypsy has his own eating utensils, towels, and soap. Other dishes and utensils are set aside for guests, and still others for pregnant women. Certain towels are for the face, and others for the nether regions—and there are different colored soaps in the sink, each with an allotted function.*

Marimé apparently originated in the early caravan period, when—for hygienic purposes—it was imperative that certain areas of the camp be set aside for cooking, cleaning, washing, taking care of body functions, and the like. Also, within the close confines of the wagons and tents, it was important that rules pertaining to sex be carefully spelled out and enforced. . . .

Gropper states that "a woman is *marimé* during and after childbirth, and during her monthly period. . . . A *marimé* woman may not cook or serve food to men. She may not step over anything belonging to a man or allow her skirts to touch his things. Women's clothing must be washed separately from men's." . . .

Marimé vs. *Melalo*

Mention should be made of the distinction between *marimé* and *melalo*. *Marimé* is pollution or defilement, as just described. *Melalo* simply means dirty, or as Lee describes it, "dirty with honest dirt." Someone who has not had a bath would be *melalo*, but not *marimé*. Hands that are dirty because of manual labor would be *melalo* rather than *marimé*—although they would be *marimé* if they had touched the genitals.

(In actual practice, Gypsies tend to wash their hands many times a day—because they may have touched any number of objects or organs that are *marimé*. Miller states that "a working Rom also washes his face and hands whenever he feels his luck leaving him during the day; he washes again upon returning from work.")

The distinction between *marimé* and *melalo* explains why a Gypsy domicile often appears dirty to a non-Gypsy—and vice versa! Some of the Romany dwelling places I have been in, for example, are anything but spic and span. Food scraps, cigarette butts, paper, wrappings—all may be thrown on the floor, presumably to be swept out later. Such a condition is not *marimé* so long as the proper rules of body hygiene, food preparation, and so forth are followed. As one writer puts it: "Americans tend to be shocked at visible dirt, but Gypsies abhor invisible pollution.". . .

Barrier to Assimilation

Do not the various rules and prohibitions involved in *marimé* impose a hardship on the Rom? In one sense, the answer is yes. The urban world, the Gypsies' major habitat, is seen as "pervasively *marimé,* filled with items and surfaces that are subject to use and re-use by careless *gadje,* polluted, diseased, and therefore dangerous." To avoid the danger, Gypsies must take any number of daily precautions—and there is no doubt that these precautions are time-consuming and burdensome. Little wonder, as Miller points out, that "the home is the final bastion of defense against defilement, and the only place that the Rom feel altogether at ease."

At the same time, *marimé* serves as an extremely effective *barrier to assimilation.* As used by sociologists, the term *assimilation* refers to the absorption of one population group by another, where the end result is a blending of culture traits. The so-called melting pot in America has boiled unevenly, some groups being assimilated much faster than others. The Gypsies, of course, would fall at the lower end of any assimilation scale—which is just where they want to be. As the Rom see it, assimilation would be tantamount to group extinction.

Nowhere is this resistance to assimilation more apparent than in their attitude toward the *gadje.* The belief that the latter are *marimé* not only serves as a barrier to assimilation but acts as an ever-present sustainer of pride and self-respect. In fact, so pervasive is their negative attitude toward the *gadje* that Gypsies will not assimilate even after death! Nemeth's analysis of cemetery plots and tombstones revealed that the Rom attempt "to maintain distance in the graveyard between themselves and non-Gypsies, and between themselves and outcasts from their own society."

One further point should be mentioned regarding resistance to assimilation. The Gypsies are a markedly low-status group, and solidifying elements such as morale and esprit de corps are extremely important.

> *Categories of thought that rank the* gadje *as inferior to the Rom in purity and health have an obvious significance to pride and self-respect. . . . Despite demeaning life circumstances, which include frequent and irritating inequalities, morale is maintained at a favorable level.*
> *When a Romni is arrested for telling fortunes, or a family is forced to move because the landlord has discovered that ten children, instead of the two or three expected, are living in one house, the aggravation can be lessened by a counter offensive of verbal abuse in Romany concerning the appearance of the skin, the odor, moral character, and personal habits of the* gadje. . . .

Marimé as Social Control

Although Gypsy leadership may or may not be an effective source of social control, *marimé* has traditionally been a powerful instrument. Indeed, it may just be the most important factor in keeping the Rom in line. The reason is not hard to find, for *marimé* is more than a simple declaration that a person or thing is polluted. A Gypsy who has been declared *marimé* is ostracized by the entire group. Other Rom will have nothing to do with him or her.

It cannot be emphasized too strongly that within the confines of their own society, Gypsies are gregarious. They are never really alone. Practically all of their waking moments are spent in the company of other Rom. Talking, laughing, working, arguing, gossiping, and most important perhaps, eating—all are considered group activities. To be declared *marimé,* therefore, effectively cuts a Gypsy off from the very roots of his existence. He brings shame not only upon himself but upon his family. . . .

Fortune-Telling

Fortune-telling is a special case, for if there is one field that has been monopolized by the Rom it is certainly fortune-telling. Indeed, the terms "Gypsy" and "fortune-teller" seem to go hand in hand—and with good reason, as the following account by Andersen indicates:

> *The little girl is expected to be a fortune-teller or reader and advisor, as early as thirteen or fourteen, and as a child she is trained for this profession. Fortune-telling as a means of livelihood is a tradition among Gypsy women, and many little girls observe their mothers, aunts, and other female relatives performing within this tradition every day.*
>
> *They are taught that they have a natural gift for the practice, that they received this gift from God, and that as fortune-tellers they will be performing a type of psychological counseling service for non-Gypsies.*

Fortune-telling is not a difficult occupation to learn, overhead expenses are negligible, and—depending on the location—business may be good. Clark cites the old Gypsy saying: "A fortune cannot be true unless silver changes hands." And there have always been enough *gadje* who believe in this aphorism to make crystal gazing, palmistry, and card reading profitable ventures. Fees typically range from $2 to $5 per session, with a surprising number of repeat customers. Most of the latter are reportedly drawn from the lower socio-economic ranks. As Bercovici puts it: "The Gypsy fortune-teller is the psychoanalyst of the poor."

Do Gypsies themselves believe in fortune-telling? The answer is yes and no. Despite what they may nominally teach their youngsters, all Rom realize that readings performed for the *gadje* are entirely fallacious. Paradoxically, however, some Gypsy women are believed to possess occult powers of prediction, at least when applied to other Gypsies. Older Gypsy women, therefore—who are felt to be experts in dream interpretation and card reading—are sometimes consulted.

The specifics of Gypsy fortune-telling vary somewhat, though not a great deal. In a few areas, Romni still travel in pairs, telling fortunes on a catch-as-catch-can basis. The most common practice, however, is to set up a fortune-telling parlor, or *ofisa,* with living quarters in the rear. The actual location of an *ofisa* depends in good part on the availability of large numbers of shoppers or passers-by, which is why resort and vacation areas are considered choice places to set up shop. . . .

The *Bujo*

On occasion, Gypsy fortune-tellers have been accused—and convicted—of flimflam, or *bujo*. The *bujo* is nothing more than a swindle, whereby a gullible customer is cheated out of a goodly portion of his or her savings. One common ruse is called "switch the bag." In this instance, a bag of fake money or cut-up paper is substituted for a bag of real cash—which the customer had brought to the *bujo* in order to have the evil spirits or curse removed. (In Romany, *bujo* means "bag.")

According to the New York police, *bujo* swindles in excess of $100,000 have occurred. And according to Mitchell, there are some Gypsy fortune-tellers "with a hundred or more arrests on their record." Obviously, these are unusual cases; in fact, many Rom frown on the *bujo* because it causes bad community relations and is likely to bring police action.

At the same time, the *bujo* has occurred often enough to cause many areas to outlaw fortune-telling. Major cities like New York and Philadelphia, as well as most Canadian regions, have banned fortune-telling. Some observers feel that the illegalization of fortune-telling may be the Gypsies' biggest problem. [Interestingly enough, in 1985 the California Supreme Court, overruling a lower court decision, found that an ordinance prohibiting fortune-telling for profit violated the constitutional right to free speech. The suit was brought by a *romni*, and the ramifications are as yet unknown. It is quite possible, however, that the case will eventually reach the United States Supreme Court!]

Legalities aside, the Rom continue to ply their trade, even though they are somewhat restricted in many areas. They often pose as "readers" and "advisors" rather than as seers. And this, in turn, may necessitate a measure of police "cooperation." But by one method or another, the Gypsies survive. Gypsies always survive.

QUESTIONS 3.1

1. See how many concepts discussed in this chapter you can find and illustrate with examples from this reading.

2. Culture, subculture, or counterculture—which most accurately describes the Gypsies? Make a case for each.

3. Social control refers to the processes by which people are made to conform to collective norms. How do the Gypsies do it?

4. Kephart found the Gypsies to be incredible. Why? What about them in particular is so different from us?

Reading 3.2

THE HIDDEN DIMENSION
Edward T. Hall

Norms indicate the appropriate behavior for a specific situation. Norms may vary from one situation to another and especially from one culture to another. Behavior that is appropriate in one culture may be unacceptable in another. This is true even in such seemingly innocent matters as talking loudly or softly, the value of smells, maintaining eye contact in conversation, or moving one's chair about. In this excerpt from his book The Hidden Dimension, *anthropologist Edward Hall shows how different cultures vary in their use of space as compared with American culture and how this affects patterns of communication.*

Germans

Germans and Europeans generally will schedule fewer events in the same time than Americans will. Americans structure time tightly and are sticklers for schedules. On the other hand, Americans treat space more casually, even wastefully, in the eyes of many Europeans. Private or personal space is very important to Germans. Prisoners of war built partitions, tiny dwelling units, so that each could have his own space. German houses with balconies are arranged so that there is visual privacy; yards tend to be well fenced. German doors are thick and substantial, and kept closed in contrast to American doors, which are more flimsy and likely to be open. Germans want to know where they stand and object strenuously to people who "get out of line" or who do not obey signs such as "keep out," or "authorized personnel only." The feeling that many Americans have that Germans are overly formal can be seen in the handling of chairs. Americans tend to move their chairs about to get closer or to get a better view. In Germany, however, it is a violation of the mores to change the position of your chair. Light furniture is disliked because it seems flimsy and because people move it and thereby destroy the order of things—they intrude on one's private space. In one instance, a German newspaper editor who had moved to the United States had his visitor's chair bolted to the floor "at the proper distance" because he couldn't tolerate the American habit of adjusting the chair to the situation.

English

In the United States we use space as a way of classifying people and activities, whereas in England it is the social system that determines who you are. In the United States, your address is an important cue to status. The Englishman, however, is born and brought up in a social system—he is Lord ____ no matter where you find him. The middle-class American feels he has a right to have his own room whereas the middle- and upper-class Englishman is brought up in a nursery shared with brothers and sisters and may never have a permanent "room of his own." Consequently, the English are puzzled by the American need for a secure place in which to work, an office. Americans working in England may become annoyed if they are not provided with what they consider appropriate work space. When the American wants to be alone he goes into a room and shuts the door—he depends on architectural features for screening. For an American to refuse to talk to someone else present in the same room, to give them the "silent treatment," is the ultimate form of rejection and a sure sign of great displeasure. The English, on the other hand, lacking rooms of their own since childhood, never developed the practice of using space as a refuge from others. Public silence in their case may represent a way of finding privacy.

Americans living in England have trouble adjusting to English ideas regarding neighbors. The fact that you live next door to a family does not entitle you to visit, borrow from, or socialize with them, or your children to play with theirs. English relationships are patterned not according to space but according to social status. In England and in Europe generally, Americans are continually accused of loud talking. Americans tend to increase volume with distance, and in many situations they do not care if they are overheard. The English do care and they have developed skills in beaming the voice toward the person they are talking to, carefully adjusting it so that it just barely overrides the background noise and distance. For the English to be overheard is to intrude on others. Even eye behavior in conversation varies: the English look straight at you and blink to let you know they have heard whereas the gaze of the American tends to wander about.

Japanese

The concept of the center that can be approached from any direction is a well-developed theme in Japanese culture. Japanese name intersections rather than the streets leading to them. How one gets from point A to point B seems almost whimsical and the "correct route" is not stressed as it is with us. Taxicab drivers have to ask local directions at police booths, not just because streets are not named but because houses are numbered in the order in which they were built. To us the walls of a house are fixed. In Japan they are semifixed. The walls are movable and rooms are multipurpose. Depending on the time of day, the room can include all outdoors or it can be shrunk in stages until all that remains is a boudoir. Some Americans feel that Japanese talk around and around the point and never get to it. Some Japanese wonder why Americans have to be so logical all the time. Crowding is distasteful to Americans, while it is less likely to bother the Japanese. Japanese feel it is congenial to sleep close together on the floor. This seems to indicate a disinterest in privacy. However, a Japanese has strong feelings against sharing a wall of his house or apartment with others. He considers his house and the zone immediately surrounding it as one structure. This free area is considered to be as much a part of the house as the roof. Traditionally, it contains a garden even though tiny, which gives the householder direct contact with nature. The Japanese garden itself, in keeping with the Japanese concept of space, is designed to be enjoyed from many points of view.

Arab World

Pushing and shoving in public places is characteristic of Middle Eastern culture. While for the American this may seem "pushy" and rude, for the Arab there is no such thing as an intrusion in public. Public means public—if A is standing on a street corner and B wants his spot, B is within his rights if he does what he can to make A uncomfortable enough to move. To the Arab good smells are pleasing and a way of being involved with each other. They consistently breathe on people when they talk. To smell one's friend is not only nice but desirable, for to deny him your breath is to act ashamed. Americans, on the other hand, trained as they are not to breathe in people's faces automatically communicate shame in trying to be polite. Arabs look each other in the eye when talking with an intensity that makes most Americans highly uncomfortable. To view the other person peripherally is regarded as impolite, and to sit or stand back-to-back is considered very rude. Privacy in a public place is foreign to the Arab. Business transactions in the bazaar are participated in by everyone. In terms of personal space, Arabs don't mind being crowded by people but hate to be hemmed in by walls. Ideally, homes must meet at least three standards: much unobstructed space without walls or partitions (so the family can be together), very high ceilings, and an unobstructed view.

In summary, views of social and personal space differ cross-culturally. By examining them it is possible to reveal hidden cultural frames that determine the structure of a given people's perceptual world. Perceiving the world differently leads to different definitions of what constitutes crowded living, different interpersonal relations, and a different approach to both local and international politics.

QUESTIONS 3.2

1. Peoples' norms concerning space determine their patterns of communication. Explain.

2. Show how the terms of concepts *ethnocentric, cultural relativism,* and *culture and personality* might be used in this article.

Reading 3.3

ONE HUNDRED PERCENT AMERICAN

Ralph Linton

The content of a culture—skills, customs, material objects, and nonmaterial ideas—comes from several sources. Some aspects of culture appear spontaneously and are developed from within the culture through the process of invention. Much more, however, is borrowed from other cultures through the usually unconscious process of diffusion. Anthropologist Ralph Linton wrote this passage some years

From *The American Mercury*, April 1937, pp. 427–429. Reprinted by permission of the publisher.

ago. It is often quoted and still remains a classic at demonstrating the importance of diffusion. Much of what we thought was "one hundred percent American" actually came long ago from societies we never heard of.

There can be no question about the average American's Americanism or his desire to preserve this precious heritage at all costs. Nevertheless, some insidious foreign ideas have already wormed their way into his civilization without his realizing what was going on. Thus dawn finds the unsuspecting patriot garbed in pajamas, a garment of East Indian origin; and lying in a bed built on a pattern which originated in either Persia or Asia Minor. He is muffled to the ears in un-American materials: cotton, first domesticated in India; linen, domesticated in the Near East; wool from an animal native to Asia Minor; or silk, whose uses were first discovered by the Chinese. All the substances have been transformed into cloth by a method invented in Southwestern Asia. If the weather is cold enough he may even be sleeping under an eiderdown quilt invented in Scandinavia.

On awakening he glances at the clock, a medieval European invention, uses one potent Latin word in abbreviated form, rises in haste, and goes to the bathroom. Here, if he stops to think about it, he must feel himself in the presence of a great American institution; he will have heard stories of both the quality and frequency of foreign plumbing and will know that in no other country does the average man perform his ablutions in the midst of such splendor. But the insidious foreign influence pursues him even here. Glass was invented by the ancient Egyptians, the use of glazed tiles for floors and walls in the Near East, porcelain in China, and the art of enameling on metal by Mediterranean artisans of the Bronze Age. Even his bathtub and toilet are but slightly modified copies of Roman originals. The only purely American contribution to the ensemble is the steam radiator.

In his bathroom the American washes with soap invented by the ancient Gauls. Next he cleans his teeth, a subversive European practice which did not invade America until the latter part of the eighteenth century. He then shaves, a masochistic rite first developed by the heathen priests of ancient Egypt and Sumer. The process is made less of a penance by the fact that his razor is of steel, an iron-carbon alloy discovered in either India or Turkestan. Lastly, he dries himself on a Turkish towel.

Returning to his bedroom, the unconscious victim of un-American practices removes his clothes from a chair, invented in the Near East, and proceeds to dress. He puts on close-fitting tailored garments whose form derives from the skin clothing of the ancient nomads of the Asiatic steppes and fastens them with buttons whose prototypes appeared in Europe at the close of the Stone Age. This costume is appropriate enough for outdoor exercise in a cold climate, but is quite unsuited to American summers, steam-heated houses, and Pullmans. Nevertheless, foreign ideas and habits hold the unfortunate man in thrall even when common sense tells him that the authentically American costume of gee string and moccasins would be far more comfortable. He puts on his feet stiff coverings made from hide prepared by a process invented in ancient Egypt and cut to a pattern which can be traced to ancient Greece, and makes sure they are properly polished, also a Greek idea. Lastly, he ties about his neck a strip of bright-colored cloth which is a vestigial survival of the shoulder shawls worn by seventeenth-century Croats. He gives himself a final appraisal in the mirror, an old Mediterranean invention, and goes downstairs to breakfast.

Here a whole new series of foreign things confront him. His food and drink are placed before him in pottery vessels, the popular name of which—china—is sufficient evidence of their origin. His fork is a medieval Italian invention and his spoon a copy of a Roman original. He will usually begin the meal with coffee, an Abyssinian plant first discovered by the Arabs. The American is quite likely to need it to dispel the morning-after effects of overindulgence in fermented drinks, invented in the Near East, or distilled ones, invented by the alchemists of medieval Europe. Whereas the Arabs took their coffee straight, he will probably sweeten it with sugar, discovered in India, and dilute it with cream, both the domestication of cattle and the technique of milking having originated in Asia Minor.

If our patriot is old-fashioned enough to adhere to the so-called American breakfast, his coffee will be accompanied by an orange, domesticated in the Mediterranean region, a cantaloupe domesticated in Persia, or grapes, domesticated in Asia Minor. He will follow this with a bowl of cereal made from grain domesticated in the Near East and prepared by methods also invented there. From this he will go on to waffles, a Scandinavian invention, with plenty of butter, originally a Near Eastern cosmetic. As a side dish he may have the egg of a bird domesticated in Southeastern Asia or strips of the flesh of an animal domesticated in the same region, which have been salted and smoked by a process invented in Northern Europe.

Breakfast over, he places upon his head a molded piece of felt, invented by the nomads of Eastern Asia, and, if it looks like rain, puts on outer shoes of rubber, discovered by the ancient Mexicans, and takes an umbrella, invented in India. He then sprints for his train—the train, not the sprinting, being an English invention. At the station he pauses for a moment to buy a newspaper, paying for it with coins invented in ancient Lydia. Once on board he settles back to inhale the fumes of a cigarette invented in Mexico, or a cigar invented in Brazil. Meanwhile, he reads the news of the day, imprinted in characters invented by the ancient Semites by a process invented in Germany upon a material invented in China. As he scans the latest editorial pointing out the dire results to our institutions of accepting foreign ideas, he will not fail to thank a Hebrew God in an Indo-European language that he is a one hundred percent (decimal system invented by the Greeks) American (from Americus Vespucci, Italian geographer).

QUESTIONS 3.3

1. Although it is true that many material objects have come to us through diffusion, most of the nonmaterial aspects of American culture are original and the products of invention. Discuss.

2. This article was written in the 1930s; the same sort of statements could not

be made today. Discuss. To prove the point, draw up a list of aspects of American culture developed through invention.

3. Analyze Linton's article, using the terms *ethnocentrism* and *cultural relativism*.

NOTES

1. Clyde Kluckhohn, "The Study of Culture," in *The Policy Sciences*, edited by Daniel Lerner and Harold D. Lasswell (Stanford, Calif.: Stanford University Press, 1951), p. 86.

2. These and other examples are contained in *Studies in Ethnomethodology*, by Harold Garfinkel (Englewood Cliffs, N.J.: Prentice-Hall, 1967).

3. The terms *folkways* and *mores* were introduced by William Graham Sumner in *Folkways* (Boston: Ginn, 1907).

4. For a more complete report of this study, see Stanley Milgram, "Nationality and Conformity," *Scientific American* 205 (December 1961), pp. 46–51.

5. This study was done at the School of Public Health at the University of California at Berkeley and was reported in newspapers and newsmagazines beginning in August 1975.

6. Robert Bierstedt, *The Social Order*, 4th ed. (New York: McGraw-Hill, 1974), ch. 9.

7. See "Social Integration and Longevity: An Event History Analysis of Women's Roles and Resilience," by Phyllis Moen, Donna Dempster-McClain, and Robin Williams, Jr., *American Sociological Review* 54 (August 1989), pp. 635–644.

PART TWO

Social Organization

In preceding chapters we have looked at basic units of analysis used by sociologists to understand, describe, and explain human behavior. In this part our view becomes more general. If we combine these basic units and move to a higher level of abstraction, we come to the concept of social organization. If, in a like manner, we were analyzing football, we could first look at the basic elements—the players, a ball, a marked-off field, and maybe some spectators. But our description is helped a great deal if we move to the next level and describe a football game; thus it is now necessary to look beyond the elements and deal with the combination of the setting, the rules, the shared expectations, and the interaction patterns of participants—in other words, the social organization that allows 22 players, 6 officials, and 20,000 observers to mutually participate in and get something out of the same event.

Just as there is an extensive degree of organization to a football game, so there is to society. The basic elements of a society are a number of people and an inhabitable geographical area. But an understanding of a society cannot be obtained until we study the patterns of interaction and organization that are characteristic of the people. We see that cities exist (why do they?), that highways link cities, that educational facilities are developed, that some individuals are more highly valued than others, and that governments run the cities. These things don't just happen by accident; they are evidence that there is organization to society. People cooperate, interact, and share expectations and mutual interests. There is a structure or system to society much as there is to a building or to a machine or to a football game: Parts link together to form a complex whole. The **social organization** *of society is frequently referred to as a* social fabric: *an integrated set of norms, roles, cultural values, and beliefs through which people interact with each other, individually and through groups. The terms* social organization *and* social structure *(which we discussed in Chapter Three) are very similar concepts and are often used interchangeably. Although* social structure *can be used to* refer more to a stable network of elements through which people relate to each other, and social organization *can be used* to refer more to the continually changing ordering and coordination of human activities, it is probably not worthwhile to try to make a distinction between the terms.

Studying various aspects of the social organization and social structure of society is the central task of sociology. We undertake this analysis in a variety of ways. It is as if we were looking at a subject, society in this case, through a number of windows. Each window is of a different shape, size, thickness, and color of glass. So, although we are looking at the same subject, each approach gives a somewhat different viewpoint, emphasizing some aspects and ignoring others. The following chapters will analyze the social organization of society through several windows. Groups will be the topic of Chapter Four, and several types of groups will be discussed in detail. Chapters Five and Six will deal with types of social differentiation and social inequality. In Chapters Seven through Ten we will discuss institutions as the sociologist sees them. And Chapter Eleven will cover population and ecology.

Groups

TWO RESEARCH QUESTIONS

1. *Is alcohol use related to either social class (one's rank in society) or religiosity (one's commitment to and involvement in religion)? If it is related, in what direction? Do higher-class people drink more or less? Do more religious people drink more or less? Which of the two—social class or religiosity—would have the greater influence on alcohol use? Why?*

2. *Does talking about one's own knowledge and experience make one more well liked and influential in small groups, or is it better to talk about others' knowledge, experience, and influence? Will the gender of the rest of the people in the group (all male, all female, or mixed) make any difference?*

A GROUP—a major unit of analysis for sociologists—is simply defined as a collection of people. Not all collections of people, however, are defined as groups. Are the people on a bus a group? How about red-haired people between the ages of 30 and 45? Or what about the students at a large university? At the same time, collectivities that are traditionally defined as groups may vary tremendously in some characteristics. Your family, my sociology department, a college football team or sorority, and the president's cabinet could probably be defined as groups, but they are quite different from each other in many dimensions: size, complexity, type of interaction, and division of labor. Much of the study of social organization, therefore, could center around several questions: What is a group? How is the group developed? What are the different types of groups?

WHY STUDY GROUPS?

Sociologists study groups for several reasons. The division of labor and the sharing of interests and jobs that occur in groups make it possible to complete tasks that would be impossible for individuals to accomplish alone. Far more important for the sociologist, however, is the role the group plays in the socialization and development of the individual. To a great degree, we feel that a person *is* the sum of the groups he or she belongs to. Transmission of culture and learning of values, attitudes, and ways of behaving and believing occur mainly in groups. As we have pointed out in Chapters Two and Three, the socialization and development of self and personality are interactive processes. Individuals interact with others and modify their own behavior according to their interpretations of the responses of others. These others are individuals: mother, father, teacher, or another significant other. These others are also groups. Most of the socialization process, both primary and secondary, takes place in *groups*. First the family, then the classroom group, the team or club, the social group, the professional organization, and numerous other voluntary and involuntary associations influence and shape the individual.

The group is more than just the sum of the people who belong to it. Perhaps a good example of this is to compare a family of five people with five people we pick at random on the street. Even if we set each "group" in front of the fire or TV or around the same dinner table and start them interacting, the differences between the two would be clear. We would shortly see that the real family group had developed a set of norms, roles, and relationships to each other that went far beyond the interactions of the five separate people. Because of the presence of this network of relationships and understandings that exists in groups, the impact or power of the group in influencing the individual is extensive.

To get a better idea of the importance of groups, imagine the effects on two people of belonging to the following contrasting groups. One person is raised in a family in which competitiveness is valued; family members are constantly competing with each other and with outsiders. The other is raised in a family in which competition is discouraged. One is raised in a family that pursues artistic endeavors, and the other in a family that prefers athletics. One is a Jehovah's Witness; the other is a member of a Unitarian church. One is a member of a rock group; the other is a member of a chamber-music group. One is on the high-school basketball team; the other is on the yearbook staff. One goes to the University of California at Berkeley; the other goes to West Point. One goes through basic training in the Marine Corps; the other joins the Peace Corps. One is raised in a modern Western family, the other in a primitive tribal family.

You're probably thinking, "Sure, the Berkeley student is a lot different from the West Point student; but they had some of those differences before, which is why one chose Berkeley and the other chose West Point." That's probably true, but those differences, too, can be explained by other earlier group memberships and experiences. Each set of group experiences prepares and inclines the person toward another set, then another set, and so on. Or, consider this example. Imagine a 46-year-old doctor at the top of his field who climbs mountains, builds sailboats, races sports cars, writes books, and is a gourmet cook. How did this superachiever get this way? Was he born this way? Certainly not, says the sociologist. Study his socialization patterns through numerous group contexts, and you will find the answer. It may not be easy, but the basic explanations are in the complicated interaction between group and individual.

GROUPS AND NONGROUPS

Sociologists say that a **group** comprises a number of people who (1) have shared or patterned interaction and (2) feel bound together by a "consciousness of kind" or a "we" feeling. "Consciousness of kind," a phrase coined by Franklin Giddings, refers to the individual's awareness of important similarities between himself or herself and certain others. This concept also refers to

Group members have common loyalties

the awareness that the individual and other group members have common loyalties, share at least some similar values, and see themselves as set apart from the rest of the world because of their memberships in this particular group. Groups vary tremendously in variety, size, and shape. A group may be as small as two people or almost infinitely large. Groups may be simple in structure or exceedingly complex; they may involve close, intimate relationships between members or more distant and infrequent personal contacts. In other words, the definition of group—patterned interaction and "we" feeling—may fit an enormous variety of situations: a family, a basketball team, a sociology class, IBM, or General Motors.

This definition of groups, imprecise as it may seem, allows us to distinguish groups from other types of collectivities of people, which we could call *nongroups*. One type of nongroup, which we will call an **aggregate** (or aggregation), consists of a number of people clustered together in one place. Examples of aggregates might be all the people in New York City, or the pedestrians at a busy intersection waiting for the light to change to "walk," or all the people in North America, or the passengers on a jet from New York to San Francisco. A second useful nongroup, called a **category,** consists of a number of people who have a particular characteristic in common. Examples of categories would be all females, or all red-haired people, or all pilots, or all teenagers, or all whites.

Although we have called them nongroups, aggregates and categories may be transformed into groups should they develop patterned interaction among members and consciousness of kind. For example, let's examine our aggregate of ten people waiting at the intersection for the light to change to "walk." Then suppose it *doesn't* change; for 5, 10, even 15 minutes the light refuses to budge from "wait." The pedestrians, strangers until now, begin talking to each other about the impossible situation. Should they race across through traffic against the light? Where's the cop?—you can never find one when you really need one. Some interaction takes place, and a consciousness of kind develops: a group of good people being victimized by a lousy, mechanical light. Or, take the passengers on the jet from New York to San Francisco, another aggregate. Somewhere over Pennsylvania the pilot says to himself, "I'm sick and tired of flying to San Francisco all the time; I guess I'll go to the North Pole." The passengers, who did not know each other before, would begin to interact, possibly in an agitated manner, and by the time they reached the Arctic Circle there would probably be highly developed interaction.

Categories may also become groups. All the red-haired people between the ages of 30 and 45 would constitute a category, as would all carpenters of Irish ancestry. But suppose the middle-aged redheads decided to get together and put out a journal telling of their common problems and aspirations. Or suppose the Carpenters of Irish Ancestry decide it's time to start an organization (CIA?), have a convention, and elect officers. In each case we might have a category developing into a group. These sound farfetched, but sociologists study both groups and nongroups (categories and aggregates), and the lines between these collectivities are somewhat fluid and easily crossed.

The study of people and society is a large task. It is made more manageable if the whole is broken down into more basic parts—parts that are smaller and have some characteristics in common. We can divide society into *categories* (males, upper-class blacks, teenage white females), *aggregates* (a crowd, a city), and *groups* (a family, a study group). In later chapters we will look at types of categories and aggregates, but the focus of this chapter will be groups. Three sections follow; first we will look at specific *types of groups* and then at groups at the two extremes in terms of size: *small groups* and *formal organizations.*

TYPES OF GROUPS

The New York Times *has described one of the most unusual clubs in the world made up of a small number of usually serious scientists stationed at the South Pole. It is called the "300 Club." Those who want to join must wait until the temperature is at least 100 degrees Fahrenheit below zero, then strip completely nude and dash 100 yards across the ice to a marker designating the South Pole and 100 yards back to the scientific hut. Anyone surviving becomes a member of this very exclusive group.*

Social groups may be classified in many ways as the following categories illustrate. In some groups, membership is automatic and the participant has no choice; in others, the option is open and individuals may or may not join as they wish. These two types are called involuntary and voluntary groups. **Involuntary groups** might include the family one is born into or the army platoon one is drafted into. **Voluntary groups** would include any of a vast number that an individual may exercise some choice in joining: lodges, fraternities, bridge clubs, student governments, or political organizations.

Reference groups are groups that serve as models for our behavior—

groups we may or may not actually belong to but whose perspectives we assume and mold our behavior after. A reference group may be made up of people one associates with or knows personally, or it may be an abstract collectivity of individuals who represent models for our behavior. Each individual has many reference groups. As a teacher I would have certain reference groups, as a sociologist others, as a husband others, and as a handball player still others.

The first research question at the beginning of this chapter asked if alcohol use was related to social class and religiosity, and if so, what the relationship was. A recent paper examined this in some detail using reference groups. There is a negative relationship between alcohol use and religiosity—that is, the more religious person drinks less; the less religious person drinks more. Religiosity was defined in terms of church attendance, strength of belief, membership in church organizations, and belief in life after death. The relationship between social class and alcohol use is less powerful but appears to be positive; that is, upper-class people drink more than lower-class people. The real emphasis of the paper, however, was on which of the two factors was more important, religiosity or social class. If the two were in conflict, which would win out? For example, what would be the alcohol use of a highly religious upper-class person or of a less religious working-class person? The answer was clear—religiosity would win out—and the explanation made sense from a reference-group viewpoint. A social class is a *category,* a number of people who have a particular characteristic in common but who otherwise don't know each other and are less likely to share common norms or similar values. However, religiosity, especially as measured here, is more likely to reflect membership in a *group.* The authors suggest that a group—people with whom one has sustained and significant interaction—reflects a "primary environment of opinion." When this is in conflict with a "secondary environment of opinion" such as a category or aggregation, the group context will have greater influence.[1]

One's **peer group** is made up of people of relatively the same age, interests, and social position with whom one has reasonably close association and contact. A peer group may consist of a class at school, a street gang, or an occupational group such as the members of a college sociology department or a group of lawyers in a law firm. Not all the members of a peer group are necessarily friends, but the peer group exercises a major role in the socialization process. During adolescence, a peer group may be *the* major socializing agent.

Groups whose members come predominantly from one social-class level are called **horizontal groups.** Examples of horizontal groups would include almost any organization formed along occupational lines: an association of doctors, carpenters, or actors. If a group includes members from a variety of social classes, it could be called a **vertical group.** Vertical groups are more difficult to find in American society because many divisions are made along social-class lines. A church congregation might constitute a vertical group, and in some cases an army platoon made up of draftees would include members from a variety of social classes. Groups are defined as open or closed according to the ease of gaining membership. A white fraternity is often a **closed group** as

"FRIENDS AND NETWORKS"—TALLY'S CORNER

More than most social worlds, perhaps, the streetcorner world takes its shape and color from the structure and character of the face-to-face relationships of the people who live in it. Unlike other areas in our society, ... resources in the streetcorner world are almost entirely given over to the construction and maintenance of personal relationships. On the streetcorner, each man has his own network of these personal relationships.... At the edges of this network are those persons with whom his relationship is effectively neutral, such as area residents whom he has "seen around" but does not know except to nod or say "hi" to as they pass on the street. In toward the center are those persons he knows and likes best, those with whom he is "uptight": his "walking buddies," "good" or "best" friends, girl friends, and sometimes real or putative kinsmen. It is with these men and women that he spends his waking, nonworking hours, drinking, dancing, engaging in sex, playing the fool or the wise

man, passing the time at the Carry-out or on the streetcorner.... So important a part of daily life are these relationships that it seems like no life at all without them. Old Mr. Jenkins climbed out of his sickbed to take up a seat on the Coca-Cola case at the Carry-out for a couple of hours. "I can't stay home and play dead," he explained, "I got to get out and see my friends."

Preston was Clarence's uncle. They lived within a block of each other and within two blocks of the Carry-out. Clarence worked on a construction job and later got Preston a job at the same place. Tally, Wee Tom, and Budder also worked at the same construction site. The five men regularly walked back from the job to the streetcorner together, usually sharing a bottle along the way. On Friday afternoons, they continued drinking together for an hour or so after returning to the streetcorner. Tally referred to the other four men as his "drinking buddies."

Tally had met Wee Tom on the job.

---

far as a black male is concerned, but the United States Army is probably a very **open group** for the same individual. Now, what type of group is the "300 Club"?

Primary and Secondary Groups

A family and a draftee's army platoon are involuntary groups, but they are quite different sorts of groups. A basketball team and the members of a large college class are voluntary groups, but again they are very different. Groups vary along a number of dimensions, and one of the most important of these is to what degree each group is primary or secondary. A family and a basketball team are more primary than are an army platoon and a large college class. Primary groups play an important role in human development. Most of the socialization process—learning society's norms and roles, and development of the self—takes place in small primary groups. The **primary group** was first described by Charles H. Cooley as referring to groups in which contacts between members are intimate, personal, and face-to-face. A great part of the individual's total life experience is bound up in the group and is known to other group members. The

Through Tally, Wee Tom joined them on the walk home, began to hang around the Carry-out, and finally moved into the neighborhood as well. Budder had been the last to join the group at the construction site. He had known Preston and Clarence all along, but not well. He first knew Tally as a neighbor. They came to be friends through Tally's visits to the girl who lived with Budder, his common-law wife, and his wife's children. When Tally took Budder onto the job with him, Budder became a co-worker and drinking buddy, too. Thus, in Tally's network, Wee Tom began as co-worker, moved up to drinking buddy, neighbor, and finally close friend; Budder from neighbor and friend to co-worker. Importantly, and irrespective of the direction in which the relationships developed, the confluence of the co-worker and especially the neighbor relationship with friendship deepened the friend relationship.

The most common form of the pseudo-kin relationship between two men is known as "going for brothers." This means, simply, that two men agree to present themselves as brothers to the outside world and to deal with one another on the same basis. Going for brothers appears as a special case of friendship in which the usual claims, obligations, expectations, and loyalties of the friend relationship are publicly declared to be at their maximum.

Sea Cat and Arthur went for brothers. Sea Cat's room was Arthur's home so far as he had one anywhere. It was there that he kept his few clothes and other belongings, and it was on Sea Cat's dresser that he placed the pictures of his girl friends (sent "with love" or "love and kisses"). Sea Cat and Arthur wore one another's clothes and, whenever possible or practical, were in one another's company. Even when not together, each usually had a good idea of where the other was or had been or when he would return. Generally, they seem to prefer going with women who were themselves friends; for a period of a month or so, they went out with two sisters.

From Tally's Corner *by Elliot Liebow. Copyright 1967 by Little, Brown and Co., Inc. Reprinted by permission of the publisher.*

---

primary type of relationship is one that involves deep and personal interaction and communication. This interaction is an end in itself: Primary groups often exist because of the value of the primary relationship rather than because of other specific goals or tasks. People conform in primary groups because of strong informal norms—for example, fear of being ostracized, scorned, or ridiculed—rather than because of any formal written rules.

A **secondary group,** on the other hand, is more impersonal. Interaction is more superficial and probably based on utilitarian goals; that is, the person is less important than a particular skill he or she may offer the group. Interaction and communication are based on the value of one's particular skill rather than on interest in one's general personal qualities. It is probably most helpful to see the concepts of primary and secondary as opposite ends of a continuum. The completely primary or completely secondary group may seldom be found. Rather, groups vary in their degree of primariness or secondariness. Moving from primary to secondary along the continuum, we might see these groups: a married couple, an extended family including parents and grandparents, a basketball team, a sorority, a professional organization or labor union, or the employees of a large corporation.

SMALL GROUPS

The study of small groups has long been a central interest of sociology and social psychology. Recently, small groups have become popular in a number of areas. They are used as vehicles for treatment: Group therapy has proven to be a useful mechanism for change in prisons, mental hospitals, and organizations dealing with alcoholism and drug usage. Sensitivity-training groups and encounter groups have become popular on college campuses and with large organizations as techniques for improving communication, interaction, and self-understanding. Likewise, the small class is believed to work better than the large class in the educational process.

Small groups combine a primary group type of interaction with a task orientation. Achievement of the task—better education, self-understanding, and improved communication and management skills—is facilitated by the primary group atmosphere. In the following paragraphs, we will discuss ways of studying small groups and some of the characteristics of small-group behavior.

Social scientists have studied groups in a variety of ways, and each of the methods they use focuses on a different aspect of small-group interaction. For example, Robert Bales developed a technique called **interaction process analysis (IPA).**[2] In this technique, a group is watched by researchers, and aspects of the group's interaction are evaluated and categorized. This act may be "showing solidarity," whereas that act may be "giving a suggestion" or "asking for an opinion." Bales's technique classifies bits of interaction into predetermined categories. He believes that groups behave in typical and predictable ways and in fact move through stages in a sort of life history.

J. L. Moreno developed a technique called **sociometry** that provides a quite different way of studying small groups. Here the focus is more on the *who* of interaction rather than on the *what*. Sociometry looks not at the nature or type of interaction, as Bales's IPA does, but at the direction of interaction—who interacts with whom. Members of the group being studied are asked questions like, "Who is your best friend in the group?" or "Who would you most like to work with on an important project?" The patterns and directions of interaction are then graphed, as shown in Figure 4.1. This figure represents a sociogram of the members of a volunteer fire department. Each member of the group was asked three questions: "Who would you most prefer to have help you fight a fire?" "Who would you most prefer to help you plan a social function?" and "Who is your best friend in the department?" The lines and arrows in the sociogram represent the resulting choices. For example, fireman 13 chose fireman 11, firemen 1 and 7 named each other, and so on. Not all the selections are shown in this sociogram to keep it from getting too complicated. Those near the center of the sociogram were chosen more often: Fireman 11 was named 28 times, fireman 9 was named 14 times, and fireman 4 not at all. Several typical sociometric patterns appear in this figure. Fireman 4 was not chosen and is therefore called an *isolate*. Firemen 5, 10, and 13 were named only once each and so are called near-isolates or *neglectees*. Fireman 11 was chosen by nearly everybody and is termed a *star*. Rank in the department and

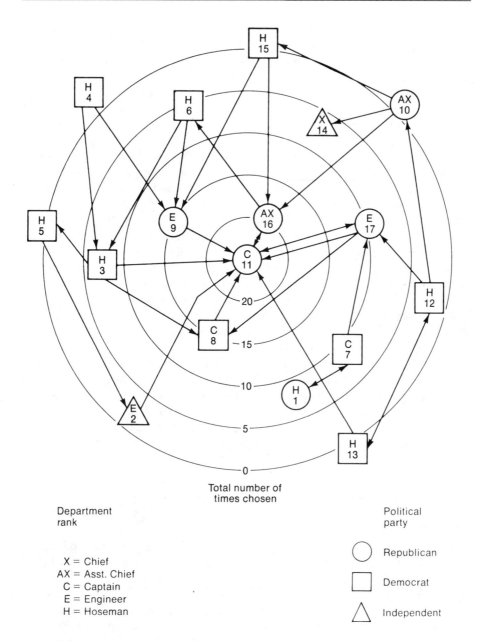

FIGURE 4.1 *Sociogram of a volunteer fire department.*

political preference are shown in the sociogram to see whether these explain patterns of selection. One of the assistant chiefs was more popular than was the chief, and the three captains are all near the center of the sociogram. Hosemen, the lowest rank, are consistently around the edge. More detail on fireman 11 indicates that, aside from being a Republican and a captain, he is one of two men in the department with a college degree, he is the third oldest member, he is one of five Catholics, he has the highest social-class standing in the department, and he participates actively in departmental activities.

As a technique for the study of groups, sociometry presents both advantages and disadvantages. On the positive side, it helps uncover informal interaction and friendship patterns and helps to identify variables associated with leadership and social isolation. On the negative side, sociometry is somewhat static and, unless repeated often, does not take into account the fluid and rapidly changing nature of group structure. It is an effective technique for mapping interaction but does not provide an in-depth analysis of that interaction.

Another way of studying small groups is by means of **participant observation.** Here, the researcher joins the group being studied and becomes involved in its activities; one learns about the group by becoming part of it. (Remember our discussion of participant observation in Chapter One—Whyte's study of Cornerville and Rosenhan's visits to mental hospitals.) Participant observation is often less controlled and provides less of the sort of detail that either IPA or sociometry gives. A talented observer, however, can use participant observation to achieve an analysis of more depth, can uncover meanings of actions, and can present a dynamic picture of a group's behavior. I have a feeling that if I were studying a group—the Hell's Angels, a campus fraternity, the San Francisco Forty-Niners, and the president's cabinet come to mind—I'd learn more through participant observation than I would through IPA or sociometry. Essentially, though, the method to use depends on the type of group being studied and on what you want to find out. In participant observation, the observer always has to be cautious so that his or her presence in the group doesn't unduly change the group's behavior and so that he or she doesn't become so much a part of the group that objectivity is lost. Both of these are crucial problems that the participant observer has to face.

Small-Group Research

The following paragraphs summarize some small-group research that I have found particularly interesting. Social psychologists have long been fascinated by conformity, and a number of studies have examined the effects of group pressure on individual judgment and behavior. (Milgram's study of conformity in France and Norway discussed in Chapter Three is an example of this.) A stationary light in a dark room appears to move; this is called the *autokinetic effect.* Muzafer Sherif asked subjects to judge how far the light moved. When small groups were tested, they quickly arrived at an agreement on the light's movement—a group norm. When the members were tested individually, they stuck to the earlier established group norm. When other subjects were tested individually *first* (no group norm available), their judgments of the light's movement were more variable and erratic. In other studies, Solomon Asch asked small groups of people to compare the lengths of lines. Some members of the group were stooges, or confederates of the researcher. The stooges were instructed to make obviously wrong selections: Two lines of different lengths were said to be the same. In studies involving eight stooges and one subject, the subject was faced with the choice of going along with the majority or making an independent judgment. Many avoided stress and simply went along with the majority.[3]

Small groups handle deviants in various ways. Laboratory experiments indicate that if one person continues to disagree with or block a position on which the rest of the group agrees, the group tends to reject the deviant. In another study of Quaker work groups and army basic-training squads, however, another pattern emerged. Robert Dentler and Kai Erikson found that in some groups deviant behavior helped maintain group equilibrium and that attempts to alienate a member whose behavior was deviant were resisted. Mentally ill (*schizophrenic*) members of basic-training squads were taken care of by the other recruits. Others performed the deviant's duties, they became very protective of him, and when he was finally hospitalized, other squad members became disturbed and angry at his removal.

Leadership and other roles in small groups follow certain predictable patterns. For example, two types of leaders that have been observed are the task leader and the social-emotional leader. The *task leader* possesses skills or characteristics that the group values highly, but he or she is not highly approachable. The *social-emotional leader* is highly approachable; he or she is the one to whom other group members can complain, show affection, or in other ways express feelings. The social-emotional leader may have some influence on the task leader. Other group roles are described by the terms *nice guys, ignored,* and *rejected*. *Nice guys* are approachable—not so much as the social-emotional leader but enough that their membership and contributions are valued. Those *ignored* are members who are usually inactive; they are an unknown quantity to the rest of the group. Those *rejected* are members perceived as unapproachable because they have no skills or characteristics valued by the group or because of hostile or confusing behavior on their part.[4]

A study of deviants in small groups was conducted to find out whether the deviant's rank in the group would affect the punishment he received. Twenty-five members of a college fraternity's pledge class were divided into four groups. Each group was told by an officer of the fraternity that another member of the fraternity had violated some rules and would have to be punished. The identity of the offender (independent variable) was changed for each group, and each group was asked to determine the appropriate punishment (dependent variable). Here are the descriptions of the four offenders:

Offender for Group 1. Highest social position, fraternity officer, very well liked, task and social-emotional leader.

Offender for Group 2. High social position, best-liked member of the sophomore class, social-emotional leader.

Offender for Group 3. Not identified by name—just as a "member of the sophomore class."

Offender for Group 4. Low social position, not very well liked.

The groups were told that the deviant had duplicated and sold fraternity files. They were given 20 minutes to discuss the case and were told to write down, individually and anonymously, the appropriate punishment. The members of Group 1 all felt the deviant should give a formal apology. Group 2 members felt the deviant should be given a one-month to one-semester suspension. Group 3 members felt a one-semester suspension to be appropriate, with one member voting for a one-year suspension. Most of the members of Group 4 felt the deviant should not be allowed to eat or sleep in the house, and one member voted for expulsion from the fraternity. We see that deviants are punished not so much according to what they have done, but according to their position or status in the group. The high-ranking individuals were lightly punished; the low-ranking person's punishment bordered on ostracism; he was physically removed from group activity. As the author of the study puts it, "The higher the position of a person within a group, the more lenient will be his punishment for deviance; the reason for this being that the higher the social position of a person, the more value he has to the group and the more they would have to lose by rejecting him."[5]

The second research question at the beginning of this chapter concerned the dynamics of small-group interaction. Talking about one's *own* knowledge, experience, or skills as a reason or explanation for behavior is called an "internal account." Talking about your own knowledge can show that you are competent and should be listened to. On the other hand, it may be seen as arrogant, especially if done by a low-status member of the group. Talking about or giving credit to *others* such as teachers, experts, and books to explain behavior is called an "external account."

Which is likely to make you more well liked and influential in a small group: giving internal accounts or external accounts? A recent paper examined this question by looking at women's influence and likability in small groups as related to the type of accounts they gave. Further, the researchers looked at three different types of small groups: all-female groups, mixed male-female groups, and male groups with one female. You can probably guess what happened. In all-female groups, the participant was most well liked and influential if she gave internal accounts. The authors of the study conclude that this reflects women's belief in their own competence and in that of their same-sex peers. In mixed male-female groups, the type of accounts given had no effect on likability or influence. Internal and external accounts worked equally. The authors conclude that in mixed groups, one's gender becomes less important and members are judged more on the quality of their contributions. If the lone female in an otherwise all-male group gave internal accounts, she was not well liked and had little influence. Only external accounts worked. The authors conclude, perhaps sadly, that in otherwise all-male groups, statements about herself made by a single low-status (female) member are seen as egotistical and self-serving.[6]

Some of the effects and constraints that small groups have on behavior can be seen in the actions of people on juries. The jury usually quickly selects a person of high status (wealth and/or education) as foreman. People of high status are more active on the jury; they talk more, have more influence, and are seen as more competent. Low-status people tend to defer to the higher-status members. Men are more active on juries, but women are better jurors;

they take more care in considering testimony. Jurors are very reluctant to come back without a verdict. The self-confidence of witnesses often has more effect on the jury than the logic or soundness of what they say. Juries often try the lawyer rather than the person he or she is representing. Finally, the mere fact of a person's being brought to trial makes him or her suspect in the eyes of the jury, in spite of the presumption of innocence in our court system.[7]

FORMAL ORGANIZATIONS

The new director of the Department of Motor Vehicles (DMV) ran head-on into a problem recently. He had DMV workers paint "no parking" in Chinese in no-parking areas around the DMV buildings. There were already signs in English and Spanish as part of a program to help people unfamiliar with English. The director was proud and the local newspaper published a picture of the new signs. But the next day they were gone, painted over. It seems that a grounds supervisor had also seen the picture in the newspaper, had checked around and couldn't find the necessary work order for having the signs painted, and had them painted over. The next day the director had the signs painted in Chinese again, but only after he had obtained the necessary work order. "It was his first real experience with bureau-cracy, and he lost it to the paper pushers," reported a spokesperson.

Large-scale formal organizations are an essential part of modern society. These organizations, often called **bureaucracies,** arise as societies' activities become increasingly planned rather than spontaneous. People discovered long ago that if several people got together and planned an activity such as building a car or educating a student, they often got the job done more rapidly and efficiently together than if each pursued the task in his or her own way, spontaneously and haphazardly. Arthur Stinchcombe defines a formal organization as any social arrangement in which the activities of some people are systematically planned by other people to achieve some special purpose. Those who plan the activities automatically have authority over the others. Formal organizations seem to arise in societies that have a money economy and that pursue complex tasks requiring the coordinated efforts of a number of people. As the size of the organization increases, administrative tasks multiply, and this encourages further bureaucratization. These organizations have an enduring quality: People leave, but like the building they worked in, their jobs and the organization remain, seemingly forever.[8]

Some years ago, sociologist Max Weber identified the characteristics of an abstract, or pure, form of bureaucracy:

1. A precise division of labor exists and therefore each individual in the organization is a specialist—an expert in performing a specific task.

2. There is a hierarchy of authority, a chain of command in which each individual is clearly in control of a particular set of people and clearly responsible to another.

3. An exhaustive and consistent system of rules has been designed to assure uniformity in the performance of every task.

4. Social relationships between individuals in the organization, especially between superiors and their subordinates, are formal and impersonal.

5. The bureaucracy is technically highly efficient, much like a machine.

6. Employees are highly trained for their tasks; they are, in turn, protected from arbitrary dismissal: Their jobs constitute a career with possibilities for advancement according to seniority and/or achievement.

Weber's characteristics described how bureaucracies *are supposed* to operate. In fact, formal organizations don't always follow Weber's description on all or even most points. In any organization, much that is unplanned occurs. The chain of command is often circumvented. Shortcuts reduce bureaucratic red tape in a variety of ways. Relationships between individuals are often personal and informal. Organizations occasionally are grossly inefficient. In fact, to counteract the somewhat negative view that many people have of the huge bureaucracy, organizations often consciously attempt to look and act less formal, less bureaucratic. Nevertheless, Weber's description provides us with a useful abstraction of the general characteristics of large organizations.

The bureaucracy I'm most familiar with is a university. If Max Weber wandered through our institution, he would probably note the following:

Division of labor, specialization: There are groundskeepers, secretaries, computer operators, librarians, janitors, and professors. The professors teach specific courses in specific areas, and these courses seldom overlap. In the sociology department I teach criminology, and another person teaches courses on Middle Eastern societies. We are both sociologists, yet he knows little about my area and I even less about his. We are all highly specialized.

Hierarchy of authority: There is an authority ladder in each section and a massive organizational chart for the college (which in fact looks like a map of the Los Angeles freeway system and I'm certain is understood by no one). Hierarchy runs from students (at the bottom) to professors to department heads to associate deans to deans to associate vice presidents to vice presidents to the president. Each must answer to the one above.

System of rules: Classes must be added or dropped by a certain day. Grades must be turned in by professors in a certain way by a certain time. Lists of rules describe how students may complain about injustices and how faculty may deal with unruly students. A faculty member wishing to leave campus to go to a professional meeting in another city must fill out a pile of forms weeks in advance of the trip. People are promoted by a mathematical model so complex that it defies comprehension.

Formal and impersonal relationships: All are handled alike; people in the same circumstances must be handled the same—the rules say so. We are all cogs in a large machine, and the machine works better if it does not have to concern itself with exceptions. People I know well become members of committees (budget, promotion, grievance), and suddenly they become very formal and impersonal.

A highly efficient bureaucracy: We have an assembly-line process with which we mass-educate thousands of students. We take flocks of uneducated high-school seniors and after only four years of subjecting them to some 40 sophisticated courses of instruction, we produce educated people.

Highly trained employees, protected from dismissal: Most of the faculty have Ph.D.'s or the equivalent in their area of specialization. After remaining with the organization for a set number of years, tenure (protection from dismissal) is granted.

But there is another side—the nonbureaucratic side of the bureaucracies, the informal organization in the formal organization. Our institution has that aspect as well. Specialization breaks down somewhat as instructors, because of changing enrollment patterns, move into new teaching areas. It is common knowledge that, contrary to the hierarchy of authority, the university is really run by the secretaries. They are here every day, they work longer hours, they have more knowledge about how the place really works, and they end up making or influencing many of the decisions. Informal small primary groups arise within the organization that counteract formality and impersonality. Often these cut across the authority structure as well. Several faculty lunch groups constantly meet to pass on information and to plan strategy. A phone call to the right person (usually a secretary) quickly cuts through the system of rules and bypasses miles of red tape. The typical organization is also so large that it cannot keep track of bits and pieces. Many professors do not turn in travel plans; they know that they could leave for months and never be missed. And efficiency? The car manufacturer proudly counts the number of units manufactured—3,000 cars today. What they don't tell us is that 2,500 broke down within a mile of the assembly line. Our institution graduates 2,000 people a year, but there is growing concern over the inability of many of the graduates to read and understand the diploma they receive.

I have perhaps exaggerated to make the point: Bureaucracies and large organizations have typical formal characteristics that develop because of their size and functions. They also have typical informal characteristics that develop because of the needs of the people who work in the organizations.

Types of Organizations

People who study organizations often come to the conclusion that organizations can be categorized. Max Weber categorized organizations as shown in Table 4.1. **Charisma,** which is a basic factor in Weber's charismatic retinue, refers to a certain superior quality an individual may have that sets him or her apart from others. This individual is viewed as superhuman and capable of exceptional acts—the sort of person you might follow to the ends of the earth if you were asked to.

Formal Organizations—Consequences and Concerns

Parkinson's Law: The Formula of 1,000
  Any enterprise with more than 1,000 employees becomes a self-perpetuating empire, creating so much internal work that it no longer needs any contact with the outside world.

Table 4.1 Weber's Types of Organizations

	Characteristics	Examples
Charismatic Retinue	Made up of the followers and disciples of a charismatic leader	Followers of civil rights leader Martin Luther King, Jr.; Reverend Moon's Unification Church; followers of a great scientist or artist
Feudal Administration	Separate from but responsible to a parent organization; has local autonomy to act in terms of own interests; is obligated to superiors, but there is no constant flow of orders or supervision	New-car dealerships; chain or franchised stores, agencies, or theaters
Modern Bureaucracy	Close to Weber's characteristics of bureaucracy; specialization, hierarchy of authority, system of rules, formality	Governmental agencies, large public utilities, military organizations
Modern Professional Organization	Bureaucratic but with more individual freedom, more decentralization of responsibility once the individual's competency is certified	Universities, large hospitals

Arthur Stinchcombe, "Formal Organizations," in Neil Smelser, ed., *Sociology: An Introduction,* 2d ed. (New York: Wiley, 1973).

A proliferation in society of large organizations and the bureaucratic values of efficiency and rationality will have consequences—some obvious, some less so. Large organizations enable the performance of tasks that could not otherwise be accomplished. Products are undoubtedly produced with greater efficiency. At the same time, **alienation** (a feeling of isolation, meaninglessness, and powerlessness) may result when workers see themselves as small cogs in a huge, impersonal machine. Stinchcombe states that in societies dominated by formal organizations, people tend to become specialists in particular activities. Educational institutions respond by encouraging students to concentrate on learning a specific technique or skill rather than attaining a broader, more general education. In societies dominated by formal organizations, a larger percentage of a person's social relationships is planned instead of spontaneous. Interaction is more superficial, and emotions play less of a part in social relationships of this type. People are judged more by what they *do* (actions) than by what they *are* (their individual human qualities). To put it another way, interaction of all types, on the job or off, is more secondary, less primary. Some analysts of modern society have commented with concern on the decline

Ron, *Toy Store Floor Manager*

I worked at Toys "R" Us in Tampa, Florida. The job was boring, but I found a number of ways to entertain myself, my fellow employees, and even the customers.

One Christmas the store had a Barbie doll house on display. Every night I would create a different scene by dressing the dolls up in strange outfits and setting them up in unusual situations. One time I dressed the Ken doll in a clown outfit, tied Barbie against a balcony, and set Ken up so he was whipping her.

One day during the summer I found a Barbie doll on the floor with half a leg chewed off. I attached a peg to her leg, resealed her in a box, and sold her as Peg-Legged Barbie. Another time I found a ripped-up Cabbage Patch Kid, and a friend and I resewed part of it and sold it as an Anatomically Correct Cabbage Patch Kid.

I had some fun with what I did. I'm not trying to sound like a jerk, but I was the best worker at the store. I kept the other employees amused. I saw it as a necessity for surviving the type of job I had.

From Sabotage in the Workplace *by Martin Sprouse. Copyright 1992 Pressure Drop Press. Reprinted by permission.*

---

in the number of primary groups with which one can affiliate. They argue that groups and contacts are becoming increasingly secondary in modern industrialized countries like the United States.

Scholars have long been interested in the contrast between primitive and modern societies or, to put it another way, the effects that modernization and industrialization have on a society. Obviously, the size of the society and the degree of modernization and development will change the social relationships within the society. Several of the typologies that have been developed to describe this process are shown in Table 4.2.

Many who feel as Cooley did—that primary group relationships are of great importance in the socialization process and in the development of the self and personality—are concerned that the increasing secondariness or *Gesellschaft*-like nature of modern societies (Table 4.2) will have serious effects on the developmental processes. Some would no doubt point to high rates of divorce, rising crime and delinquency, increased mental illness, and other individual and group pathologies as consequences of restricted primary-group contacts. Others would argue that these conditions are merely the consequences of any modern, complex society. Moreover, the huge, complicated, multilayered bureaucracies that perform tasks for us today are highly efficient even though somewhat impersonal. To idealize the *Gemeinschaft* society, they would argue, is unrealistic and shows a lack of understanding of the positive aspects of the large organizations of today. They would conclude that there is not really more pathology, such as crime, marital breakup, and mental illness; our efficient bureaucracies just do a better job of ferreting it out and letting us know about it.

In the early part of this century, a German sociologist named Robert Mi-

Table 4.2 Types of Societies

Author	Smaller, Less-Industrialized Societies	Larger, More-Industrialized Societies
Robert Redfield	**Folk** Isolated villages, small, homogeneous, self-contained, subsistence activities	**Urban** Large, industrialized nations, highly developed division of labor
Emile Durkheim	**Mechanical Solidarity** Homogeneous, preindustrial, uniformity among people, lack of differentiation or specialization	**Organic Solidarity** Complex societies, greater specialization and division of labor, increasing individualism
Ferdinand Toennies	**Gemeinschaft** Primary, close-knit, relationships are personal and informational, people are committed to community	**Gesellschaft** Secondary, impersonal, utilitarian, rational thought rather than emotion, contractual arrangements, well-developed division of labor

chels described a curious and perhaps threatening characteristic of large organizations. He used the phrase "Iron Law of Oligarchy" to describe what happens. Michels felt that **oligarchy**—that is, rule by a few—is inevitable because it is difficult for a large number of people to arrive at a decision efficiently. So, experts emerge to run things, and they eventually become indispensable. Once they are in power, they spend more and more of their time and energy maintaining their positions, rather than working toward the goals of the organization. You would think that the rest of the workers would get upset at this, but on the contrary, according to Michels, they are very happy to have somebody to look after things. The elite, as a consequence of working to stay in power, becomes increasingly conservative, and its interests become predominant over those of the masses they are supposedly serving. Michels concludes that in large organizations, oligarchy is inevitable and abuse of power follows.[9]

Another critique of organizations appeared some 40 years ago when W. H. Whyte, Jr., wrote *The Organization Man*. He painted a rather gloomy picture of how large organizations were affecting people's lives. Americans have long been guided by what has been called the Protestant Ethic—the idea that pursuit of individual salvation through hard work, thrift, and competitive struggle is the heart of the American achievement. According to Whyte, a new ethic has emerged as a consequence of the proliferation of large organizations. Now the young person going out into the world believes that to make a living he or she must do what somebody else tells him or her to do. The idea of individual achievement and the encouragement of personal creativity is disappearing. Instead, the guiding values of the organization person include a belief in the group as source of creativity, in belongingness and togetherness, and in the value of fitting in. These phrases are part of the organizational ethic: "Be a compromiser." "Maintain group solidarity at all costs." "Don't rock the boat." "The unorthodox can be dangerous." The "organization man" is conforming and "well-rounded"; he looks at his company this way:

> *Be loyal to the company and the company will be loyal to you. After all, if you do a good job for the organization, it is only good sense for the organization to be good to you, because that will be best for everybody. There are a bunch of real people around here. Tell them what you think and they will respect you for it. They don't want a man to fret and stew about his work. It won't happen to me. A man who gets ulcers probably shouldn't be in business anyway.*[10]

The organizational ethic maintains that the goals of the organization and the goals of the individual are the same. People don't buck the system; they cooperate with it.

Where did the "organization man" come from? He was produced by a society that is now geared to the large organization. The educational system, which is increasingly influenced by business, is the first step. The university curriculum has become more practical, vocational, and job-oriented, with a declining emphasis on liberal arts and the humanities. And what has become of the "organization man" now? He lives an outgoing, friendly life of inconspicuous consumption in suburbia, where he joins clubs, has kids, and plays golf, and where all the houses, cars, appliances, lawns, and even people tend to look alike.

And if the individual messes up, can't hack it, gets disgusted, and drops out—whose fault is it? The organization's? Certainly not. It's the *individual's* fault; he or she doesn't fit in. Our worship of organizations bothers Whyte, and he objects to the emerging values and behaviors that result.

Is it really as bad as he says? It's difficult to calculate the effects of formal organizations accurately because they are so much a part of our lives; but remember, he wrote *The Organization Man* many years ago (in 1956), and we are certainly a far more bureaucratized society now.

Can Formal Organizations Change?

In sometimes interesting ways, formal organizations modify their forms and structures and change their philosophies. For example, bureaucracies have long held a basic belief in specialization: The simpler the task, the better and more efficiently it can be performed. (See Weber's first characteristic of bureaucracy earlier in this chapter.) Break the job down into its basic parts. Make it simple, precise, and repetitive, and the worker can't possibly mess it up. Well, this sort of approach to tasks—three turns on the same nut with the same wrench 3,000 times a day—may in fact be more efficient, but people don't like it. They get bored, job satisfaction goes down, and absenteeism goes up. One popular management approach is to reverse the cherished specialization ethic and to *enrich* jobs by making them more interesting, more varied, and not so repetitive. This view has been supported widely by industry and by business graduate schools. Jobs should be meaningful, self-fulfillment is a worthwhile goal, and work groups are important for improving job satisfaction. Workers will take more pride in the product if they can follow it all the way through, performing a variety of tasks toward its completion.

They may be right, and it certainly sounds like a more enlightened approach to the work situation. But a skeptic might inquire about management's reasons for this new approach. Companies like to feel they are progressive. And, of course, it's good for the image. Ads have portrayed the happy Swedish automobile workers who start working on a car and then carry the job through the whole construction process. You just *know* that's a better car. Industry can be humanistic after all. But the clincher for the organization is the economic payoff. Improved worker morale means less absenteeism, less employee turnover, less sabotage, and improved quality of product. The best of all possible worlds exists if organizations can improve their images and cut costs at the same time. Imagine the company's dilemma, however, if the new program increases worker morale, happiness, and self-fulfillment—but efficiency and output are unchanged or reduced.

Weber's general characteristics of bureaucracy should not blind us to the fact that individual differences in organizations, as in people, are vast. Some are more affected by and responsive to change than others. For example, educational institutions are probably more likely to react to demands for change than are military organizations. Bureaucracies may be responsive to change in one area but resist it in another. The auto industry changes styles yearly, but its responses to suggestions of environmentalists come more slowly. And finally, organizations traditionally have been very slow to change. Problems inherent in industrial bureaucracies have been discussed for well over 100 years, yet little has changed. Today we are more bureaucratized than ever; large formal organizations are the handmaidens of mass society.

THEORY AND RESEARCH: A REVIEW

All three of the major theoretical perspectives introduced in Chapter One are included in this chapter, but *functional analysis* again, as in Chapter Three, is the major viewpoint. The characteristic emphasis on structure and function is seen clearly in the discussion of groups, types of groups, small groups, and large organizations. (See, for example, Weber's characteristics of bureaucracy and the analysis of types of organizations in Table 4.1.) *Symbolic interaction* is often helpful in understanding what happens in small groups, especially in participant observation and perhaps sociometric research situations. *Conflict theorists* would enjoy interpreting the findings of the study of fraternity deviants and also the description of jury behavior. Of the research efforts described in the chapter, Sherif's and Asch's classic studies as well as the study of fraternity deviants are classified as *experiments*. *Surveys* (on jury behavior by Dentler and Erikson) and *participant observation* (the sociometric study of the volunteer fire department conducted by a member of that group) are also shown.

SUMMARY

The social organization, or *social fabric*, of a society is woven of cultural norms, values, beliefs, and patterns of behavior through which people interact with each other. The analysis and description of this social organization is a major area of study for sociologists. The analysis of the social organization of a society can be approached from a number of viewpoints. The viewpoint we have taken in this chapter introduces the concepts of group, category, and aggregation. A *group* exists when a number of people experience shared or patterned interaction and feel bound together by a consciousness of a kind. A collectivity of people lacking these traits is called a *category* (if they have a particular characteristic in common) or an *aggregate* (if they are located in a specific area).

Sociologists categorize groups in a variety of ways. Of special importance are primary groups, small groups, secondary groups, and formal organizations. Ways of studying small groups were described in this chapter, as were some of the effects small groups have on their members. Characteristics of past, present, and future bureaucracies were compared. By categorizing and analyzing specific types of groups, we are better able to understand and explain the variety, importance, and effect of the group affiliations individuals maintain. However, any one group will cut across types and carry several labels. I work for a large formal organization in which I am a member of a group—a sociology department—that could be described as a reference group, a peer group, a voluntary group, a horizontal group, a closed group (depending on the job market and how many professors are seeking the California climate), and sometimes a primary group—sometimes a secondary group.

The first of the readings that follow is an excerpt from Christopher Edwards's book, *Crazy for God,* in which he describes how group pressure can be used to manipulate the individual. The second reading, by Barbara Garson, shows how automation and computerization can affect the everyday work situation.

TERMS FOR STUDY

aggregate (113)

alienation (126)

bureaucracy (123)

category (113)

charisma (125)

closed group (115)

folk society (128)

Gemeinschaft (128)

Gesellschaft (128)

group (112)

horizontal group (115)

interaction process analysis (118)

involuntary group (114)

mechanical solidarity (128)

oligarchy (128)

open group (116)

organic solidarity (128)

participant observation (120)

peer group (115)

primary group (116)

reference group (114)

secondary group (117)

social organization (108)

sociometry (118)

urban society (128)

vertical group (115)

voluntary group (114)

Reading 4.1

CRAZY FOR GOD

Christopher Edwards

Groups are important for the ways they affect and shape the individual. We don't generally think of the coercive power of groups or of how they indoctrinate members, but that is also part of group life. Group pressure is a powerful and very effective instrument for producing individual change, as can be seen in this excerpt from Christopher Edwards's book, Crazy for God. *After his graduation from Yale, Edwards traveled to Berkeley, where he met some young people who invited him to a training camp for a "fun weekend." This led to his joining the Reverend Moon's Unification Church. He spent seven months as a "Moonie" until he was kidnapped by his father and deprogrammed.*

After a quick lunch on the lawn, Family members jumped to their feet, took their guests by the hands, and led them to a nearby field to play dodge ball. As I kicked at the ground, sending up clouds of dust, I savored schoolboy pleasures revisited. I looked around at the usual sea of smiling faces. But now deep in the Family members' eyes I glimpsed a chilling single-mindedness, an intent that both bewildered and fascinated me. What lay behind this supposedly innocent game?

Each group divided up into one of two teams. Each team was appointed a captain who suggested a cheer and team chant. During the entire game our team chanted loudly, "Bomb with Love," "Blast with Love," as the soft, round balls volleyed back and forth. Again I felt lost and confused, angry, remote and helpless, for the game had started without an explanation of the rules. The guests were being moved around the field like robots on roller skates.

"Listen, Chris," Jacob called from the sidelines. "If you don't understand the rules, just chant or cheer as loudly as possible. The important thing is to do whatever a Family member tells you. Remember, unity is everything here."

I dutifully started shouting.

"Louder, louder, Chris. That's it. Just follow me." He placed me in a new position and I clapped mechanically. A few minutes later, he moved me again.

I noticed how aggressively the Family members played, how they constantly eyed their guests, coaxing them to chant out loud. As I clapped and shouted, I could feel my tension slipping away, my sense of involvement growing. In spite of myself, I felt a desire to merge into this Family, this group, this game, to become a part of this vibrant, loving circle. "Give in, Chris," urged a voice within me. "Just be a child and obey. It's fun. It's trusting. Isn't this the innocence, the purity of love, you've been searching for?"

The game ended before I had figured out the rules. . . .

Davey read the work assignments for each group. I was assigned to the trail crew, a group that was clearing a path up to the hills so that Family members could conduct spiritual hikes at sunset. After being given our picks and axes in the barn we proceeded to the trail. As we started to work, Edie shouted:

"Now, remember, first-weekers can't talk to first-weekers. Second-weekers can't talk to second-weekers. Promise you'll obey that!"

I resented this kind of regimentation, but I kept my mouth shut.

Seeing how awkwardly I was handling my pick, Scott stepped over to show me how to use it. As we were chatting, Jim suddenly appeared at our side.

"Hey, you're both first-weekers! You aren't allowed to talk to each other. If you want to know something or need help, just ask an older brother or sister. That's what we're here for. We've all been first-weekers. We know it's tough."

Annoyed at the interruption, I muttered "Okay," and picked away silently. Apparently sensing my anger, Jim suggested:

"How 'bout a song?" Did he really think he could dissipate my negative feelings that easily?

Jim distributed songbooks to the "new members," as first-weekers were called. All afternoon, we took turns singing religious and patriotic solos as everybody else busily picked and shoveled away. The hours passed quickly. My initial resentment toward Jim gradually melted away in the mass effect. . . .

The fifty people crammed in the little trailer suddenly whistled and burst into simultaneous applause. Everyone was smiling, grinning from cheek to cheek, delighted by Heavenly Father's power. Marilyn sat down, beaming. We all dug into our squash, bean, and salad dinner as Moses, at the center of the circle, called on another person to give testimony.

I could feel myself getting high, high from this smiling group, this happy Family surrounding me. Whatever my doubts might be, it seemed that this loving circle must have an element of goodness. And if there is goodness, then mustn't it be true?

"Individual Entertainment" followed. Various members of the audience volunteered to sing selections out of the songbook on an impromptu stage. What innocent joy, this entertainment was, I thought, a great way for guests to release their inhibitions and prove themselves to older members. Mary stood up and sang "Go Down, Moses." Keith, as an older member, sang a more militant song of self-righteousness, "Marching on, Heavenly Soldiers." The crowd cheered frantically. Edie, who was keeping close tabs on me as usual, elbowed me.

"What, Edie?"

"Well, aren't you going to sing? You're the only second-weeker who hasn't sung."

"I am?"

"Sure. Now, go sing."

"No thanks. I don't care to."

"Go. Go do what you're told."

"No. C'mon, Edie, I don't want to."

Phil, overhearing the conversation, collapsed in prayer behind me to ask for the salvation of my soul. I could feel my face grow hot. No one had ever prayed for me before, let alone for God to forgive me for not singing. Maybe I just had a lot to learn about the way in which God works. . . .

"Don't be such a spoiled child, Chris," Edie persisted. "Now, come on, do what I tell you. It's for your own good. And remember, if you want to grow in the Family you have to follow center. Come on, are you afraid of what people will think of you? Are you really afraid?"

Me—afraid? I'd never thought of it that way. It looked like I'd have to prove to her that I wasn't and that I wanted to be part of this Family.

"What should I sing?"

"Sing something that indicates your rebirth as a heavenly child. Sing something you'd never sing in Old Life."

As she said Old Life, Edie wrinkled her nose. She paged through the songbook, paused to think, tugged my shirt. "Sing 'God Bless America.' Go on!"

No, I could never sing that song. My God, I had been so ashamed of my country, a country consumed by worship of the dollar and aggrandizement of the self. What would my leftist college friends think if they saw me now? But since that was the most difficult song to sing of all, I decided I'd have to do it to prove to myself that I could, that I could even become a heavenly child if I wished. Edie raised my hand, and Moses called on me. I stood up and walked over to the stage.

"God bless America—land that I love . . . ," I began in a choked voice. An image flashed across my mind. I was nine years old, at a baseball game with my brother and Dad. A scratchy record was playing the National Anthem as the Yankees paused from their pregame warm-up. I felt the same lump in my throat that I was feeling now.

As I continued, singing louder, my voice cracked with emotion. My hands trembled. Everyone was staring at me, trying to establish eye contact, looking completely trans-fixed in typical Family fashion. My voice trembled and I began to choke again as I spilled out the faintly remembered words.

"God bless America—my home sweet home!"

The audience rose in unison, breaking into frenetic applause. Two brothers picked me up, led me around the room, and dropped me off next to an ecstatic Edie. Tears streamed down my face, tears of joy or sadness, or God knows what in this chaotic caldron of unbridled emotion.

I was a little child once again, the child who cried when Tarzan slipped from his vine or Lois Lane got captured by crooks. The Family kept jostling me, elbowing me, shaking my hand. Nobody had ever loved me so much, nobody had really cared. . . . This was where I wanted to stay, where I could be loved and accepted. This was the place where God wanted me to be. . . .

Remembering how Jacob had handled me that first weekend, I set out to do the same with Bill, to establish a good "give and take," as the lectures called it. It was essential for him to trust me, to consider me his friend, as I led him through the group rituals, deliberately creating a new reality for him, the reality of the group. One of my aims was to "love-bomb" him, to appeal to his ego in every way possible. I wanted him to feel as positive about himself as possible, for he had to feel accepted enough to want to stay before we could remold him, reshape him into the heavenly child we wanted him to be. I couldn't love him for what he was, for according to the Principle he was a product of Satan's world; I loved him for what I wanted him to *become,* an obedient follower of Reverend Moon, the mysterious Messiah behind this movement.

After two hour-long lectures, I took Bill aside and asked him, "What do you think of the Farm, Bill?"

"It's really amazing. The group's sincerity and dedication are fantastic. But I have so many questions about the lectures and about the social service projects in Berkeley, which nobody will talk about in any detail. And I can't get accustomed to the group life in our circles here. You know, how everything comes from or goes to Edie. Is this true in all the groups?"

"Well, yes. We feel that elder members have a special relationship to God, so the elder members direct the activities and mediate the conversations."

"I find that hard to accept, Chris."

"But you get used to it. You will learn that everything works better that way. And you'll find that the older members have a great deal of wisdom."

"Maybe so, but Edie can't or won't answer all my questions."

"Don't worry. They'll all be answered in time. You have to have faith, brother. Just have faith."

I patted him on the back and hugged him.

Our group walked out to the field for the usual weekend dodge ball game. Edie was the team captain for the four groups that composed our team.

"Okay, brothers and sisters. We have to think up a name for ourselves. How about the Heavenly Hustlers?"

"How 'bout the Boonville Bombers?"

"How's that, everybody?" Edie cried.

"Great!"

"Great! Just great!"

"Great!"

"Great!"

As I said "Great," I could hear the emptiness in my voice; I could feel a haunting hollowness in my head. The word echoed over and over after it spilled automatically from my lips. Over the weeks I learned that this word was an instant Family turn-on, as well as a term of obedience. I had learned to say "great" whenever somebody asked me if everything was all right, when they spoke in a tone that indicated that he or she demanded acknowledgment of the God-given authority of their orders.

I had noticed older Family members use the cue phrase in answer to their central figures even in the moments of depression, hopelessness, and despair they experienced while under heavenly subjugation. Their faces looked as rigid as mine, and their cheeks must have hurt as much as mine often did, the muscular pain of constant smiling, hour after hour, day after day.

So it would be the Boonville Bombers vs. Father's Flying Forces. Edie quickly explained the rules of the game, too quickly for the newcomers to understand, leaving them dependent on their spiritual parents, as I had been the first week. I marveled at the beauty of the system.

"Just follow me. I'll help guide you," I explained to Bill. Each team formed three lines, the front line always throwing the ball and back lines moving forward as the front-liners were knocked out of the game.

I realized by now that this was a particularly violent game, one in which a person could unleash his aggressions, his frustrations, at being controlled and manipulated. The older Family members played it extremely well, and I found myself getting better at playing it each weekend, enjoying this rare moment of individual expression. As the game progressed, Bill looked toward me, helplessly, for instructions. I told him where to stand, how to clap and chant, how to dodge the ball.

Edie took me aside. "How's Bill doing?"

"Okay so far. I've been establishing common bases and we're having good give-and-take. I think that he is beginning to look up to me."

"Good! Is he with us yet?"

"Not completely, but I'll see what I can do at lunch."

"Good going. You're quite a heavenly child, Chris." She gave me a broad smile and whispered in my ear: "Your Heavenly Father loves you!" I grinned. We did a "choo-choo-choo," and I fell back into line, into the game.

Bill was obviously enjoying the game now, happy to be part of a team. I knew that teaching teamwork, teaching mutual interdependence was the first step on the road to indoctrination. Edie had explained that you have to consider yourself to be part of the team before you can accept the ideology of the team. This is a most important point, for it means that group affiliation is a stronger force than ideology, which only justifies and reinforces the affiliation.

As the game progressed, I walked around Bill, encouraging him to clap and chant. The group got very, very high off this chanting. Angry faces, excited faces, joyous faces, all united under one common objective. The game ended when the last person was knocked out of the roped-off square. The team huddled momentarily and Edie shouted:

"What are you going to do?"

"Face the ball!" we shouted.

"What else?"

"Catch the ball!"

"What else?"

"Throw the ball!"

"What else?"

"Win the game! Yay, yay!" we all shouted as the new game began. Bill looked confused.

"Where do I stand?" he pleaded once again.

"Right here." By the hand I led the Harvard grad student to his position. He passively followed.

"Now, catch the ball, face the ball, throw the ball. Win the game! Let's have a choo-choo."

Dazed, Bill chanted with me as I grabbed his hands and shook them up and down in ritual.

"Choo-choo-choo. Choo-choo-choo. Choo-choo-choo. Yay, yay—Pow!" He giggled, and I smiled, reinforcing his childish behavior. Recently divorced, confused about the meaning of life, this lonely academic, starved for affection, was finding the simple love and acceptance he sought in this classless universe of giggling children.

Bill had the final shot in the game, smashing a fat little girl in the knee with the flexible rubber ball, winning a victory for the Boonville Bombers. Everybody crowded around him, mussing his hair, shaking his hand, patting his bottom. He was all smiles, feeling the love radiating from each member of the team—purposeful affection that older Family members could turn on and off at will. Everybody was high once again.

"How did you like the game, Bill?"

"Great, just great!" he said, parroting those around him. . . .

"I'd like to be excused to go to the bathroom," Bill announced. Edie look startled.

"Can't you wait until the group is finished?" she asked sharply.

"No, I want to go now."

"You must obey the rules of the group, Bill. That was assumed when you came here."

"Sorry, but I have to go *now.*"

"Okay, Chris will go with you."

We trotted off together. As Bill entered the stall, he said "Gee, I don't know why you had to come along."

"It's part of the Heavenly life, brother. We have to be so close that we'd follow each other anywhere." I didn't tell him that the real reason I had followed him was the Family's belief that Satan would attack him if he was left alone. Satan lurked everywhere, even in those stalls. Reverend Moon himself had been quoted many times as saying: "Satan is everywhere, and you are vulnerable to his attack!"

As we left the brothers' bathroom, I reached for Bill's hand. This time he instinctively pulled back. Looking him straight in the eyes, I said sternly, "You want unconditional love, don't you?"

"Why, yes, but . . ."

"No buts. If you want unconditional love, you have to obey the rules. That's all there is to it. As the principles say, Truth is Truth."

Bill reluctantly took my hand, and we returned to our circle, where Edie was leading a sensitivity game designed to get the group accustomed to taking orders from their leader, their center person.

"Now, I want everybody to lie down in the circle and put your feet in the center of the circle. Wiggle your toes. That's it. Now, wiggle your toes to say hello to the partners across from you. Good."

I admired the Family's genius, for I realized that this seemingly innocent exercise was, in fact, part of a complex and powerful mechanism for gaining control. I was to learn over the next few months that it worked as follows: You were cajoled to give up control to a person for five minutes, and that person structured your environment for that time. Then you gave up control for another twenty minutes, following the wave of group singing. Then you listened to lectures, giving up your critical control, since control in the discussion groups was contingent upon accepting the ideology of the lectures. You actually begin to listen to the lectures *only* to gain an awareness of what the group leaders would do to you and how they would justify it, for at that point it becomes a matter of survival.

Bill seemed to be responding well. When Edie asked for his inspiration, he replied, "I'm really impressed by the people here. I can't believe there's a community like this, smack in the center of California, a community of such loving, growing people—a real family. It's just tremendous. It's . . ."

QUESTIONS 4.1

1. Outline the techniques the group in this selection used to manipulate the individual. Do groups generally use these same sorts of techniques to persuade group members to conform? If yes, show how; if no, are you sure? And why don't they?

2. Why are these group pressure techniques so effective? Discuss in terms of primary group, reference group, socialization, resocialization, and so on.

3. What was the point of age-segregating (first-weekers and second-weekers) the prospective members?

Reading 4.2

THE ELECTRONIC SWEATSHOP

Barbara Garson

Barbara Garson believes that "the underlying premise of modern automation is a profound distrust of thinking human beings." She feels that computers and automation can be useful, but when they define and control the whole work experience, they can turn work, which should be a pleasant and creative experience, into an "electronic sweatshop."

In this excerpt from her book, Garson describes jobs in three workplaces (a griddleman at McDonald's, a reservations agent for American Airlines, and two social workers at the NTW [Not the Worst] welfare center in Boston). She focuses on how jobs have been changed by the emphasis on automation and computers.

Jason Pratt

"They call us the Green Machine," says Jason Pratt, recently retired McDonald's griddleman, "'cause the crew had green uniforms then. And that's what it is, a machine. You don't have to know how to cook, you don't have to know how to think. There's a procedure for everything and you just follow the procedures."

"Like?" I asked. I was interviewing Jason in the Pizza Hut across from his old McDonald's.

"Like, uh," the wiry teenager searched for a way to describe the all-encompassing procedures. "O.K., we'll start you off on something simple. You're on the ten-in-one grill, ten patties in a pound. Your basic burger. The guy on the bin calls, 'Six hamburgers.' So you lay your six pieces of meat on the grill and set the timer." Before my eyes Jason conjures up the gleaming, mechanized McDonald's kitchen. "Beep-beep, beep-beep, beep-beep. That's the beeper to sear 'em. It goes off in twenty seconds. Sup, sup, sup, sup, sup, sup." He presses each of the six patties down on the sizzling grill with an imaginary silver disk. "Now you turn off the sear beeper, put the buns in the oven, set the oven timer and then the next beeper is to turn the meat. This one goes beep-beep-beep, beep-beep-beep. So you turn your patties, and then you drop your re-cons on the meat, t-con, t-con, t-con." Here Jason takes two imaginary handfuls of reconstituted onions out of water and sets them out, two blops at a time, on top of the six patties he's arranged in two neat rows on our grill. "Now the bun oven buzzes [there are over a half dozen different timers with distinct beeps and buzzes in a McDonald's kitchen]. "This one turns itself off when you open the oven door so you just take out your crowns, line 'em up and give 'em each a squirt of mustard and a squirt of ketchup." With mustard in his right hand and ketchup in his left, Jason wields the dispensers like a pair of six-shooters up and down the lines of buns. Each dispenser has two triggers. One fires the premeasured squirt for ten-in-ones—the second is set for quarter-pounders.

"Now," says Jason, slowing down, "now you get to put on the pickles. Two if they're regular, three if they're small. That's the creative part. Then the lettuce, then you ask for a cheese count ('cheese on four please'). Finally the last beep goes off and you lay your burger on the crowns."

"On the *crown* of the buns?" I ask, unable to visualize. "On top?"

"Yeah, you dress 'em upside down. Put 'em in the box upside down too. They flip 'em over when they serve 'em."

"Oh, I think I see."

"Then scoop up the heels [the bun bottoms] which are on top of the bun warmer, take the heels with one hand and push the tray out from underneath and they land (plip) one on each burger, right on top of the re-cons, neat and perfect. [The official time allotted by Hamburger Central, the McDonald's headquarters in Oak Brook, Ill., is ninety seconds to prepare and serve a burger.] It's like I told you. The procedures make the burgers. You don't have to know a thing."

"I would never go back to McDonald's," says Jason. "Not even as a manager." Jason is enrolled at the local junior college. "I'd like to run a real restaurant someday, but I'm taking data processing to fall back on." He's had many part-time jobs, the highest-paid at a hospital ($4.00 an hour), but that didn't last, and now dishwashing (at the $3.35 minimum). "Same as McDonald's. But I would never go back there. You're a complete robot."

"It seems like you can improvise a little with the onions," I suggested. "They're not premeasured." Indeed, the reconstituted onion shreds grabbed out of a container by the unscientific-looking wet handful struck me as oddly out of character in the McDonald's kitchen.

"There's supposed to be twelve onion bits per patty," Jason informed me. "They spot check."

"Oh come on."

"You think I'm kiddin'. They lift your heels and they say, 'You got too many onions.' It's portion control."

"Is there any freedom anywhere in the process?" I asked.

"Lettuce. They'll leave you alone as long as it's neat."

"So lettuce is freedom; pickles is judgment?"

"Yeah but you don't have time to play around with your pickles. They're never gonna say just six pickles except on the disk. [Each store has video disks to train the crew for each of about twenty work stations, like fries, register, lobby, quarter-pounder grill.] What you'll hear in real life is 'twelve and six on a turn-lay.' The first number is your hamburgers, the second is your Big Macs. On a turn-lay means you lay the first twelve, then you put down the second batch after you turn the first. So you got twenty-four burgers on the grill, in shifts. It's what they call a production mode. And remember you also got your fillets, your McNuggets. . . ."

"Wait, slow down." By then I was losing track of the patties on our imaginary grill. "I don't understand this turn-lay thing."

"Don't worry, you don't have to understand. You follow the beepers, you follow the buzzers and you turn your meat as fast as you can. It's like I told you, to work at McDonald's you don't need a face, you don't need a brain. You need to have two hands and two legs and move 'em as fast as you can. That's the whole system. I wouldn't go back there again for anything."

Kenny

Until recently, airline reservation agents have been high-paid, long-term employees. They were valued because they had to learn all the company's routes, fares and policies. Then they had to apply this knowledge while responding on the telephone to the thousands of turns that even the simplest conversation can take. It seemed that this would always involve a great deal of personal judgment. After all, a two-way conversation isn't made out of interchangeable parts.

But in a feat of standardization even more phenomenal than McDonald's fry vat computer, the airlines have found ways to break down human conversation into predictable modules that can be handled almost as routinely as a bolt or a burger.

This has already downgraded reservation work. And as I listened to airline reservation agents I slowly realized that something more than their job was being simplified.

I met young people like Kenny at American Airlines, who explained the dozens of ways in which the two-minute reservation conversation was now timed and graded. Kenny, a new and enthusiastic employee, told me that he liked the work, despite its stresses, because he liked being "involved with people." This modern young man seemed to have no idea that involvement had once meant something far more spontaneous than what he now did.

I also met many reservation agents (most of them older) who understood how their own responses were being streamlined, and they resisted. They would have followed company instructions for assembling a car, I believe, but assembling a conversation was somehow different.

"There's AHU, that After Hang Up time. It's supposed to be fourteen seconds. It just came down to thirteen. But my average is five seconds AHU, because I do most of the work while the customer's still on the phone. There's your talk time, your availability, your occupancy—that's the percent of time you're plugged in, which is supposed to be 98 percent. That's not the same as your availability. There's bookings. You're supposed to book 26 percent of your calls, which is very low. I've averaged 37 percent in the five months I've been here."

Kenny was explaining the computer-collected statistics on which his first raise would be based. Raises currently ranged from 2 to 7 percent. The exact rate depended on how many of the numerical standards, for AHU, talk time, occupancy and so on, you met or exceeded.

"One of the big numbers up to now is conversion rate." I'd never heard of a conversion rate.

"Every call is either action or potential," Kenny explained.

"If someone calls up and says, 'Hi, I want to go to L.A. first class,' that's action. There's nothing I have to do but get the information.

"But the fare shopping! [He held his hands up in horror.] Where they want to know is it cheaper if I go in the middle of the week or at night or after the holiday."

"You mean like I want to visit my mother in Florida, so I listen to all the rates and then say, 'Thanks, I'll call back after I talk to my mother'?"

"Right. Now if someone were listening in, I would be graded on how I probed for the business."

"O.K.," I said, becoming the customer. "Thank you. I'll have to call my mother and see what time is most convenient for her."

Kenny played the reservation agent: "And do you think you'll be flying toward the beginning or the middle of the week?"

"Well, uh, I have to be back by the following weekend so I guess the beginning."

"Tuesday we have a flight leaving at 10 and at 4. Which would you prefer? (See, I'm trying to get you to make a tentative booking.) Ten?"

"But I don't really know yet. I'll call back after I . . . "

"So I give you a sales pitch. 'It costs nothing to save a seat. There's limited seating at the SuperSaver fare.' I could give them every sales pitch in the world and they still say [now he imitates me, repeating like a parrot], 'I have to call my mother, I have to call my mother, I have to call my mother.'

"If we're going round and round and you keep resisting, I move into what we call a close. 'I still have room on the flight leaving at 10 on Tuesday. Let's hold the seat for you tentatively. What's your last name?' And if you say, 'My last name is Garson,' then I know I've booked it. I've converted a potential!"

"Congratulations." I shook his hand across the table.

"But if you still say, 'Dah dah dah dah,' then I could go back to a sales pitch. 'We're coming on the holidays; summer is a hard time to fly; it doesn't cost you anything to hold the booking. Now of the 10 o'clock and the 4 o'clock, which do you prefer?'"

"You're closing again," I said.

"No," he explained, "'Which do you prefer' is called a probe. 'What's your last name' is a close. Now if I come to the close and they don't book I could start around again. I could go around and around for two hours. It's always a pull between keeping your talk time down and getting your conversion rate up. There's no actual rule on it. Generally, I say if you come to close twice and they don't book, let them go."

I stared at Kenny. "That's amazing." I knew, of course, that good salesmen had always analyzed every phase of a contact. But it had never occurred to me that my two-minute reservation conversation could be broken down into opening, sales pitch, probe and close. When Kenny is monitored live, by his supervisor, he'll also be graded for such subjective qualities as tone of voice and effective use of name, as in, "If you like, Mrs. Garson, I can book you on the . . ."

"How does the monitoring work?" I asked. "I mean the human monitoring, what does it feel like?"

"We don't know when we're being listened to. You know afterwards because normally you get calls one after another without any let up. Suddenly there's a break, you look around and you see there's calls coming in to the other agents but you're not getting any. Then you hear a voice in your ear, 'Hello.' It's the supervisor from a listening post. Mine is usually on another floor. There are listening posts all over the office.

"You never know when it will happen. I haven't been monitored yet this month so I figure some time this week. It's usually once a month."

"What do they say?"

"Let me try to remember how it was last month. She said, 'Hello, I've been listening to your calls for an hour and you've been doing everything fine. Keep it up.'"

"What would a criticism be like?"

"Usually it's, 'You didn't ask for the business.' 'You should have closed.'

"Sometimes they also listen from the corporate headquarters in Dallas. It's called trunk monitoring. But *they* never come in on the line. You never know about Dallas."

Eddy Malloy

When I was in college, many people studied social work with the expectation of a civil service job in some do-good agency, the largest of which was Welfare. They knew they'd never get rich as social workers, but, like teaching, it was a respectable profession that paid them for helping others. Besides, it was a secure job—for we have the poor always with us.

The main reward of these jobs was a work style (lifestyle was a later invention) that allowed social workers to interview, visit, fill out forms (alas), make evaluations and decisions and generally function as free human beings while doing worthwhile work.

CPAs deal with numbers. Lawyers deal with contracts. Social workers deal with people. It was the right kind of work for my fellow Berkeley students who pinned IBM registration cards onto their chests to declare *Do not fold, spindle or mutilate.* They demanded to be treated as individuals, not numbers. So they chose to work with individuals, not numbers.

Today I don't meet many undergraduates going into social work. In part this is a matter of fashion. Poor people are out of style. Teaching similarly declines in prestige when children decline in prestige.

But by now social work isn't just a less-prestigious or lower-paid profession; it's hardly a profession at all.

Of course the poor are still with us. So are the welfare centers. But these centers are no longer staffed by autonomous professionals. They're staffed by increasingly regimented clerks. These automated welfare workers have no more chance to deal with people than do airline reservation agents.

How can the "helping professions" be clericalized? How can you eliminate the listening, talking, asking and advising from a job intended to help? It sounds impossible to make social work as physically and mentally constrained as an assembly line job, but that's exactly what's being done.

"You mean that for every form you fill out, you have to fill out another form saying you filled out a form?"

"You'll never be able to get this system straight," Eddy expressed his confidence in me. For the entire lunch hour he'd been trying to explain the point system that had evolved from the time studies.

It went roughly like this:

Time standards, in tenths of an hour, had been established for each function that an FAW [Financial Assistance Worker] like Eddy performs. For instance:

Issue food stamp I.D. card _____ .3
    [meaning three-tenths of an hour, or 18 minutes]
Authorize funeral and burial expenses _____ .7
Replace lost or stolen check _____ .4
Complete redetermination _____ 1.8

The full list contained over sixty functions and was frequently updated.

Each function, like "update medical report" encompassed many "actions," such as going downstairs to see what the client wants, pulling the file, filling out several forms, distributing the forms to supervisors or the Electrical Data Processing center (EDP), returning the case file. The original time studies had identified and assigned time standards for over 1000 actions.

But how did they keep track of it all?

When a worker completed a function—when, for instance, he handed his supervisor the forms necessary to record a change of address or a change of food stamp benefits—he filled out an additional form called a "control slip." At the end of the month the supervisor used these control slips to count the number of points (tenths of hours) each worker had earned.

It was remarkably like the piece-work system in a mink factory I had visited. There, as you finished cleaning each pelt you tore off a tag and tossed it into a cigar box for the supervisor to count. Here, as you finished adding a baby you handed in a control slip. (At the welfare center the control slip was sometimes called a "penny slip," just as it is in a garment shop.)

At the end of the month the supervisor counted the worker's points, his tenths of hours, and calculated how close he came to using 100 percent of his available time. Each worker received a monthly report card with a percentage grade. So far a worker wasn't disciplined unless he fell below 70 percent. That was the system after the time study but before full computerization.

"So these control slips mean you fill out a form for every form you fill out?" I asked incredulously.

"You'll never get it down straight," Eddy assured me. "They keep changing everything anyway."

Daniel Sheridan

Several people had described Daniel Sheridan as the fairest and most efficient supervisor at NTW. "Straight arrow" was the phrase that kept coming up.

. . . When we were seated and served, I posed my first question as open-endedly as I could. "What do you think of the time standards and the point system?"

"I blame the union for the way it's operating," Daniel answered concisely.

"You mean because they're sabotaging it?"

"No, because they're not sabotaging it."

"What do you mean?"

"If they followed the rules as the department issued them, this system would have collapsed in three months."

"You want it to collapse?"

"If I were a worker and a union activist," Daniel said, "the first time I did 100 percent in the first three weeks of the month I'd stop work. And if they tried to make me do anything over 100 percent, I'd fill out an overtime form.

"The problem," he declared, "is that all the workers have developed systems of their own to get the points they need and still deliver a timely service. That's what keeps this place going."

QUESTIONS 4.2

1. See how many of Max Weber's characteristics of bureaucracy you can find in this reading.

2. Find examples of the other side—the nonbureaucratic side, the informal organization within the formal organization.

3. Look at the criticisms of bureaucracy under "Formal Organizations—Consequences and Concerns" in this chapter and describe which ones you think fit here. Or, with respect to the jobs described in this reading, analyze the statement: "It may seem cold and impersonal, but it's the fastest and most efficient way to get things done."

4. Make a list of five other jobs (perhaps several at your college or university) and show how they have become automated and computerized. Do you think this makes them less pleasant and creative?

NOTES

1. See "Religiosity, Social Class and Alcohol Use: An Application of Reference Theory," by Leslie Clark, Leonard Beeghley, and John Cochran, *Sociological Perspectives* 33 (Summer 1990), pp. 201–218.

2. Bales's major work is *Interaction Process Analysis* (Cambridge, Mass.: Addison-Wesley, 1950). Shepherd's summary of Bales's work has been especially helpful; see Clovis Shepherd, *Small Groups* (San Francisco: Chandler, 1964), pp. 27–36. Also see Michael Olmsted, *The Small Group* (New York: Random House, 1959), pp. 117–132, and *The Small Group*, 2d ed., by Michael Olmsted and Paul Hare (New York: Random House, 1978).

3. Muzafer Sherif, *The Psychology of Social Norms* (New York: Harper, 1936); Solomon Asch, "Effects of Group Pressure on the Modification and Distortion of Judgment," in *Groups, Leadership and Men*, edited by M. H. Guetzkow (Pittsburgh: Carnegie, 1951); and S. Asch, "Studies of Independence and Conformity: A Minority of One Against a Unanimous Majority," *Psychological Monographs* 70, no. 9 (1956), whole no. 416.

4. The studies discussed in this and the previous paragraph are summarized by Shepherd in *Small Groups*, pp. 69 and 74–79 (see Note 2).

5. Lowell W. Gerson, "Punishment and Position: The Sanctioning of Deviants in Small Groups," *Case Western Reserve Journal of Sociology*, vol. 1 (1967), pp. 54–62.

6. See "Gender Composition, Attributional Accounts, and Women's Influence and Likability in Task Groups," by Judith Taps and Patricia Martin, *Small Group Research* 21 (November 1990), pp. 471–491.

7. For more detail on these and other studies of jury behavior, see Hans Toch, ed., *Legal and Criminal Psychology* (New York: Holt, Rinehart & Winston, 1961).

8. See Arthur Stinchcombe, "Formal Organizations" in *Sociology: An Introduction*, 2d ed., edited by Neil Smelser (New York: Wiley, 1973). Also see Peter Blau and Marshall Meyer, *Bureaucracy in Modern Society*, 2d ed. (New York: Random House, 1971) and H. H. Gerth and C. W. Mills, *From Max Weber: Essays in Sociology* (New York: Oxford University Press, 1958).

9. The major source for this section is *Elites and Masses* by Martin Marger (New York: Van Nostrand, 1981). Especially see pp. 70–73.

10. William H. Whyte, Jr., *The Organization Man* (Garden City, N.Y.: Doubleday, 1956), p. 143.

CHAPTER FIVE

Social Inequality: Social Class

TWO RESEARCH QUESTIONS

1. *Are wealthy women more likely than poor women to have childbirth by Cesarean section? Or would it be the other way around, or no difference? Why?*

2. *Are women on welfare more likely to give birth to children than women not on welfare? How does length of time on welfare affect the likelihood of giving birth?*

SOCIOLOGISTS STUDY GROUPS and nongroups to understand the social organization of society. The previous chapter dealt with groups—specifically, small primary groups and large formal organizations. In this chapter our attention turns to nongroups. Many sociologists spend much of their time specializing in the study of *categories* of people—people who have a particular characteristic in common but who do not constitute a group. Those who specialize in the study of older people, an age category, are called gerontologists. Others study race, ethnic, religious, social-class, and gender categories.

The process of defining, describing, and distinguishing among different categories of people is called **social differentiation.** People differ across a range of variables, and we find ourselves automatically in some of these categories, such as age, gender, and race. Other categories, such as social class and religion, have a greater degree of flexibility, and to a certain extent one's position

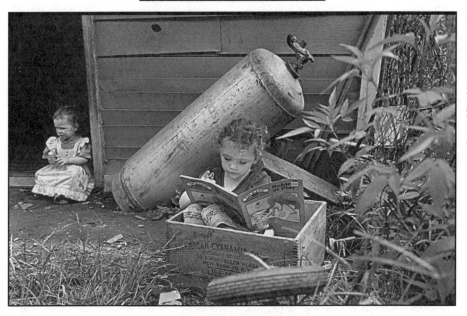

Leonard Freed/Magnum Photos, Inc.

within them can be changed. (Recall the discussion of ascribed and achieved status in Chapter Three.)

It is perhaps an unfortunate fact of life that social differentiation leads to **social inequality.** We differentiate among people, and very often the distinctions we make lead to judgments of inequality. In a sense, we "rank" categories of people at different levels. These are social definitions, and they may be different from one society to the next, but they have important social meaning. Whether one has brown or blue eyes is probably of little significance to his or her future. However, think of the social meaning attached to being black or white, male or female, old or young, rich or poor. Among the questions that sociologists are interested in are: *What* distinctions are made, *why* are these distinctions made, and what are the *consequences* of these distinctions for the people involved?

In this chapter and the next, we will focus on several important categories—social class, race, ethnicity, age, and gender—all of which involve judgments of social inequality.

SOCIAL STRATIFICATION

Semper Anglia—A dignified English solicitor-widower with a considerable income had long dreamed of playing Sandringham, one of Great Britain's most exclusive golf courses, and one day he made up his mind to chance it when he was traveling in the area.

>   *Entering the clubhouse, he asked at the desk if he might play the course. The club secretary inquired, "Member?" "No, sir." "Guest of a member?" "No, sir." "Sorry."*
>
>   *As he turned to leave, the lawyer spotted a slightly familiar figure seated in the lounge, reading the London Times. It was Lord Parham. He approached and, bowing low, said, "I beg your pardon, your Lordship, but my name is Higginbotham of the London solicitors Higginbotham, Willingby, and Barclay. I should like to crave your Lordship's indulgence. Might I play this beautiful course as your guest?"*
>
>   *His Lordship gave Higginbotham a long look, put down his paper and asked, "Church?" "Church of England, sir, as was my late wife." "Education?" "Eton, sir, and Oxford." "Sport?" "Rugby, sir, a spot of tennis and a number four on the crew that beat Cambridge." "Service?" "Brigadier, sir, Coldstream Guards, Victoria Cross, and Knight of the Garter." "Campaigns?" "Dunkirk, El Alamein, and Normandy, sir." "Languages?" "Private tutor in French, fluent German, and a bit of Greek."*
>
>   *His Lordship considered briefly, then nodded to the club secretary and said, "Nine holes."*

The study of **social stratification** arises out of the recognition that people are ranked or evaluated at a number of levels. As the geologist finds layers when examining a cross section of the earth, so the sociologist finds layers in the social world. Societies today are stratified into many layers, but this was not always the case. Hunting and gathering societies, the oldest known, were very egalitarian. Members of these small (50 people or less) societies spent all their time finding food; there was little, if any, division of labor and little social differentiation. The horticultural societies that followed were larger (from 200 to 4 million people). They were able to cultivate domestic plants and thereby produce a surplus, which led to activities beyond pure survival. Jobs aside from food gathering emerged and accumulation of wealth began. Specialization, division of labor, and presence of a surplus produced the beginnings of social stratification.

This trend continued in agrarian societies. Improved tools—the plow replaced the digging sticks and hoes of the horticultural societies—meant more efficient and productive use of the land and greater surpluses. This meant that a smaller proportion of the population remained involved in food production while more people occupied themselves as makers of tools and weapons, and as miners, merchants, traders, and soldiers. The division of labor and the greater surpluses in agrarian societies brought pronounced social stratification and social-class differences. These trends have continued and probably accelerated in industrialized societies. The division of labor is greater and stratification is marked.[1]

We find in societies today that people are *unequal*; they are constantly evaluated and ranked by others on a number of criteria. The idea exists that we are all created equal, but what is really meant is that we all have an equal chance to become unequal. Even that is not true, though, because some people have an obvious head start and others are blocked from the time they are born. For example, imagine two boys, Billy and Tommy. Billy is the one-year-old son of black parents in rural Mississippi. His parents are tenant farmers, the family income is less than $3,000 a year, he has six brothers and sisters, and neither of his parents went beyond the eighth grade in school. Tommy is the one-year-old son of white parents in Scarsdale, New York. His father is a surgeon, his mother is an author, the family income is over $150,000 a year, he has one brother, and both his parents are college graduates. Now map out each boy's future. What are the chances of each living to the age of six? Of graduating from high school? Of going to college? Of going to an Ivy League school? Of receiving regular medical and dental care? Of living to the age of 65? Of getting married while still in his teens? Of having many children? Of having an arrest record? Of going to prison? Of having his name in *Who's Who in America?* These are two healthy, happy, bouncing babies, but their life chances are vastly different.

PARALLEL STRATIFICATION SYSTEMS

Most societies have more than one system of stratification. These parallel stratification systems are based on different factors. The system we usually use when discussing social class is based on occupation, education, wealth, power, and prestige. We could refer to this as socioeconomic status. Another stratification system might be based on race: Individuals are ranked high or low purely on the basis of racial characteristics. We see this kind of stratification in the United States and to a greater extent in South Africa. Age, gender, and religion may be the bases for other stratification systems. In some societies, older people automatically have higher status than do younger people. Northern Ireland is religiously stratified—the battles between Protestants and Catholics are often in the news.

When there are a number of parallel stratification systems, an individual will be ranked in several different systems at the same time. One may be ranked in terms of race, then in terms of socioeconomic factors, then in terms of age, and so on. And though the stratification systems are parallel, they are not necessarily equal. The status of an American black who is upper-class in terms of wealth and occupation may be roughly equivalent to the status of a white who is middle-class in wealth and occupation. The discrepancy is explained by the existence of two stratification systems—one that ranks a person according to socioeconomic status, a second according to race.

Our society, then, is stratified in numerous ways. In the remainder of this chapter and in the next chapter we will examine several ways by which people are categorized and stratified. The subject of this chapter will be socioeconomic stratification. The following chapter will deal with race, ethnic, gender, and age categories.

SOCIOECONOMIC STATUS AND SOCIAL CLASS

Status based on socioeconomic factors represents one of the major systems of stratification. Following the ideas of Max Weber, socioeconomic status is usually determined by wealth, power, and prestige.[2]

GOLD—WHO'S GOT IT?

Who makes the most money in American society? Is it people in business, entertainment, or sports? Well, people at the top in any of these fields do well, as can be seen below. The figures show 1993 income (in *millions* of dollars) of the top five earners in each of the three areas as reported by *Forbes* magazine.

Sports

Who	How Much
M. Jordan, basketball	$36
R. Bowe, boxing	25
A. Senna, auto racing	19
A. Prost, auto racing	16
G. Foreman, boxing	16

Business

Who	How Much
T. First, HCA Hospital	$127
L. Hirsch, U.S. Surgical	75
S. Weill, Primerica	67
C. Lazurus, Toys "R" Us	64
A. O'Reilly, H.J. Heinz	36

Entertainment

Who	How Much
O. Winfrey	$52
S. Spielberg	42
Guns N' Roses	27
K. Costner	27
B. Cosby	26

From Forbes, *May 24, September 27, and December 20, 1993.*

Wealth

Generally, when comparing and evaluating people, we rank higher those who have the greater *wealth* and store of material possessions—type and size of house, area of residence, make and number of cars, quality of clothes, and so on. In a society that places value on wealth and material possessions, it becomes important to allow others to find out how well-off one really is, and so we have **status symbols.** If a person drives a Rolls Royce, Lincoln Continental, or exotic foreign car with a low-numbered license plate; lives in Beverly Hills or on Park Avenue; is mentioned in the society pages; and winters in Palm Springs and summers in Switzerland, then we have a pretty good idea of where he or she belongs on the social-class ladder. The game becomes complicated, however, when status symbols change or become generally available. Then it becomes more difficult to tell who ranks where. Only the Internal Revenue Service knows for sure.

Wealth is strongly correlated with education, income, and occupation, and when socioeconomic status is measured, these other factors are usually included. Income refers to current earning capacity, whereas wealth refers more to an accumulation of money and property over time. The relationship between income and education is shown in Figure 5.1.

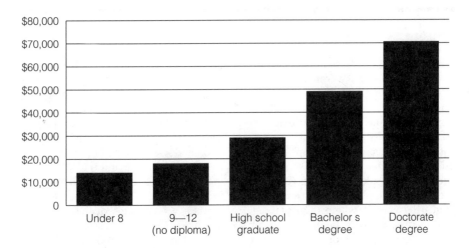

FIGURE 5.1   *Median income and school years completed, 1991.*

United States Bureau of the Census, *Statistical Abstracts of the United States, 1993* (113th ed.), Washington, D.C.

Power

People are also ranked according to the amount of power we think they have. **Power** refers to the ability of one party (either an individual or a group) to affect the behavior of another party. Individuals who run things, who make the important decisions at a city or national level, whether they are formally elected or informally appointed, are accorded high status by those who know of the power they exercise. Elected government officials, advisors to presidents, and consultants in high, mysterious-sounding positions are automatically given high status because we know that they are at or near the centers of power.

Power is a fascinating variable that stimulates numerous questions: Who has it? How did they get it? How is it used? How is power distributed in society? Some analysts of American society believe that power is concentrated in the hands of a few people who are not directly responsible to anyone else. Others believe that power is fairly equally spread across a number of interest groups who respond to the will of the people. (These issues are hotly debated, and we will return to them in more detail in Chapter Nine.)

Some theorists suggest that it is basically wealth, accompanied by the related variables of education and occupation, that establishes a person's position in society and that this wealth *leads* to power. Wealth is not shared equally in our society and never has been. In 1990 the IRS reported that in the most recent year available (1986), the wealthiest 1.6 percent of the U.S. population—those with assets of $500,000 or more—controlled 29 percent the wealth of the country. In fact, their holdings exceeded the entire gross national product of the

POWER WATCHES

Telling time is the least important function of a watch. What a watch really does is tell the world about the wearer. If you work it right, your watch can tell the world what you want it to. You're powerful. A contender. Someone who's taken out a lifetime lease on The Top.

Power watches are all Swiss watches. The two best sales areas are California and Texas, both places with strong entrepreneurial energy, where money, image, and keeping up with others is very important. Rolex ($8,800 for the President in gold) is currently the best-selling prestige watch in the United States. But according to one expert, Rolex is for new money. Old money has different rankings. The real power watch is the Patek Philippe. This company made the first watch. Albert Einstein, Queen Victoria, and Leo Tolstoy each wore a Patek Philippe. It is a basic round watch with only an hour and minute hand; it sells for $3,950 and is so understated that it looks like it costs $25. Very powerful. The Vacheron Constantine and the Audemars Piguet are also very high status. They are very conservative in look, nothing flashy. Old money doesn't need to impress. The Piaget Polo ($13,500 in gold) is the preferred power watch of the young jet set.

And the cost? "I had one client," reports a power watch dealer, ". . . a commodities broker. He came in one afternoon and bought a Rolex President, and had to use three credit cards to make the down payment. I asked him if he really could afford it. He said, 'I can't afford *not* to buy it. In this business I'm in, if I'm not wearing a power watch, I'm dead.'"

From Robert Ferrigno in Orange County Register. *Reprinted in* San Francisco Examiner, *"This World," December 16, 1984.*

Table 5.1 Proportion of Income Received by Each Fifth of the Population

Population Group and 1991 Income	1950	1991
Wealthiest fifth (more than $62,992)	43%	44%
Second fifth ($43,001–$62,991)	23%	24%
Third fifth ($29,112–$43,000)	17%	17%
Fourth fifth ($17,001–$29,111)	12%	11%
Poorest fifth (under $17,000)	5%	5%

United States Bureau of the Census, *Statistical Abstracts of the United States, 1993* (113th ed.), Washington, D.C.

country for that year! As the data in Table 5.1 indicate, the top fifth of the population controls far more than its share of national income, and the poorest fifth has only 5 percent of the national income. To put it another way, one fifth of the population in the United States has nearly half of the income, and this relationship has not changed at all in more than 40 years.

Prestige

Prestige means distinction or reputation and refers to how we subjectively evaluate others. It is usually connected to the position one holds in society.

STATUS SYMBOLS

If you live in Hong Kong, driving a Mercedes without an "appropriate" license plate is as awkward as wearing a tuxedo with sneakers. Chinese are known throughout Asia for their obsession with "lucky" numbers, and Hong Kong, where enormous amounts of money are paid to get the right license plate, is a remarkable example. Wealthy Chinese will pay thousands and sometimes millions of dollars for the right plate. A local business tycoon recently set a record by paying $3 million for the single-digit registration number nine, or "gaau," which sounds similar to the Cantonese word for longevity. Another unusual plate—AA4444—sold for $14,900. This is odd because four (a homonym of the word for death in Chinese) is usually considered unlucky, but apparently rarity outweighed bad luck. The government accommodates the situation by holding public auctions of special numbers—between 1973 and 1994 over 200 auctions have been held, raising about $66 million.

Some positions have high prestige (doctor, Supreme Court justice); others carry low prestige (shoeshiner, garbage collector). Prestige—distinction in the eyes of others—is less concrete than are wealth and power. It is no less important, however. In fact, if they had the opportunity to choose, many would probably rather have prestige than wealth.

In 1947, Cecil North and Paul Hatt asked a national cross section of people to evaluate a series of occupations as "excellent," "good," "average," "somewhat below average," or "poor." This resulted in an occupational prestige scale that has been widely used by sociologists.[3] The study has been repeated and updated numerous times since 1947, and the occupational prestige scores have remained amazingly consistent. Some of the findings of one of these replications are shown in Table 5.2. Look at the occupations and prestige scores in Table 5.2 and see if you can figure out what leads people to rank occupations in particular ways. Is it pay, or power, or education, or government service, or what?

Kingsley Davis and Wilbert Moore believe that the prestige and status accorded one's position are determined by its scarcity or importance to society and by the amount of training or talent needed to obtain the position.[4] Logically, we would expect that one holding a position high in prestige would receive greater financial rewards than one would in a position of less prestige. This is usually true, but not always. A carpenter or plumber often makes more money than does a teacher, but the teacher, whose skill requires more training, has higher occupational prestige; this may make the teacher *feel* better, but it won't pay the bills.

People in the situation just described—material wealth but low prestige, or vice versa—are victims of **status inconsistency.** They are status-inconsistent because the factors that determine their ranks in society are not consistent with each other. If we recall that there are many stratification systems—race, religion,

Table 5.2 Occupational Prestige Scores

Occupation	Prestige Score
U.S. Supreme Court justice	94
Physician	93
Nuclear physicist	92
Cabinet member, federal government	90
College professor	90
Chemist	89
Lawyer	89
Dentist	88
Civil engineer	86
Banker	85
Public school teacher	81
Building contractor	80
Railroad engineer	76
Electrician	76
Trained machinist	75
Farm owner and operator	74
Welfare worker	74
Policeman	72
Carpenter	68
Mail carrier	66
Traveling salesman	66
Plumber	65
Automobile repairman	64
Barber	63
Truck driver	59
Clerk in a store	56
Dockworker	50
Night watchman	50
Farmhand	48
Janitor	48
Garbage collector	39

Robert Hodge, Donald Treiman, and Peter Rossi, *Status and Power* (New York: Free Press, 1966), pp. 424–425.

age, gender, as well as socioeconomic factors—it is clear that there are many possibilities for status inconsistency. Examples of status-inconsistent people in the United States would include a college-educated carpenter, a wealthy business executive with an eighth-grade education, or a person of recent wealth but no prestige. The problem for the status-inconsistent person is that he or she behaves toward others in terms of his or her high rank (wealthy business executive), but others tend to behave toward him or her in terms of his or her low rank (eighth-grade education). Some sociologists report that status inconsistency may lead to other outcomes: stress, political liberalism, involvement in social movements, withdrawal, or psychosomatic illness.[5]

How Many Social Classes?

A variety of factors, then, produce a layered society. The question of just exactly *how many* layers or social classes we have has long been of interest to sociologists. Karl Marx was an economic determinist. He believed that a person's position in the economic structure of society determined his or her lifestyle, values, beliefs, and behavior. In his writings in the mid-19th century, Marx described societies as being composed of two classes: the *bourgeoisie,* or capitalists, and the *proletariat,* or workers. The bourgeoisie controls the capital, the means of production. The proletariat owns only its own labor. Marx envisioned a continuing struggle between these two classes, leading to the overthrow of the bourgeoisie by the proletariat and eventually the emergence of a classless society.

Studies of social class in the United States have produced various results. Lloyd Warner studied a New England community and discovered six social classes. August Hollingshead, using a technique similar to Warner's in a midwestern town, found five social classes. Dennis Gilbert and Joseph Kahl have devised a system of six social classes that they believe provides a reasonable picture of class structure in the contemporary United States. They see class structure as growing out of the economic system and placement as based on source of income (either capitalist property, labor-force participation, or governmental transfers such as social security, welfare, and so on), type of occupation, and educational preparation. Table 5.3 summarizes their model.

Other students of stratification believe that one distorts the picture to speak in terms of specific social classes, whether it might be two, three, five, six, nine, or whatever. One is more accurate, they feel, to speak of a stratification continuum consisting of many rankings with small gradations between them (much like a thermometer) and with no readily discernible social-class categories. Logically, this approach makes sense and may be the most accurate description of the way things really are. However, because people seem to be-

Table 5.3 Model of the American Class Structure

Proportion of Households	Class	Education	Occupation of Family Head	Family Income, 1990
1%	Capitalist	Prestige university	Investors, heirs, executives	Over $750,000, mostly from assets
14%	Upper middle	College, often with postgraduate study	Upper managers and professionals; medium business owners	$70,000 or more
60%	Middle	At least high school; often some college or apprenticeship	Lower managers; semiprofessionals; nonretail sales; craftspeople, foremen	About $40,000
	Working	High school	Operatives; low-paid craftspeople; clerical workers; retail sales workers	About $25,000
25%	Working poor	Some high school	Service workers; laborers; low-paid operatives and clericals	Below $20,000
	Underclass	Some high school	Unemployed or part-time; many welfare recipients	Below $13,000

From Dennis Gilbert and Joseph Kahl, *The American Class Structure: A New Synthesis,* 4th ed. (Belmont, Calif.: Wadsworth, 1993), pp. 305–323.

lieve in social classes and because teachers and researchers often find it easier to make distinctions among several broad categories than among numerous slightly differing gradations, we usually view stratification in terms of specific classes.[6]

Determining Social Class

Suppose you want to know what social class you belong to. There are several ways of determining one's rank. These methods include the subjective, reputational, and objective techniques. In the *subjective* method, individuals are asked what social class they think they belong to. This seems a rather naïve method, but many argue, as Marx did, that the world is divided along class lines and that a person's social-class identification is very much a part of his or her self-image; it is much of what he or she *is*. This view holds that social class is important to people, that these people therefore have a high degree of **class consciousness,** and that they should be able to place themselves accurately in the social-class spectrum.

The *reputational* method involves finding out what others think. People from the community are selected to act as prestige judges, and these judges in turn evaluate others in the community. Social class becomes what selected people say it is. This method is similar to the subjective technique, and again it hinges on community members' class consciousness.

The *objective* method differs from other methods in that individuals are evaluated in terms of specific factors that sociologists assume are related to social class. For example, Warner examined four factors when he used the objective technique: the individual's occupation, his or her source of income, the type of house, and the area of the city in which he or she lives. By knowing these facts about a person, Warner held that he could objectively place him or her in a specific social class. Other objective factors that have been used include years of education, amount of income, type of possessions, and even type and quality of home furnishings.

Using any of these methods, especially the objective method, may give the impression that social class is a very clear-cut phenomenon and that each person can be adequately plugged in. This, of course, is an oversimplification. The categories are not precise and are not always related one to another. For example, in what social class do we put the status-inconsistent person whose occupational prestige is low, income is high, who lives in an apartment in a slum, and is a college graduate—and when we ask her, she says she doesn't believe in social class?

Social Class and Behavior

When the Titanic sank in 1912, only 3 percent of the women in first class died, but 45 percent of the women in third class perished. It was supposed to be women and children into the lifeboats first, but the death rate was higher for children in third class than for men in first class.[7] We saw near the beginning of this chapter that the life chances of Tommy and Billy were affected drastically by their social class. It is clear that social class is related to numerous patterns of behavior.

Social-class differences in nutrition and health care lead to different disease patterns among the classes. A recent study comparing death rates in 1960 and 1986 found that although death rates had declined overall, poor and poorly educated people had higher death rates than people with higher incomes and better educations. The differences were greater in 1986 than they had been in 1960. Studies of social class and mental illness have found that psychoses, particularly schizophrenia and manic-depressive disorders, are more prevalent among lower-class than middle-class people. Middle-class people, on the other hand, are more likely to be neurotic, or so diagnosed by psychiatrists, than are lower-class people. There are class differences in treatment as well. Middle-class people are likely to have psychotherapy, highly qualified practitioners, and more thorough treatment of mental illness. Lower-class people are more often treated by interns and less qualified practitioners, and they are more likely to be subjected to questionable procedures, such as shock treatments, lobotomies, and drugs.[8]

The first research question at the beginning of this chapter asked whether there might be a difference between rich and poor women in their rates of childbirth involving Cesarean section. A University of California, Berkeley, study in 1989 reported that affluent women are nearly twice as likely as poor women to have Cesarean sections. Several reasons are suggested. Cesarean sections cost more than natural childbirth and rich people have the money, so physicians wanting to make more money may suggest the procedure. Or perhaps the wealthy are more likely to have Cesarean sections if the birth looks difficult because physicians fear that these patients will sue them for malpractice if problems occur. Also, poor patients are more likely to give birth at large teaching hospitals staffed by young resident physicians who are closely supervised and limit Cesarean sections to those cases in which they are truly necessary. Ironically, because Cesarean sections are more dangerous than natural childbirth, poor women may be getting better care than richer women.[9]

Voting behavior is affected by many factors, and social class is certainly one of them. Lower-class people are more likely to vote Democratic, and middle- and upper-class people are more likely to vote Republican. In addition, middle-class people are more likely to vote than are lower-class people. Turning to deviant behavior, we find that homicide is more prevalent among lower-class people than among middle-class people in the United States and throughout the world. Assault is also a lower-class phenomenon.[10]

Leisure-time activities are often class-related as well. Interest and participation in sports and other recreational activities are shown in Table 5.4. In some activities like backpacking, fishing, and exercise walking, high- and low-income households participate equally. Higher-income households are more involved in aerobic exercising, bicycling, golf, skiing, swimming, and tennis. Can you give some reasons to explain the differences?

In terms of the way they raise their children, middle-class parents are probably more permissive, lower-class parents more rigid. Melvin Kohn reports that middle-class mothers value self-control, dependability, and consideration,

Table 5.4 Participation in Sports Activities, 1991

	Percentage Who Participate	
Activity	Household Income Under $15,000	Household Income $50,000 and Over
Aerobic exercise	15	24
Backpacking	19	18
Bicycling	16	23
Exercise walking	18	22
Fishing	19	18
Golf	9	30
Downhill skiing	6	37
Swimming	15	24
Tennis	11	27

United States Bureau of the Census, *Statistical Abstracts of the United States, 1993* (113th ed.), Washington, D.C.

whereas lower-class mothers value obedience and the ability in a boy to defend himself. Vanfossen suggests that the way children are raised reflects their parents' work situation. The middle-class child is raised in an atmosphere in which independence, achievement, curiosity, autonomy, and *self*-control are valued. The lower-class child is raised in an atmosphere in which the immediate and concrete rather than the new and unfamiliar are emphasized; the child is taught to focus on getting by, and he or she learns obedience, conformity, propriety, and control by *others*.[11]

Many of the current theories attempting to explain delinquent and criminal behavior are based on social class. Walter Miller, for example, argues that lower-class juveniles who become delinquents do so because of the lower-class value system. Values and beliefs important to lower-class youth include trouble, toughness, smartness, excitement, fate, and autonomy (resistance to being bossed or controlled). Miller believes that the greater presence of these values in the lower class than in other social classes makes it inevitable that many lower-class children will run afoul of the law. He seems to be saying that we have two different cultures—middle-class and lower-class; but the *laws* in United States society emerge solely from the middle-class culture, and therefore one should expect that most illegal behavior will come from the lower-class culture. The rather pessimistic inference from this theory is that the only way to make it is to become middle class.

Several other theories examine the social-class–linked motivations for delinquency in a slightly different way. Albert Cohen, Richard Cloward, and Lloyd Ohlin agree that ours is a middle-class–based society, but they do not believe that a lower-class value system or culture leads one to delinquency. Rather, they feel that the basic problem arises out of the lower-class individual's attempt to move into the middle class. Cohen holds that most who try to make it will not. Anticipating failure, their reaction is to invert the middle-class system—to do just the opposite of what middle-class people say is correct. The result is malicious, nonutilitarian, and frequently criminal behavior. Cloward and Ohlin agree with Cohen up to the point of the lower-class individual's reaction of inverting the values of the middle class to spite them. Cloward and Ohlin believe that lower-class individuals still want to make it and that when they see they cannot succeed legally, they decide to try it illegally. They become involved, therefore, in utilitarian property crime, which allows them to collect the good things, the symbols of status associated with middle-class culture. Regardless of whether the theorists agree on particulars, the important factor in all of these theories is that they are based on social class. The social class to which people belong apparently provides them with a characteristic view of the world that does much to determine their behavior.

The theories discussed above deal only with lower-class crime and may leave one with the impression that all crime is committed by lower-class people. This is a common error that is made if one looks only at arrest figures, which *are* higher among the lower class. Middle-class people commit different types of crimes—fraud, embezzlement, white-collar offenses—that are less easily detected and punished. Also, middle-class people have the financial resources to deal with the law, to avoid arrest to begin with, and to obtain better legal defense should arrest occur. Typically, the middle-class juvenile delinquent is informally apprehended and returned to his or her parents, whereas the lower-class delinquent is formally arrested and sent to a juvenile hall. The actual extent of hidden middle-class crime is difficult to assess. Perhaps if all offenses could be known, the differences among social classes would be only in types of offenses, not in numbers.[12]

Social Mobility

Social mobility refers to movement within the social-class structure. This movement can be up, down, or sideways. **Horizontal social mobility** refers to movement from one occupation to another within the same social class. If an architect becomes a minister or if a mail carrier becomes a carpenter, we would say that horizontal social mobility has occurred—occupational change but no change in social class. The term is also used by sociologists to describe geographical mobility, moving one's home from one place to another. Vertical social mobility is more relevant to our discussion here, however. **Vertical social mobility** refers to movement up or down the social-class ladder—movement from one rank to another.

Societies with unlimited possibilities for vertical social mobility are described as *open*. Those with no possibilities for mobility are called *closed*. In reality, all societies rank somewhere between the extremes of completely open and completely closed. For example, India's system of stratification—a **caste system**—is nearly closed. Lines between levels, or castes, are often firmly drawn; moving from the caste into which one is born may be difficult if not impossible. Some have suggested that we have a type of caste system in the United States relative to blacks and whites; that is, there is a firm line between black and white stratification systems, and although a black may rise within the black class system, he or she is either blocked from the white system or is automatically assigned low rank in the white system. In societies with stratification systems that are more open, mobility from one level to another may be both possible and frequent.

Examining mobility in a given country is a complicated procedure. Americans, for example, have traditionally considered their system to be completely open. Mobility and achievement are supposedly available to anyone who works hard and has led a clean life. We *know* that the rags-to-riches Horatio Alger story is true and that we can make it too. The truth is, however, that mobility is very much the same in all modern industrialized countries. The dramatic leap from rags to riches is a rare occurrence indeed. Mobility, when it occurs, usually happens in a series of small steps and possibly over several generations. Pure luck, of course, can increase one's opportunities, as can the needs of technology—the growth of the computer industry, for example. We find more mobility *within* classes (horizontal mobility) than *between* classes (vertical mobility). Even though there seems in general to be more upward vertical mobility than downward and many individuals are mobile, few move very far.

A study of the mobility of men and women in the United States reported an interesting finding: Women are more mobile, both upward and downward, through marriage than men are through occupations. Men are much more likely to inherit their fathers' statuses than are women. Because our society lacks mechanisms for women to directly inherit occupations, they are freer to marry into statuses both above and below their origins.[13]

If I have convinced you that it is difficult to become upwardly mobile but you still want to pursue this aim without relying completely on luck, there are a few general rules to follow: Defer gratification, marry late, and have a small family if you must have one. If you live in a small town, leave. Vertical mobility is easier in medium-sized and large cities. If you are an immigrant to the United States, be Japanese, Jewish, or Scottish. These groups are more mobile than are other immigrant groups, who more often than not come in at the bottom of the social-class ladder and stay there. In modern societies in which skills and knowledge are increasingly important, education is essential for upward mobility. In the United States a college degree helps. For some not only the degree, but the right school—Harvard, Yale, Stanford, Princeton—is essential to mobility. Finally, if all else fails in your quest for upward mobility, marry someone who has followed the above rules.

But before you decide to move up, beware of the consequences. Sociologists have been studying the social-psychological effects of upward mobility. One result of uprooting oneself from the past and moving to a new level is that it becomes difficult to form satisfying relationships. One is alone, a **marginal person** between two worlds and a member of neither. Another view agrees that socially mobile people are isolates but argues that they were isolates to begin with. They were unhappy and isolated, and this led to attempts at mobility, so they are really no worse off. A third view of the situation, and the one that is most accurate according to current research on social mobility, is that socially mobile people have few problems because they are so anxious to get to the next level that they adopt the patterns of behavior of that level long before they ever get there; this behavior is called *anticipatory socialization*. Consequently, they are accepted and much more at home when they finally arrive.[14]

In a society that is upwardly oriented, downward mobility is hard to take. Studies cite relationships between downward mobility and both suicide and mental illness, especially schizophrenia. The same factors that are related to upward mobility are conversely related to downward mobility. Marrying early, having a large family, or failing to get an education may do it. Personal factors may help: business failure, sickness, or rejection of the ethic that says that climbing the social ladder is important.

POVERTY AND WELFARE

Earlier in this chapter we learned that wealth is an important variable in determining social class and that wealth is not equally shared in American society. The top 20 percent of the population controls 44 percent of the national income, but the poorest 20 percent have only 5 percent of the national income (see Table 5.1). This calls our attention to one of this country's major social issues: poverty. Poverty is a major social issue because of the links between it and crime, delinquency, self-esteem, mental illness, educational opportunities, physical health, life expectancy, infant mortality, and numerous other factors.

The United States government has developed a definition of poverty that is based on the amount of money needed to buy a nutritionally adequate diet, assuming that no more than one third of the family income is used for food. The poverty line varies according to family size, location, and the value of the dollar. In 1991, for example, the poverty line for a nonfarm family of four was $13,924. During that year, 36 million people (roughly 14 percent of the population) were below the poverty line. Large as that number is, it hardly represents the whole picture of poverty in this country. The poverty line refers to a so-called survival income, and many people who are above the poverty line don't have a normal or acceptable standard of living. They may live in substandard dwellings without electricity or running water, and although they may be able to survive, they know that most of the rest of the people in the country live far better than they do. It is difficult to arrive at an accurate estimate of how many poor and near-poor there are, but the bottom 20 percent of the population is the figure often used. Roughly 50 million people would fall into this category, and as we can see from Table 5.1, they were no better off in 1991 than they were in 1950. Each year they had only 5 percent of the country's total yearly income, although they represented 20 percent of the population.

Unemployment, inflation, and recession are part of the poverty picture. Although family income tends to increase each year because inflating dollars buy less, the "real" family income often *drops.* This forces still more Americans below the poverty level. In 1992, 9.4 million workers, or 7.4 percent of the labor force, were unemployed.

But who exactly are the poor? Obviously poverty doesn't strike all equally. The characteristics of the poor (see Table 5.5) show them to be the young members of ethnic and racial minorities, the illiterate and unemployable, and the disabled and handicapped. Poverty is higher in certain parts of the country, primarily in rural areas and in the South.

How does a society that sees its country as a land of opportunity deal with

Table 5.5   Characteristics of Families and Persons Below the Poverty Level, 1991

Characteristic	Percentage of Population
All persons	14
Blacks	33
Hispanic origin	29
Age 15–24	36
Age 65 and over, living alone	25
No high school diploma	24
Female householder, no spouse present	36
7 persons or more in family	33

Current Population Reports, Series P-60, No. 181, *Poverty in the United States: 1991.*

the existence of poverty? In curious and interesting ways. One approach is to hide it. The poor are located in areas we don't see and consciously avoid. The sick and aged are almost by definition invisible. The subject is not popular and won't be found on prime-time television except for an occasional documentary. The poor have no economic or political clout. The poor are a large but essentially anonymous and hidden group; we try to deny their existence.

Another way of dealing with an unwelcome problem is to blame the victim. William Ryan begins his book *Blaming the Victim* with the story of actor Zero Mostel impersonating a pompous senator investigating the origins of World War II. The climax of the senator's investigation comes when he booms out with triumph and suspicion, "What was Pearl Harbor *doing* in the Pacific?" Exactly. Thus, according to this view, the poor are poor because they are lazy and unintelligent, they lack motivation, they tend to drink and have loose morals, they can't manage money, and they don't want to work anyway. Blaming the victim releases the nonvictims from responsibility for the victims' plight. Recent studies have found that many Americans (often, a majority of those asked) believe that (1) most people on welfare who can work don't try to find jobs, (2) many people on welfare are not honest about their need, and (3) many mothers on welfare have babies to increase their welfare payments. In fact, studies of those on welfare show that all of these beliefs are *incorrect*.

The second research question asked at the beginning of this chapter concerned the relationship between welfare and fertility—whether women on welfare are more likely to give birth than those who are not. A 1989 study reports that welfare recipients have a low fertility rate, considerably below that of women in the general population. The study found further that the longer a woman remains on welfare, the less likely she is to give birth. Interviews with women of childbearing age who were on welfare showed that the economic, social, and psychological pressures of being on welfare were not conducive to wanting more children. Having more children was clearly seen as making the situation worse, not better. Nearly all the women interviewed wanted to get off public assistance, and having more children was seen as a major hindrance to doing that.[15]

Why do we have these beliefs about the poor? Probably because Americans fervently believe in the work ethic. All able-bodied people should work, their income should come from work, and the more hard-working they are the better people they are. Work is good—even if it is low-paying, exhausting, and degrading temporary work. There is also a welfare ethic, which emphasizes that no one should accept money unless absolutely necessary. This leads to suspicions regarding welfare recipients, more extensive monitoring to find suspected welfare cheaters, and confirmation of the view that people who take charity are bad people to begin with.

Finally, we should recognize that poverty is *functional*. The existence of the poor and their exploitation benefits other Americans in many ways. The poor do society's dirty work—the dirty, dangerous, undignified, and menial jobs. They work for low wages, and as domestics, for example, they make life easier for the middle and upper classes. The existence of poverty creates all sorts of jobs and occupations to service the poor: social workers, police officers, ministers, pawnshop operators, and the peacetime army (which recruits heavily from the poor). The poor help the economy by purchasing goods that the more affluent don't want: day-old food, secondhand clothes, and deteriorating merchandise. In many ways, then, the existence of poverty is advantageous to the nonpoor segment of American society.[16]

Pictures of starving people in Bangladesh, Ethiopia, and other countries should continually remind us that poverty is a worldwide problem. Wealth discrepancies between rich and poor countries are vast, and they represent a political powder keg and a disaster for humanity. Table 5.6 shows per capita gross national product figures for 20 countries. Think of the amount of poverty and the plight of the poor in this country, and then imagine the situation for a child born in a country in which the per capita income is a small fraction of what it is in the United States. What resources would such a country have to deal with poverty, starvation, or disaster?

The Welfare System

The welfare system was devised ostensibly to assist the victims of poverty—to help them fare well in society. The idea of welfare has been around for a long time. Originally, churches were the primary agencies for charity and relief giving, but as the urban labor class grew, attempts to control the poor emerged. In 1351 under King Edward III of England, a law was passed that established maximum wages, required the unemployed to work for whomever demanded labor, restricted the travel of workers, and forbade charity to the able-bodied. In large part, these have become the basic features of welfare in the United States today.[17]

Public-program social-welfare expenditures in 1990 were more than $1,045 billion. These programs included aid for families with dependent children (AFDC); Medicaid; aid for the aged, the blind, and the disabled; food stamps; veterans aid; workmen's compensation; black lung benefits; and so on. The

Table 5.6 Per Capita Gross National Product (GNP) in Selected Countries (in Dollars), 1989

Country	Per Capita GNP	Country	Per Capita GNP
Argentina	1,694	Mexico	2,170
Austria	16,430	Nepal	150
Bangladesh	180	Pakistan	330
Belgium	15,620	South Africa	2,253
Canada	20,240	Soviet Union	9,226
Egypt	1,342	Switzerland	27,510
Ethiopia	120	United Kingdom	14,580
India	321	United States	20,910
Indonesia	479	West Germany	19,520
Japan	22,900	Zaire	258

United States Bureau of the Census, *Statistical Abstracts of the United States, 1993* (113th ed.), Washington, D.C.

list of programs is long, and the amount of money spent is substantial and increasing sharply.

Critics of the welfare system, however, believe that in spite of the huge amounts of money spent, the system actually operates to *perpetuate* poverty. Those statutes that Edward III established over 600 years ago helped him *control* the poor. Piven and Cloward argue that our current system has the same intent: to regulate and exploit the poor. During periods when high unemployment leads to public unrest, welfare programs are expanded to absorb and pacify the masses. At times of political and economic stability and low unemployment, welfare rolls are narrowed, forcing people into low-paying jobs in the labor market.[18]

Turner and Starnes suggest that we really have two systems operating, a *wealthfare system* and a *welfare system*. The **wealthfare system** enables the middle and especially the upper class to maintain their privileged position by allowing them to avoid taxation and by giving them and the businesses they own and operate direct cash payments and subsidies. The wealthfare system is an example of the exercise of power, and it works through the intricacies of the federal tax system.

The **welfare system** exists, according to Turner and Starnes, because it appears to be a humanitarian effort to keep people from starving; it is politically necessary because, due to their large numbers, the poor have some power and they may revolt; and this system is a nice balance to and keeps attention diverted from the far more expensive wealthfare system. The welfare system perpetuates poverty by keeping payments low, by forcing people to work in any available job, and by diffusing any collective power the poor may have.[19]

What is the answer? We must recognize that poverty is a problem we want to solve and try to redistribute wealth and resources, perhaps through a guaranteed annual income; possibly in these ways could inequality be reduced. But we cannot be hopeful about the prospect. If our analysis is accurate, the middle and upper classes benefit from the system as it exists. Poverty is unpleasant, but the alternative to it—massive changes in existing education, government, and economic institutions—and the threat that change suggests are more unpleasant for the affluent to consider, and power resides with privilege. Again we must consider power—who has it, how it is used, and for whose benefit.

THEORY AND RESEARCH: A REVIEW

This chapter represents an excellent opportunity to compare two of the theories that were introduced in Chapter One; in the area of social stratification, functional analysis and conflict theory come head to head, presenting quite contrasting analyses of this concept. And it is here, perhaps, that the differences between the two theories become most evident. Proponents of functional analysis hold that societies are made up of complex and interconnected parts, each of which performs specific activities, or functions, to maintain the existence of the society as a whole. *Functionalists* like Parsons, Davis, and Moore suggest that stratification systems fit the pattern well. Their argument goes something like this: Societies that are at all complex must have an extensive division of labor to accomplish tasks and to assign workers to positions; this will guarantee smooth functioning and maintain stability. But the positions are not all equally important. Some (the physician, for example) require greater skill or responsibility than do others. To guarantee that the most qualified people, the ones with special talents, are encouraged to seek these jobs, greater rewards must be offered. Differential rewards result in the development of different classes of people. Every society must have a system that ensures that the most important positions will be filled by the most qualified people. Hence, stratification is a functional necessity.

Conflict theorists see it differently. Whereas functionalists suggest that stratification ensures the smooth and efficient operation of society, conflict theorists find that stratification is caused by the domination and exploitation of one group by another. Conflict theorists believe that society is best understood as a struggle (a conflict) for scarce resources such as food, land, wealth, or power. Some groups of people, perhaps because of inheritance or use of force, obtain more of these resources and are able to acquire a privileged position. People who have gained high positions will inevitably act in terms of self-interest and will continually strive to maintain and increase their influence.

The key for conflict theorists is control of scarce goods and services; those who can gain this control will hold power and prestige. Some occupations are rewarded more than others, not because of the importance of the task (the functionalist argument), but because of the bargaining power of the group. For example, if an occupation, by requiring apprenticeship or complicated certification, limits membership to a very few, then the services of those few will be in greater demand. Perhaps physicians are highly paid not because they have some special skill but rather because admission to medical school is artificially limited. Or a group may enhance its position by creating demand where none existed. The fashion industry, with its designer jeans and running shoes, may be an example.

The functionalists see society as a smooth-running machine with the parts fitting nicely. The conflict theorists see instead an exploitive adversary system involving coercion and force, with dominant classes manipulating and controlling subordinate classes. It seems difficult to believe that these two groups of theorists are talking about the same thing. They are presenting contrary but valid interpretations of the same set of observations.

The data used in this chapter are descriptive and are for the most part census data obtained by *questionnaire* or *interview surveys*. The North-Hatt study and the sports-activities survey (Table 5.4) are also examples of *reactive surveys* that obtain information through mailed *questionnaires*. The data described in the section on social class and behavior were obtained in general through *reactive surveys* using *questionnaires* and *interviews*.

SUMMARY

Sociologists study the social organization of society by studying groups and nongroups. The subject of the previous chapter was groups, and the characteristics and the variations in behavior in small and large groups were discussed.

This chapter focused on social differentiation and social inequality. People are distinguished from each other by a number of categories: age, gender, race, social class, and religion. These are not groups as we discussed them earlier; they are nongroups, or categories, describing very large collectivities of people. A central point for us here is that when people are separated into categories, they are *ranked.* People regard some categories as better than others. In this chapter we examined the idea of social stratification, the ranking produced by socioeconomic status.

The important variables in socioeconomic status are *wealth, power,* and *prestige.* Wealth, income, and education are separate but closely related variables. Several interesting debates run through the discussion of social class. Who has the power in the United States? What does the American class structure look like? It depends on whom you ask. Some see a continuum with no specific classes; others see three, four, five, six, or even more reasonably well defined classes.

We found that social class is related to numerous aspects of behavior: voting, mental illness, recreational patterns, homicide, child-rearing patterns, and delinquency. *Vertical social mobility* refers to movement up or down the social-class ladder. Analysis of social mobility in the United States indicates that Horatio Alger doesn't live here anymore; that is, few people move very far up or down the ladder.

Nearly 50 million people are poor or nearly poor in this country. Their poverty and the ways society deals with it are a major social issue for Americans. Analysis of the welfare system suggests that although it was established to help the poor, it may actually operate to perpetuate poverty.

In the next chapter, this analysis of social differentiation will continue as we direct our attention to other categories: race, ethnicity, gender, and age.

In the first of the readings that follow, William Ryan examines several conflicting models in an attempt to understand what we really mean by equality. The second reading, by Kristin Luker, is a study of women's opposing views on abortion. This reading suggests that the women's social-class differences relate in part to their belief differences.

TERMS FOR STUDY

caste system (160)

class consciousness (156)

horizontal social mobility (160)

marginal person (161)

power (151)

prestige (152)

social differentiation (146)

social inequality (147)

social mobility (160)

social stratification (148)

status inconsistency (153)

status symbol (150)

vertical social mobility (160)

wealthfare system (165)

welfare system (165)

Reading 5.1

THE EQUALITY DILEMMA

William Ryan

In this excerpt from his book Equality, *William Ryan (a psychology professor at Boston College) examines two conflicting models of equality: Fair Shares and Fair Play. Ryan suggests that we face the following dilemma in American society: Although the Fair Play model is favored by most Americans, it is based on faulty assumptions; perhaps the Fair Shares model is more appropriate and closer to what is really meant by equality.*

Thinking about equality makes people fidgety. Insert the topic into a conversation and listen: voices rise, friends interrupt each other, utter conviction mingles with absolute confusion. The word slips out of our grasp as we try to define it. How can one say who is equal to whom and in what way? Are not some persons clearly superior to others, at least in some respects? Can superiority coexist with equality, or must the one demolish the other? Most important, are the existing inequalities such that they should be redressed? If they are, what precisely should be equalized? And by whom?

We feel constrained by the very word—to deny equality is almost to blaspheme. Yet, at the same time, something about the idea of equality is dimly sinful, subject to some obscure judgment looming above us. It is not really adequate to be "as good as." We should be "better than." And the striving for superiority fills much of the space in our lives, even filtering into radio commercials. Listen to a chorus of little boys singing, "My dog's better than your dog; my dog's better than yours." They are selling dog food, it turns out, and apparently successfully, which must mean that a lot of people want very much to have their dogs be better than your dogs. Their *dogs!*

This passing example suggests how far it is possible to carry the competition about who and whose is better. The young voices sing fearlessly. There is little danger that some fanatic will leap up and denounce the idea of one dog's being superior to another or quote some sacred text that asserts the equality of all dogs. In regard to people, however, we do hesitate to claim superiority or to imply inferiority. No commercials announce, "My son's better than your son" or "My wife's better than yours" or, most directly, "I'm better than you." That we remain reticent about flaunting such sentiments and yet devote ourselves to striving for superiority signals the clash of intensely contradictory beliefs about equality and inequality.

We all know, uneasily but without doubt, that our nation rests on a foundation of documents that contain mysterious assertions like "all men are created equal," and we comprehend that such phrases have become inseparable components of our license to nationhood. Our legitimacy as a particular society is rooted in them, and they bound and limit our behavior. We are thus obliged to agree that all men are created equal—whatever that might mean to us today.

But our lives are saturated with reminders about whose dog is better. In almost all our daily deeds, we silently pledge allegiance to inequality, insisting on the continual labeling of winners and losers, of Phi Beta Kappas and flunkouts, and on an order in which a few get much and the rest get little.

We re-create the ambiguity in the minds of our children as we teach them both sides of the contradiction. "No one is better than anyone else," we warn. "Don't act snotty and superior." At the same time we teach them that all are obliged to get ahead, to compete, to achieve. Everyone is equal? Yes. Everyone must try to be superior? Again, yes. The question reverberates. . . .

It should not surprise us, then, that the clause "all men are created equal" can be interpreted in quite different ways. Today, I would like to suggest, there are two major lines of interpretation: one, which I will call the "Fair Play" perspective, stresses the individual's right to pursue happiness and obtain resources; the other, which I will call the "Fair Shares" viewpoint, emphasizes the right of access to resources as a necessary condition for equal rights to life, liberty, and happiness.

Almost from the beginning, and most apparently during the past century or so, the Fair Play viewpoint has been dominant in America. This way of looking at the problem of equality stresses that each person should be equally free from all but the most minimal necessary interference with his right to "pursue happiness." It is frequently stressed that all are equally free to *pursue,* but have no guarantee of *attaining,* happiness.

The Fair Shares perspective, as compared with the Fair Play idea, concerns itself much more with equality of rights and of access, particularly the implicit rights to a reasonable share of society's resources, sufficient to sustain life at a decent standard of humanity and to preserve liberty and freedom from compulsion. Rather than focusing on the individual's pursuit of his own happiness, the advocate of Fair Shares is more committed to the principle that all members of the society obtain a reasonable portion of the goods that society produces. From his vantage point, the overzealous pursuit of private goals on the part of some individuals might even have to be bridled. From this it follows, too, that the proponent of Fair Shares has a different view of what constitutes fairness and justice, namely, an appropriate distribution throughout society of sufficient means for sustaining life and preserving liberty.

So the equality dilemma is built in to everyday life and thought in America; it comes with the territory. Rights, equality of rights—or at least interpretations of them—clash. The conflict between Fair Play and Fair Shares is real, deep, and serious, and it cannot be easily resolved. Some calculus of priorities must be established. Rules must be agreed upon. It is possible to imagine an almost endless number of such rules.

- Fair Shares until everyone has enough; Fair Play for the surplus

- Fair Shares in winter; Fair Play in summer

- Fair Play until the end of a specified "round," then "divvy up" Fair Shares, and start Fair Play all over again (like a series of Monopoly games)

- Fair Shares for white men; Fair Play for blacks and women

- Fair Play all the way, except that no one may actually be allowed to starve to death

The last rule is, I would argue, a perhaps bitter parody of the prevailing one in the United States. Equality of opportunity and the principle of meritocracy are the clearly dominant interpretation of "all men are created equal," mitigated by the principle (usually defined as charity rather than equality) that the weak, the helpless, the deficient will be more or less guaranteed a sufficient share to meet their minimal requirements for sustaining life.

The Fair Play concept is dominant in America partly because it puts forth two most compelling ideas; the time-honored principle of distributive justice and the cherished image of America as the land of opportunity. At least since Aristotle, the principle that rewards should accrue to each person in proportion to his worth or merit has seemed to many persons one that warrants intuitive acceptance. The more meritorious person— merit being some combination of ability and constructive effort—*deserves* a greater reward. From this perspective it is perfectly consistent to suppose that *unequal* shares could well be *fair* shares; moreover, within such a framework, it is very unlikely indeed that equal shares could be fair shares, since individuals are not equally meritorious.

The picture of America as the land of opportunity is also very appealing. The idea of a completely open society, where each person is entirely free to advance in his or her particular fashion, to become whatever he or she is inherently capable of becoming, with the sky the limit, is a universally inspiring one. This is a picture that makes most Americans proud.

But is it an accurate picture? Are these two connected ideas—unlimited opportunity and differential rewards fairly distributed according to differences in individual merit— congruent with the facts of life? The answer, of course, is yes and no. Yes, we see some vague congruence here and there—some evidence of upward mobility, some kinds of inequalities that can appear to be justifiable. But looking at the larger picture, we must answer with an unequivocal "No!" The fairness of unequal shares and the reality of equal opportunity are wishes and dreams, resting on a mushy, floating, purely imaginary foundation.

Income in the United States is concentrated in the hands of a few: one-fifth of the population gets close to half of all the income, and the top five percent of this segment gets almost one-fifth of it. The bottom three-fifths of the population—that is, the majority of us—receive not much more than one-third of all income. Giving a speech at a banquet, a friend of mine, James Breeden, described the distribution of income in terms of the dinners being served. It was a striking image, which I will try to reproduce here.

Imagine one hundred people at the banquet, seated at six tables. At the far right is a table set with English china and real silver, where five people sit comfortably. Next to them is another table, nicely set but nowhere near as fancy, where fifteen people sit. At each of the four remaining tables twenty people sit—the one on the far left has a stained paper tablecloth and plastic knives and forks. This arrangement is analogous to the spread of income groups—from the richest five percent at the right to the poorest twenty percent at the left.

Twenty waiters and waitresses come in, carrying one hundred delicious-looking dinners, just enough, one would suppose, for each of the one hundred guests. But, amazingly, four of the waiters bring twenty dinners to the five people at the fancy table on the right. There's hardly room for all the food. (If you go over and look a little closer, you will notice that two of the waiters are obsequiously fussing and trying to arrange ten dinners in front of just one of those five.) At the next-fanciest table, with the fifteen people, five waiters bring another twenty-five dinners. The twenty people at the third table get twenty-five dinners, fifteen go to the fourth table, and ten to the fifth. To the twenty people at the last table (the one with the paper tablecloth) a rude and clumsy waiter brings only five dinners. At the top table there are four dinners for each person; at the bottom table, four persons for each dinner. That's approximately the way income is distributed in America—fewer than half the people get even one dinner apiece.

When we move from income to wealth—from what you *get* to what you *own*—the degree of concentration makes the income distribution look almost fair by comparison. About one out of every four Americans owns *nothing.* Nothing! In fact, many of them *owe* more than they have. Their "wealth" is actually negative. The persons in the next quarter own about five percent of all personal assets. In other words, half of us own five percent, the other half own ninety-five percent. But it gets worse as you go up the scale. Those in the top six percent own half of all the wealth. Those in the top one percent own one-fourth of all the wealth.

These dramatically *unequal* shares are—it seems to me—clearly *unfair* shares. Twenty million people are desperately poor, an additional forty million don't get enough income to meet the minimal requirements for a decent life, the great majority are just scraping by, a small minority are at least temporarily comfortable, and a tiny handful of persons live at levels of affluence and luxury that most persons cannot even imagine.

The central problem of inequality in America—the concentration of wealth and power in the hands of a tiny minority—cannot, then, be solved by any schemes that rest on the process of long division. We need, rather, to accustom ourselves to a different method of holding resources, namely, holding them in common, to be *shared* amongst us all—not divided up and parceled out, but shared. That is the basic principle of Fair Shares, and it is not at all foreign to our daily experience. To cite a banal example, we share the air we breathe, although some breathe in penthouses or sparsely settled suburbs and others in crowded slums. In a similar fashion, we share such resources as public parks and beaches, although, again, we cannot overlook the gross contrast between the size of vast private waterfront holdings and the tiny outlets to the oceans that are available to the public. No one in command of his senses would go to a public beach, count the number of people there, and suggest subdividing the beach into thirty-two-by-twenty-six-foot lots, one for each person. Such division would not only be unnecessary, it would ruin our enjoyment. If I were assigned to Lot No. 123, instead of enjoying the sun and going for a swim, I might sit and watch that sneaky little kid with the tin shovel to make sure he did not extend his sand castle onto my beach. So, we don't divide up the beach; we own it in common; it's *public;* and we just plain *share* it.

We use this mode of owning and sharing all the time and never give it a second thought. We share public schools, streets, libraries, sewers, and other public property and services, and we even think of them as being "free" (many libraries even have the word in their names). Nor do we need the "There's no such thing as a free lunch" folks reminding us that they're not really free; everyone is quite aware that taxes support them. We don't feel any need to divide up all the books in the library among all the citizens. And there's no sensible way of looking at the use of libraries in terms of "equal opportunity" as opposed to "equal results." Looking at the public library as a tiny example of what Fair Shares equality is all about, we note that it satisfies the principle of equal access if no one is *excluded* from the library on the irrelevant grounds of not owning enough or of having spent twelve years in school learning how not to read. And "equal results" is clearly quite meaningless. Some will withdraw many books; some, only a few; some will be so unwise as to never even use the facility.

The *idea* of sharing, then, which is the basic idea of equality, and the *practice* of sharing, which is the basic methodology of Fair Shares equality, are obviously quite familiar and acceptable to the American people in many areas of life. There are many institutions, activities, and services that the great majority believe should be located in the public sector, collectively owned and paid for, and equally accessible to everyone. We run into trouble when we start proposing the same system of ownership for the resources that the wealthy have corralled for themselves. It is then that the servants of the wealthy, the propagandists of Fair Play, get out their megaphones and yell at every-body that it's time to line up for the hundred-yard dash.

One can think of many similes and metaphors for life other than the footrace. Life is like a collection of craftsmen working together to construct a sturdy and beautiful build-ing. Or, a bit more fancifully, it is like an orchestra—imagine a hundred members of a symphony orchestra racing to see who can finish first! When we experience a moving performance of the *Eroica,* how do we judge who the winners are? Is it the second violins, the horns, the conductor, Beethoven, the audience? Does it not make sense to say that, in this context, there are no losers, that all may be considered winners?

Most of the good things of life have either been provided free by God (nature, if you prefer) or have been produced by the combined efforts of many persons, sometimes of many generations. As all share in the making, so all should share in the use and the enjoyment. This may help convey a bit of what the Fair Shares idea of equality is all about.

QUESTIONS 5.1

1. Distinguish between the Fair Shares and Fair Play views of equality. What are the problems or dilemmas with each?

2. Would functional theorists and con-flict theorists respond differently to this selection? If so, how? If not, why not?

3. Is William Ryan a functionalist or a conflict theorist? Explain your choice.

4. Select the definition of equality that makes most sense to you and defend it. Briefly describe a program to make it work in the United States.

Reading 5.2

PRO-LIFE VERSUS PRO-CHOICE WOMEN—A STUDY IN CLASS CONTRASTS

Kristin Luker

In this excerpt from her book Abortion and the Politics of Motherhood, *Kristin Luker profiles the characteristics of women who are in favor of and opposed to abortion. She illustrates the point, a central one in sociology, that one's attitudes, values, and beliefs are a reflection of one's position in the social structure. In this case, peoples' social class and economic well-being affect how they respond to certain social issues.*

Ever since abortion became a divisive issue, many people have sensed something about the women actively involved on opposite sides. That something was difficult to articulate yet relevant to the bitterness of the controversy. Pro-choice women, it seemed, were *different* from pro-life women, and vice versa.

That perception, as it turns out, is accurate. A five-year study of a representative sample of these women activists demonstrates that the two groups are separated by far more than their disagreement over whether abortion is murder. By social class, by economic status, by educational attainment, they *are* different. And that is only the beginning. Their beliefs about abortion are simply the tip of the iceberg, reflecting attitudes that are also at odds on all the most central aspects of life: work, women's roles, marriage, parenthood, sex, and family.

Moreover, those attitudes were not arrived at by some intellectual process. If that were the case, there might be more hope of reversing the process and bridging the gap. Rather, their cherished beliefs flow out of realities in their own daily lives that are not easily changed.

This is why the debate is so passionate and uncompromising. The women who are most centrally involved in it have little basis for a dialogue and, in fact, few incentives to enter into one.

To explore what it is that makes the abortion debate so heated, a research team at the University of California at San Diego of which I was a part conducted interviews over a five-year period (1977–83) with a sample of nearly two hundred California women who spend at least five hours a week working for advocacy groups on that issue.

The typical pro-choice activist in our study was a forty-four-year-old married woman whose father was a college graduate. She married at age twenty-two or older, has one or two children and had some graduate or professional training after college. (Thirty-seven percent of all pro-choice women in this study have received at least some postbaccalaureate training.) She is married to a professional man, is employed and has a family income of more than $50,000 a year. She attends church rarely, if at all; indeed, religion is not particularly important to her.

The average pro-life activist is also forty-four and married. However, she wed at seventeen and has three or more children. (Sixteen percent of the pro-life women in the study have seven or more children.) Her father graduated from high school but not college, and only sixty percent of pro-life activists (compared with ninety percent of pro-choice activists) have a baccalaureate degree.

This typical pro-life activist is not employed, and is married to a small businessman or a lower-income white-collar worker; her family income is less than $30,000 a year. Her religion is one of the most important aspects of her life; she attends church at least once a week. She is probably a Catholic, but may be a convert to Catholicism. (Almost eighty percent of the pro-life activists in this study were Catholics at the time of the study, but only fifty-eight percent had been raised as Catholics.)

One small example points up the way these differences shape attitudes. A number of pro-life activists emphatically reject an expression that pro-choice women tend to use almost unthinkably—the expression "unwanted pregnancy." Pro-life women argue forcefully that a better term would be "surprise pregnancy," reflecting their view that although a pregnancy may be momentarily unwanted, the child that results from the pregnancy almost never is.

As our profile of the average pro-life activist makes clear, a woman who is not employed, who does not have an advanced degree, whose religion is important to her, and who has already committed herself wholeheartedly to marriage and a large family is well equipped to believe that an unanticipated pregnancy usually becomes a beloved child. Her life is so arranged that for her this belief is true, and her world view leads her to believe that everyone else can "make room for one more" as easily as she can, and that, therefore, abortion is cruel, wicked, and self-indulgent.

By the same token, our profile of the average pro-choice woman—a woman who is employed full-time, has an affluent lifestyle that depends in part on her contribution to the family income and expects to give her child a life at least as educationally, socially, and economically advantaged as her own—makes it evident that she draws on a different reality that makes her skeptical about the ability of the average person to transform unwanted pregnancies into well-loved and well-cared-for children.

Two prominent groups of women—blacks and working-class whites—are hardly represented at all among the ranks of abortion activists in our study. Many working-class women, often stereotyped as conservative on social issues, actually are pro-choice but seldom have the time or resources to be active. Blacks tend to be conservative on the abortion issue, but blacks often focus on issues perceived to be of more immediate concern to the black community, such as education.

Nevertheless, it is the activists who shape the debate, and it is the sharp cleavage among them that makes political compromise and dialogue so unlikely.

The gulf separating these two groups of women in our study is evident in the fact that few women on either side of the issue have any friends or even acquaintances who disagree with them about abortion. In fact, a number of pro-life women spontaneously declared during the interviews that they would end a friendship if they discovered that the friend did not share their views on abortion.

That is hardly surprising given the opposing views held by pro-choice and pro-life activists on so many issues. Consider, for example, attitudes about the roles of men and women. Pro-life activists believe that men and women are intrinsically different, and that this is both a cause and a product of different "natural" roles in life. As one said:

"Men and women were created differently, and were meant to complement each other, and when you get away from our proper roles as such, you start obscuring them. That's another part of the confusion going on now; people don't know where they stand, they don't know how to act, they don't know where they're coming from, so your psychiatrists' couches are filled with lost souls, with people who have gradually been led into confusion and don't even know it."

Men, the pro-life activists believe, are best suited to the public world of work and women to the private world of rearing children, managing homes, and caring for husbands.

Pro-choice activists reject this notion. They believe that men and women are fundamentally equal, by which they mean substantially similar, at least as regards rights and responsibilities. As a result, they see women's reproductive and family roles not as a natural niche, but as a potential barrier to full equality. So long as society is organized to maintain motherhood as an involuntary activity, they argue, "women's sphere" connotes a potentially low-status, unrewarded role to which women can be relegated at any time.

"I just feel that one of the main reasons women have been in a secondary position culturally is because of the natural way things happen. Women would bear children because they had no way to prevent it, except by having no sexual involvement. And that was not practical down through the years; so without knowing what to do to prevent it, women would continually have children. And then if they were the ones bearing the child, nursing the child, it just made sense for them to be the ones to rear the child. I think that was the natural order."

"When we advanced and found that we could control our reproduction, we could choose the size of our families, or whether we wanted families. But that changed the whole role for women in society. . . . It allowed us to be more than just the bearers of children, the homemakers. That's not to say that we shouldn't continue in that role. It's a good role, but it's not the only role for women."

For the pro-life people we talked to, the primary purpose of sexuality is procreation. It is not surprising, given this view, that the pro-life activists in this study are opposed to most contraceptives. Although they are careful to point out that the pro-life movement is officially neutral on the topic, most are confident that any law outlawing abortion would also outlaw The Pill and the I.U.D., a result they favor. A substantial number of pro-life activists use periodic abstinence, or natural family planning, as their only form of fertility control.

In contrast, pro-choice people generally focused on sexuality as an emotional outlet. They argued that the main purpose of sex is not to produce children but to afford pleasure, human contact and, perhaps most important, intimacy. But since such closeness requires trust, familiarity, and security, practice is required. As a result, contraception, which allows people to focus on the emotional aspects of sex without worrying about its procreative aspects, is a good thing, according to pro-choice activists.

A general theme in the interviews with pro-life activists (many of whom have large families, it will be recalled) is that there is an anti-child sentiment abroad in American society, as exemplified by the strong pressures to have only two children:

"My husband, being a scientist, gets a lot of questions. You know, having a large family, it's just for the poor, uneducated person, but if you have a doctor's degree and you have a large family, what's wrong with you?"

The values of pro-life activists about parenthood follow from their views on gender, sex, and contraception. Since the purpose of sex is procreation, they believe, married couples should be willing to have whatever number of children come, at whatever time they are conceived. Second, since motherhood is a natural role, one should not try to plan carefully for it through contraceptive use, and one need not prepare for it. Finally, since motherhood is the most satisfying and meaningful role for a woman, it is incomprehensible to pro-life people that a woman might want to postpone or avoid pregnancy in favor of something else, such as work, education, or worldly success.

Because pro-choice people see raising children as requiring financial resources, interpersonal and social skills, and emotional maturity, they often worry about how easy it is to have children. In their view, parenthood is far from natural; they feel that too many people stumble into it without appreciating what it takes:

"I would say that the tip of the iceberg is purposeful parenthood. I think life is too cheap. I think we're too easy-going. We assume that everyone will be a mother. . . . Hell, it's a privilege; it's not special enough. The contraceptive agent affords us the opportunity to make motherhood really special."

Since pro-choice activists think that in the long run abortion will enhance the quality of parenting by making it optional, they see themselves as being on the *side* of children when they advocate abortion.

As one pro-choice woman said, "Many people don't want the child when they find out they're pregnant, but they resolve negative feelings and by the time the child is born they do want it. But if they don't want the child enough to seek an abortion, then probably they shouldn't have a child. I'm as concerned with the rights of the child as the rights of the woman."

Pro-life women have built their lives on the premise that reproduction is a *resource,* and they resist all those cultural values—small families, contraception, abortion, non-family roles for women, day care—which diminish the value of children or dilute the unique value of motherhood.

In the same way, pro-choice women resist those values which suggest that motherhood is a natural, primary or inevitable role for a woman. Pro-choice activists believe that men and women are equal because, in their own lives, men and women have substantially the same kinds of experiences. The pro-choice women in this study have had approximately the same education as their husbands, and many of them have the same kinds of jobs—they are lawyers, accountants, pharmacists, ministers, and physicians. Even those who do not work in traditionally male occupations have salaried jobs and thus share common experiences.

Women who have "male" resources want to see motherhood recognized as discretionary. Women who have few of these resources, and limited opportunities, have incentives to motherhood recognized as the most important role a woman can have.

QUESTIONS 5.2

1. What type of research is this? Categorize using terms from Chapter One (reactive, nonreactive, and so on).

2. The point of this reading is that social class and economic well-being help mold beliefs. How and why did this happen? What is the logical connection

between social-class position and beliefs about abortion?

3. What other aspects of these women's backgrounds aside from social class seem to have contributed to their viewpoints? What was the most important factor?

NOTES

1. The major source for this section is *The Structure of Social Inequality* by Beth E. Vanfossen (Boston: Little, Brown & Company, 1979), ch. 3.

2. H. H. Gerth and C. W. Mills, *From Max Weber: Essays in Sociology* (New York: Oxford University Press, 1958), ch. 7.

3. Cecil North and Paul Hatt, "Jobs and Occupations: A Popular Evaluation," *Opinion News* 9 (September 1947), pp. 3–13.

4. Kingsley Davis and Wilbert Moore, "Some Principles of Stratification," *American Sociological Review* 10 (April 1945), pp. 242–249.

5. See Gerhard Lenski, "Status Crystallization: A Non-Vertical Dimension," *American Sociological Review* 19 (August 1954), pp. 405–413; Gerhard Lenski, "Status Inconsistency and the Vote: A Four Nation Test," *American Journal of Sociology* 32 (April 1967), pp. 298–302; and "Status Consistency and Right-Wing Extremism," *American Sociological Review* 32 (February 1967), pp. 86–92.

6. In this section we have directly or indirectly referred to several classic works in stratification. Lloyd Warner's study of Newburyport, Massachusetts, was first described in Lloyd Warner and Paul S. Lunt, *The Social Life of a Modern Community* (New Haven, Conn.: Yale University Press, 1941). Robert S. Lynd and Helen M. Lynd described stratification in a midwestern city in *Middletown* (New York: Harcourt, 1959) and *Middletown in Transition* (New York: Harcourt, 1963). So did August Hollingshead in *Elmtown's Youth,* first published in 1949 and now available as

Elmtown's Youth and Elmtown Revisited (New York: Wiley, 1975). Stratification along a continuum is discussed by John Cuber and William Kenkel in *Social Stratification in the United States* (New York: Appleton-Century-Crofts, 1954).

7. Walter Lord, *A Night to Remember* (New York: Holt, Rinehart & Winston, 1955).

8. The mortality data come from Gregory Pappas, Susan Queen, Wilbur Hadden, and Gail Fisher, "The Increasing Disparity in Mortality Between Socioeconomic Groups in the United States, 1960 and 1986," *The New England Journal of Medicine* 329 (July 8, 1993), pp. 103–109. Social class and mental illness data are from A. B. Hollingshead and F. C. Redlich, *Social Class and Mental Illness: A Community Study* (New York: Wiley, 1958).

9. See Jeffrey Gould, Becky Davey, and Randall Stafford, "Socioeconomic Differences in Rates of Cesarean Section," *The New England Journal of Medicine* 321 (July 27, 1989), pp. 233–239.

10. Marshall Clinard and Robert Meier discuss the data on homicide in *Sociology of Deviant Behavior,* 5th ed. (New York: Holt, Rinehart & Winston, 1979), p. 199.

11. See Melvin Kohn's book, *Class and Conformity* (Homewood, Ill.: Dorsey Press, 1969) and the discussion by Vanfossen in *The Structure of Social Inequality,* pp. 272–273 (see Note 1).

12. The original works referred to in this section include William C. Kvaraceus and Walter B. Miller, *Delinquent Behavior: Culture and the Individual* (Washington, D.C.: National Ed-

ucation Association, 1959); Walter Miller, "Lower-Class Culture as a Generating Milieu of Gang Delinquency," *Journal of Social Issues* 14 (1958), pp. 5–19; Albert K. Cohen, *Delinquent Boys: The Culture of the Gang* (New York: Free Press, 1955); and Richard Cloward and Lloyd Ohlin, *Delinquency and Opportunity* (New York: Free Press, 1960). Also see "The Myth of Social Class and Criminality: An Empirical Assessment of the Empirical Evidence" by Charles Tittle, Wayne Villemez, and Douglas Smith, *American Sociological Review* 43 (October 1978), pp. 643–656.

13. Ivan Chase, "A Comparison of Men's and Women's Intergenerational Mobility in the United States," *American Sociological Review* 40 (August 1975), pp. 483–505.

14. See the discussion of social mobility and its effects in *Social Stratification in America* by Leonard Beeghley (Santa Monica, Calif.: Goodyear, 1978), pp. 311–318.

15. The major sources I am using in this section on poverty are Beeghley, *Social Stratification in America*, ch. 8; Jonathan Turner and Charles Starnes, *Inequality: Privilege and Poverty in America* (Pacific Palisades, Calif.: Goodyear, 1976); Jerome Skolnick and Elliott Currie, *Crisis in American Institutions*, 5th ed. (Boston: Little, Brown, 1982); and William Ryan, *Blaming the Victim*, rev. ed. (New York: Random House, 1976). Facts on money and numbers come from *Statistical Abstracts of the United States, 1993* (113th ed.), Washington, D.C. The study of welfare beliefs and facts is discussed in Beeghley, pp. 135–138. The interview study of the fertility rates of welfare women is from Mark Rank, "Fertility Among Women on Welfare: Incidence and Determinants," *American Sociological Review* 54 (April 1989), pp. 296–304.

16. Herbert Gans, "The Positive Functions of Poverty," *American Journal of Sociology* 78 (September 1972), pp. 275–289; and Beeghley, *Social Stratification in America*, p. 140.

17. Turner and Starnes, *Inequality: Privilege and Poverty in America*, p. 122.

18. Frances Piven and Richard Cloward, "The Relief of Welfare," in Skolnick and Currie, *Crisis in American Institutions*, 5th ed. (see Note 15). Also see Beeghley, *Social Stratification in America*, pp. 139–140.

19. Turner and Starnes, *Inequality: Privilege and Poverty in America*, pp. 62–63, 93.

CHAPTER SIX

Social Inequality: Race, Ethnicity, Gender, Age

TWO RESEARCH QUESTIONS

1. Race is viewed differently in the United States than it is in the Caribbean. How is it different? What happens when Caribbean Hispanics settle in the United States—do they adopt U.S. patterns or maintain their old beliefs?

2. In a group discussion involving men and women, who is more likely to interrupt whom, or is there any consistent pattern? Does the composition of the group affect interruption patterns?

THE PREVIOUS CHAPTER introduced the concepts of social differentiation and social inequality. People fall into categories on the basis of characteristics such as age, gender, race, and social class. Further, the elements that make up these categories are ranked—people in them are not seen or treated equally. The study of differentiation and inequality is very helpful because it gives us a more precise picture of society. It also helps us understand the forces that bind people together and the characteristics of inequality.

In Chapter Five we discussed the inequality produced by socioeconomic factors. Socioeconomic status can be described as an achieved status because there is some flexibility in the ranking; one can perhaps move up or down the

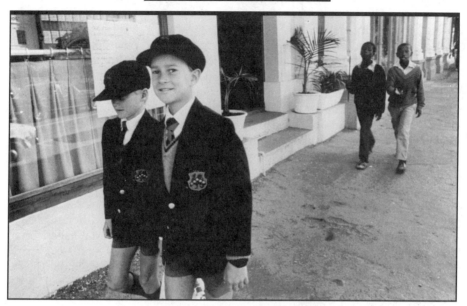

social-class ladder. In this chapter we look at categories that are more firmly established. They are automatically conferred, and there is little an individual can do to change them. We call them ascribed statuses.

MINORITY STATUS

Minority status is an important ascribed status and can be helpful in explaining how people interact with one another. Minority status not only names the group or category to which one belongs, but it also describes one's social position—how one is likely to be treated by others.

We will find in this chapter that groups of minority status may be considered lower in social standing than the rest of society; they may be subject to domination and may be denied specific rights and privileges.

It is important to understand that minority status is a social condition, not necessarily a statistical one. A minority group could represent a majority of a society's population but hold minority status because of the way it is viewed and treated by other segments of society. For example, women in the United States are a numerical majority, but they have minority status because they are not treated on a level of equality with men. For years the black population in South Africa has held minority status even though it greatly outnumbers the white population, which has been dominant. In most instances, minority

groups do represent a smaller proportion of the population—blacks and Hispanics in the United States, for example—but it is important to recall the *social*
nature of minority status.

Where do definitions of minority status come from? Who decides? A number of factors are probably at work. Some decisions about minority status arrive through custom and tradition. People who are successful are sometimes
proprietary about their good fortune and look down on those who aspire to
the same position. Some groups blend in more easily or appear to. If their skin
color and language are the same and their cultural backgrounds are similar to
those of the society they are entering, their differences will be less apparent,
and they will be more likely to escape minority status. We also learned in the
previous chapter of the importance of wealth and power. Assume that there
is a limited supply of the good things in life; it follows, then, that those with
wealth and power will want to protect their positions by denying access to
them by others. By labeling and treating categories of people differently, as if
of lower quality, they are effectively removed as threats to established positions of privilege. The creation and maintenance of minority status can work
much the same way as the treatment of the poor (discussed in Chapter Five).

RACE AND ETHNICITY

We have learned that people differ and are ranked by such socioeconomic
factors as wealth, education, occupation, power, and prestige. People also differ and are ranked by race and ethnicity. A person's race and ethnic affiliation
confer on him or her a status or position in society, and sometimes this is a
minority status.

Race is a vague and ambiguous term. It is generally defined as a group of
people bound together by hereditary physical features. The difficulty with the
notion of race is that practically no pure races exist. Substantial biological mixing has blurred boundaries to the point that it is difficult to define uniform
hereditary physical features. Scientists attempting to define race have named
anywhere from three racial categories to more than 30. Three definitions of
race—biological, legal, and social—seem to emerge.[1] A *biological* definition is
based on observable physical features such as skin color, hair texture, and eye
color, or on differences in gene frequencies. A *legal* definition is that incorporated into the laws of states or nations. For example, the law in one state has
defined a black as a person who has "one-eighth or more Negro blood,"
whereas in other states a black is a person with "any ascertainable trace of
Negro blood." A *social* definition refers to what members of society consider
the important distinctions about race. Brewton Berry has suggested that in the
United States the social definition of a black is anyone who identifies himself
as black or who has any *known* trace of black ancestry. It can be seen from this
that the term *race* is not a precise term, but it is nevertheless widely used.

A sociology class is shown slides of people who identify themselves as
blacks but who, because of an absence of typical racial characteristics, don't
look like blacks. The class (which is predominantly white) is asked to identify
the race or nationality of the light-skinned, straight-haired people on the

slides, and the guesses range across the globe. Finally the class is told, "They are all black Americans." The reaction is astonishment, of course, because "they don't look like blacks." This illustrates several things, among them the difficulty of defining race, the blurring of racial boundaries, and the importance of the social definition. Once the class is told that the people on the slides are identified as blacks (probably because of some black ancestry), the class believes it. No question about it, regardless of how they look.

When race is discussed, the question of racial differences inevitably comes up. That is, do races biologically differ in such matters as I.Q., achievement, or susceptibility to diseases? The haziness in definitions of race supplies part of the answer: If the boundaries of race are unclear, how can any statements about racial differences be made? Nevertheless they are, as witness the controversy over inherent racial I.Q. differences. Most social scientists conclude that there are no significant differences caused by race. There are numerous differences among people caused by cultural, social, and geographic factors, however, and these are often thought to be racial differences. For example, some gene frequencies and blood characteristics have developed in particular areas of the world. Population groups migrating from those areas, regardless of race, carry these characteristics to other parts of the world. Likewise, differences in attitude, achievement, and perception are explained by cultural differences: One culture emphasizes achievement, another tranquility; one is aggressive, another encourages passivity; and so on. In short, social scientists argue from a cultural-determinist viewpoint and reject the idea of racially caused differences. Many people remain unconvinced, however, and the argument will undoubtedly continue.

Ethnicity or *ethnic group* are more useful terms than *race*, at least from a sociological viewpoint. Members of an ethnic group are bound together by cultural ties, which may have several origins. When groups migrate to a new area, their ties to a previous culture may remain strong. Religious beliefs may also provide a basis for ethnicity, as in the case of the Jews, the Amish, and the Hutterites. The term *ethnicity* is useful because it appropriately brings to focus cultural similarities, which to the social scientist are more important and have more explanatory power than racial similarities.

The first research question at the beginning of this chapter concerned definitions of race in the Caribbean and in the United States. Race in the United States is an either-or concept—one is either black or white. In the Caribbean, however, race is a multicategory continuum, with many people falling in between black and white. Caribbean Hispanics are an interesting case in that although they are racially diverse—a mixture of black, white, Asian, and American Indian—they share a common ethnicity: their use of Spanish, their Spanish-colonial heritage, and their Catholic religion. In other words, their ethnic-group identity is more important than their racial identity.

Now, what happens when Caribbean Hispanics settle in the United States? Do they maintain their view of race and ethnicity, or do they change? A recent paper found that Caribbean Hispanics coming to the United States are torn between conforming to the American views of race and holding on to their own multiracial/Hispanic ethnic identity. The outcome is interesting but perhaps predictable. Caribbean Hispanics (especially Puerto Ricans) reject the bi-polar racial ideas of the United States—they consider themselves to be racially intermediate between black and white. They continue to preserve an ethnic identity—"Spanish race"—that draws them together. However, census data suggest patterns of segregation similar to those in American society. Carib-bean Hispanics are not segregated from white Hispanics, but they are highly segregated from both Hispanic and non-Hispanic blacks. They are also highly segregated from Euro-Americans. It appears that people of mixed racial an-cestry are accepted by white Hispanics on the basis of shared ethnicity but are rejected by Euro-Americans on the basis of race. The authors conclude that although both race and ethnicity are important in American life, race is especially so.[2]

Prejudice and Discrimination

Prejudice has been defined as a fixed attitude—favorable or unfavorable—toward a person or thing, probably not based on actual experience. A preju-diced person ignores the individual and his or her particular qualities or char-acteristics, and groups the person with others who happen to have the same skin color (brown or yellow), or speak with the same accent (New England), or have the same type of name (Cohen, Greenberg), or come from the same part of the country (the South).

Prejudice tends to be generalized. People who are prejudiced against one group will probably be prejudiced against others. Some years ago E. L. Hartley asked college students what they thought of Canadians, Turks, Jews, atheists, Irish, Negroes, Wallonians, Pireneans, Danireans, and other groups. He found that people who were prejudiced against blacks were also likely to be prejudiced against Jews, Wallonians, and Danireans. Now there are about 17 million Jews in the world, but no Wallonians or Danireans. Hartley used these fictitious groups to see how people would react to groups they knew nothing about. He found that people tend to generalize their prejudices.

Discrimination is actual behavior unfavorable to a specific individual or group. When people are denied equal treatment, they are being discriminated against. Discrimination occurs when a person is denied a desired position or right because of irrelevant factors—for example, when skin color is used to determine eligibility to vote or when religious affiliation is used to determine one's place of residence.

A particular type of prejudice and discrimination is described by the term *racism* (and by the related terms *sexism* and *ageism*). **Racism** is the belief that people are divided into distinct hereditary groups that are innately different in their social behavior and mental capacities and that, because of these differ-ences, they can be ranked as superior and inferior. This hereditary ranking then "justifies" differential treatment in terms of access to society's resources such as jobs, living conditions, wealth, and power.[3] You can see that racism is involved in some of the definitions of race discussed earlier (see p. 182).

Prejudice and discrimination are usually associated but not always. The situation becomes complicated when people think one way but behave another way. Robert Merton attempted to describe what happens by suggesting four categories of people:[4]

Unprejudiced nondiscriminator or all-weather liberal: Confirmed and consistent liberal; not prejudiced; doesn't discriminate; believes in American creed of justice, freedom, equality of opportunity, and dignity of the individual.

Unprejudiced discriminator or fair-weather liberal: Thinks in terms of expediency; not prejudiced but keeps quiet when bigots are about; will discriminate for fear that to do otherwise would "hurt business"; will make concessions to the intolerant.

Prejudiced nondiscriminator or fair-weather illiberal: Timid bigot; doesn't believe in the American creed but conforms to it when slightest pressure is applied; hates minorities but hires them to obey the law.

Prejudiced discriminator or all-weather illiberal: A bigot, pure and unashamed; doesn't believe in the American creed and doesn't hesitate to tell that to others; believes it is right and proper to discriminate, so does.

The origins of prejudice are explained in various ways. Some psychologists believe that there is a certain type of personality called an **authoritarian personality** that is especially prone to prejudice.[5] The authoritarian personality is also ethnocentric, is rigidly conformist, and worships authority and strength. According to psychologists, the authoritarian personality can most often be traced to faulty emotional development born of harsh discipline and lack of affection and love from parents during childhood. Most sociologists believe that prejudice, like other attitudes and behaviors, is learned in interaction with others, mainly in the family, and that personality is not as important as is the social or cultural situation in which one interacts. We learn race prejudice in the same way we learn how to eat with a fork, to study for a test, or to drive a car. Also, some life situations affect how this learning takes place. For example, people who are downwardly mobile—moving from middle to lower class—show more intense feelings of prejudice toward blacks and Jews; and in times of rapid social change, prejudice and discrimination may become more intense and more generalized.

Sociologists also associate stereotyping with prejudice. In **stereotyping,** we apply a common label and set of characteristics to all people in a certain category, even though none or only some of the people in that category fit the description. The stereotyped male college professor is absentminded, smokes a pipe, is an extreme liberal, wears horn-rimmed glasses and a tweed jacket with elbow patches, and generally is not much in touch with the real world. This image is applied to everyone in the male college-professor category, even though most do not fit the part. Professors will probably survive their stereotype without damage because the characteristics of the stereotype are not negative. Besides, a person *chooses* as an adult to become a college teacher and, therefore, to enter a stereotyped category.

Someone of minority status, on the other hand, who is stereotyped from birth as ignorant, lazy, dirty, happy-go-lucky, morally primitive, emotionally unstable, and fit only for menial work can be psychologically damaged by the effects of the lifelong stereotype. When others react to us on the basis of such stereotypes, it strongly affects our self-concepts—the way we see ourselves. This again is the self-fulfilling prophecy, which we discussed several times earlier. Remember what happened to the "bright" rats and "dull" rats? Or what happened when teachers thought some of their students were "academic spurters"?

A classic example of the effect of minority status on how people see themselves was provided by Professor Kenneth Clark when he asked 253 black children aged three to seven to choose between black and white dolls. Two thirds of the children preferred *white* dolls! This research, done in the 1940s, was cited by the Supreme Court in its 1954 school-desegregation decision. Those who hoped that recent gains in black achievements and renewed emphasis on black pride have improved images of self-worth were disappointed by several studies reported in 1987. Psychologist Darlene Powell-Hopson gave 155 children aged three to six a set of black and white Cabbage Patch Dolls and asked the children to give her the doll that "you want to be, you want to play with, is a nice color and would take home if you could." Sixty-five percent of the black children selected white dolls. From the results and from students' comments, Powell-Hopson concluded that black children are getting the message that it's preferable to be a member of another race. In a similar study by Sharon Gopaul-McNicol of 144 preschoolers in Trinidad, 74 percent of the black children chose white dolls. She suggests that although this Caribbean island has a black government and many successful blacks in business, they apparently are not enough to overcome the legacy of white supremacy passed on by 400 years of British rule and the influence of North American and European television.[6]

Several theories have been offered to explain prejudice and discrimination. One such theory explains conflict between unlike groups in terms of scapegoating. If attention can be focused on some out-group, this may strengthen the boundaries and unity of the in-group. It is suggested that the Nazis' attacks on the Jews before and during World War II were so motivated. Marxist theory holds that economic competition is the best explanation for prejudice and discrimination. When access to desired goods or valued positions in society is limited, discrimination against specific categories of people helps ensure that others can more easily dominate and obtain their goals. By reducing the status of others to second-class citizens and by eliminating some people from highly valued jobs, education, and access to wealth, we can guarantee that our own path is free of obstacles. Some have said that the conflict that arises when unlike groups meet may be best and most simply explained as resulting from a struggle for status among competing groups.

The common realms of discrimination in the United States include employment, education, housing, and, to a lesser extent, voting. In the following paragraphs we will take a brief look at some patterns of prejudice and discrimination as they affect blacks, Native Americans, Hispanic Americans, and Jews.

Bypass Surgery—Who Gets It?

A study of all Medicare files in 1986 discovered that American whites receive heart bypass surgery at a rate four times that of blacks (in Alabama, the ratio is eight to one). There is no evidence that whites are significantly more likely to *need* such operations than blacks. How, then, to explain the difference? Several factors appear to contribute—greater poverty of blacks, a reluctance among blacks to have surgery, physician racial bias, lack of trust in the system, and treatment by less experienced doctors in public hospitals.

Sacramento Bee, *March 18, 1992*

Blacks

There were 31.4 million blacks in the United States in 1992, or roughly 12 percent of the population. Blacks have been a part of American history from its beginning; the economy of the earliest days of our settlement was based on slavery. We are still trying to overcome the effects that the culture of slavery had on blacks and whites in America. Specific dates stand out:

1619 Twenty blacks purchased from a Dutch ship

1644 First slaves imported directly from Africa

1861–1865 Civil War with emancipation as a major issue

1865 Legal abolition of slavery by the Thirteenth Amendment to the Constitution

1896 *Plessy v. Ferguson* decision by the Supreme Court that supports "separate but equal" doctrine

1918–1960s Black migration from the rural South to jobs in the urban North

1954 *Brown v. Board of Education* decision by the Supreme Court stating that "separate facilities are inherently unequal"

1956 Montgomery bus boycott led by Martin Luther King, Jr.

1960–1965 Passive resistance, sit-ins, demonstrations

1965–1970 Urban riots, beginning of black militancy and "Black Power"

1970–1980s Prison riots, widespread disagreement over busing to achieve school desegregation, affirmative action, court suits over "reverse discrimination"

1990s Continued urban unrest, spotlight on the criminal justice system with the Rodney King case and the Los Angeles riots

The Kerner (riot commission) Report in 1968 stated that "our nation is moving toward two societies, one black, one white—separate and unequal." Some examples of how separate and unequal status affects other conditions of life are shown in Table 6.1.

Table 6.1   Comparison by Race on Selected Characteristics

		White	Black
Birth rate per 1,000 people (1990)		15.5	22.4
Life expectancy (1991)			
	Males	73	66
	Females	80	74
Infant deaths per 1,000 live births (1990)		8	17
Unmarried mothers percentage of all births (1990)		20	65
Homicide victims per 100,000 people (1990)			
	Males	9	69
	Females	3	14
Suicide victims per 100,000 people (1990)			
	Males	20	12
	Females	5	2
Percentage unemployed (1992)		7	14
Percentage below poverty line (1991)		9	30
Percentage of families earning under $10,000 (1991)		7	26
Percentage of families earning over $50,000 (1991)		34	15
Percentage completing four years of college (1992)		22	12

United States Bureau of the Census, *Statistical Abstracts of the United States, 1993* (113th ed.), Washington, D.C.

FIGURE 6.1   *Median family income (1991 dollars) by race and Hispanic origin.*
United States Bureau of the Census, *Statistical Abstracts of the United States, 1993* (113th ed.), Washington, D.C.

Jobs, Income, Unemployment   Median income for white families in 1991 was $31,569; for black families it was $18,807 (see Figure 6.1). The unemployment rate for blacks was more than double that of whites (14.1 percent compared with 6.5 percent for 1992). The jobless rate for white teenagers was 17 percent, but the comparable figure for black teenagers was 40 percent in 1992. Jobs available to blacks show a similar pattern. In 1992, although blacks made up 12 percent of the nation's population, they made up 3 percent of the nation's physicians, 1 percent of the dentists, 3 percent of the lawyers and judges, 5 percent of the college and university teachers, 27 percent of the postal clerks, 19 percent of the short-order cooks, 31 percent of nursing aides and orderlies, and 27 percent of the maids and housemen.

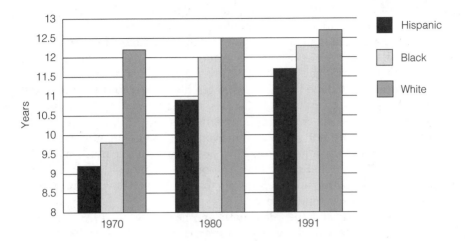

FIGURE 6.2 *Median school years completed, by race and Hispanic origin.*
United States Bureau of the Census, *Current Population Reports,* Series P-20, no. 462.

Table 6.2 Level of Education for People 25 Years and Older, 1992

Years of Schooling	Percentage of the Population				
	Whites	Blacks	Mexican Americans	Puerto Ricans	Cubans
Less than five	1	4	15	7	7
High school or more	84	68	45	61	62

United States Bureau of the Census, *Statistical Abstracts of the United States, 1993* (113th ed.), Washington, D.C., and *Current Population Reports,* Series P-20, no. 465.

Housing Finding adequate housing is difficult for blacks. First there were restrictive laws, and now there are tacit agreements to keep blacks out of the better districts in most cities and suburbs of the North, West, and South. Because middle-class whites are leaving the cities for the suburbs, *de facto* segregation is turning the inner cities into black ghettos. Blacks isolated in these areas find their housing, their public facilities, and their schools automatically segregated and made inadequate by the lowered tax base, the backwardness (and often helplessness) of city government, and the exploitation or apathy of absentee landlords and local businesspeople.

Education Blacks receive less education than do whites in the United States, as Figure 6.2 and Table 6.2 indicate. The gap is closing slowly. The United States Supreme Court outlawed segregated schools in *Brown v. Board of Education* in 1954, but integration of schools did not proceed very rapidly. For some years now, the nation has struggled with ways to achieve school integration in the face of housing segregation. One answer is court-ordered busing of students from one area to another to maintain a racial balance. The move in this direction was encouraged by a report by James S. Coleman in 1966, which concluded that deprived students performed better scholastically when their classmates came from backgrounds emphasizing educational achievement. It was concluded that the educational chances of blacks could be improved by moving them from poor school districts to higher-quality (predominantly white, middle-class) school districts. Busing has moved the cause of school integration along, but not without controversy and some unanticipated consequences. Riots and demonstrations have occurred. White parents have generally opposed busing, and many have responded by moving their children to private schools or by moving their family to areas not affected by busing, both of which may *increase* segregation.

Coleman complicated the issue with another report in 1975, which suggested that perhaps schools weren't so important as family background was in determining educational achievement; maybe there wasn't any point in busing after all. Then, another Coleman report in 1981 concluded that private schools do a better job than public schools even after family background factors are controlled—in other words, that schools *do* make a difference. Specific educational policies and techniques such as more homework, more-difficult subjects, patterns of discipline, and teacher interest in students can lead to higher achievement, regardless of family background. How can this be of help to minority families who cannot afford private schools? Coleman suggests tuition tax credits, which would probably increase the private-school enrollment of blacks and Hispanics more than it would that of whites.[7] The equal education issue is a difficult one complicated by the interaction of several important factors—minority status, social class, family background, and educational policies—all playing against the backdrop of racism.

The picture is brighter at colleges and universities. College enrollment increased more rapidly for blacks than for whites between 1970 and 1980. A gap remains, however. The percentages of black high-school graduates enrolled (32 percent) was still below that of white high-school graduates (42 percent) in 1991.

Voting and Elected Officials Blacks had been denied the vote for years in the South. However, implementation of the Civil Rights Act, calling for federal voting registrars in the South, seemed to change this form of discrimination. Voter-registration figures in 11 southern states in 1960 indicated that 61 percent of the eligible whites were registered, as compared with 29 percent of the eligible blacks. By 1992, however, 70 percent of the whites and 64 percent of the blacks were registered. In 1970, there were fewer than 1,500 black officeholders. In January 1992, 7,517 blacks were holding office in 46 states and the District of Columbia. The increase in the number of elected officials has been most evident in the South. However, blacks still account for fewer than 1 percent of all elected officials. In 1993 there were 38 blacks in the House of Representatives out of 435 individuals, and one black in the Senate.

Native Americans

In the middle- and late-19th century, Americans believed in the doctrine of Manifest Destiny, which held that it was the destiny of the United States to expand its territory and enhance its political, economic, and social influence over the whole of North America. And so, the original occupants of the land, the American Indians, were killed, captured, displaced, and placed on reservations as settlers moved west. It was too bad, as one Army general who supervised the removal of the Indians put it; they were a brave and proud people but they finally realized that it was their destiny to "give way to the insatiable progress of [the white] race. . . ."[8]

In his book, *Custer Died for Your Sins*, Vine Deloria, Jr., discusses an interesting paradox in American society. Americans are told not to trust Communist countries, for they do not keep treaty commitments. We fought in Vietnam for years and thousands of people lost their lives ostensibly because the United States had to keep its commitments in Southeast Asia. The paradox is this: At the very time we were spending $100 billion keeping commitments in Vietnam, the United States government was also breaking the oldest Indian treaty, the Pickering Treaty of 1794 with the Seneca tribe of the Iroquois nation. In fact, as far as breaking treaties is concerned, it is doubtful that anybody can beat our record. Deloria reports that "America has yet to keep one Indian treaty or agreement despite the fact that the United States government signed over four hundred such treaties with Indian tribes."[9]

Desirable land has been systematically taken from Indians, who are then relocated on reservations in less desirable areas. Indians held 138 million acres in 1887; this was reduced to 55 million acres by 1970 and to 52 million by 1982. If formerly undesirable land becomes desirable because of the discovery of oil, or if it contains a river needed for power or irrigation, or if it is considered to be a prime area for development of any kind, further relocation and manipulation of Indian rights and lands are sure to follow.

The 1990 census reported that there were approximately 2 million American Indians in the United States. In general, living conditions on reservations have long been substandard. Median family income is low, and unemployment and the number of families below the poverty level is high. Educational attainment and life expectancy are well below that of whites. Death statistics provide further evidence of the effects of minority status. Indians are less likely to die of heart disease and cancer than are other Americans, but they are about three times as likely to die of accidents. Indians have a high birth rate (about two and a half times that of whites), but Indian deaths in infancy and early childhood are about double that of other Americans. In 1990, 54 percent of all American Indian births were to unmarried mothers. Indian deaths by cirrhosis of the liver are more than double the national average, by homicide more than triple, and by tuberculosis four times the national average. Indian alcoholism and suicide rates are also very high; suicide is especially high for Indians in their late teens and twenties. Dr. Bertram Brown, a director of the National Institute of Mental Health, has stated that Indians suffer greatly the effects of

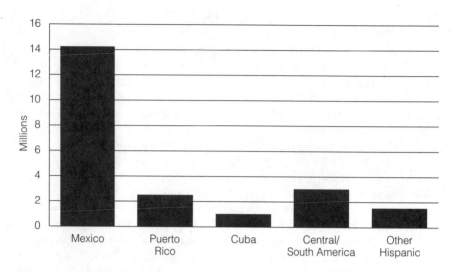

FIGURE 6.3   *Origins of Hispanic population, 1992.*
United States Bureau of the Census, *Statistical Abstracts of the United States, 1993* (113th ed.), Washington, D.C.

isolation and alienation from society brought on by the general neglect of many generations. Racism and bigotry have an adverse effect on mental health, and the American Indian has been the victim of this abuse longer than has any other American.

Hispanic Americans

Americans of Spanish origin have become this country's fastest growing minority. Census figures put the number of Hispanics in the United States in March 1992 at 22.1 million (9 percent of the population). The regions and countries that Hispanics come from are shown in Figure 6.3. Population projections suggest that the Hispanic population may outnumber that of blacks within the next decade.

The major unifying factors for Hispanic Americans are religion (Roman Catholicism) and language, but many diversities exist as well. Most of the Hispanics of Mexican origin have settled in the West and Southwest (about 3 million in the Los Angeles metropolitan area alone). More than half of those from Cuba have settled in the Miami area, and more than 1 million Hispanics from Puerto Rico are living in the New York City area.

Many Hispanic Americans face the conditions that seem to accompany minority status. Their median family income in 1991 was $22,691 (compared with $30,706 for non-Hispanics), and 28 percent were below the poverty level (41 percent of Puerto Ricans). Hispanics face health difficulties because, although their employment rates are high, they are severely underinsured (often being self-employed or working for small businesses). They may delay doctor visits because of lack of insurance as well as suspicion of the medical profession. Many Hispanics in the West are migrant farm workers laboring in the fields of California and Oregon. Farm labor is seasonal, uncertain, underpaid, and subject to difficult working conditions. Hispanic Americans are also underrepresented in the political process. In 1992, only 35 percent were registered to vote compared with 66 percent of non-Hispanics. They hold a low percentage of jobs in the federal government. In 1992, 4,994 Hispanics were holding office in 35 states. (Over half of these were in two states, Texas and New Mexico.) In 1993, there were 17 Hispanics in the House of Representatives and no Hispanic senators.

Perhaps the most serious problem for Hispanic Americans is in the realm of education. They have a high dropout rate, as can be seen in Figure 6.2 and Table 6.2. One of the central issues is language and the necessity for bilingual education. Some 11 million people in the United States speak Spanish as their first language. Fifteen percent of these who are under 18 and nearly 30 percent who are 18 and over speak English "not well" or "not at all." Schools, however, tend to communicate with parents in English only. Exclusion of their language leads to disintegration of their family heritage and discourages participation in school affairs. Children in school have to learn a new language in addition to other material assigned. They read and perform below grade level and ultimately may drop out of school early.

Jews

There are approximately 17 million people of the Jewish faith in the world. There are 6 million Jews in the United States, or 2.5 percent of our population in 1990. Religion is the foundation, but being Jewish also involves an ethnic and cultural identity. Unfortunately, persecution has been a consistent factor in Jewish cultural history. Jews were forced off their land and driven into other areas as early as the sixth century B.C. Especially sharp conflict between Christians and Jews emerged after the third century A.D. Starting around the tenth century, Jews were segregated into ghettos, at first voluntarily, then on a compulsory basis. In 1555, the Pope decreed that Jews must be segregated strictly in their own quarter of the city surrounded by a high wall with gates closed at night. Jews were forced out of certain occupations and denied citizenship. They were used as scapegoats to distract attention from other issues and to promote national unity. This was played to its most horrible extreme during World War II with the extermination of 6 million Jews in Germany.

The consequences of minority status for Jews in the United States do not seem to be as severe as they are for other groups we have considered. This is probably because Jews are less visible than are most minority groups and because they have been more upwardly mobile through their emphasis on education and participation in the professions. Jews, however, have faced certain types of prejudice and discrimination in the United States. For a time, certain jobs were not open to Jews. Colleges and universities, especially in the eastern United States, had quotas covering the percentage of Jews they would accept. In 1958, 33 percent of the real-estate agents in a large midwestern city reported they did not want to rent to Jews, and a study of 3,000 resorts across the country in the same year found that 22 percent discriminated against Jews. In the late 1960s and 1970s, such overt discrimination against Jews probably declined, although anti-Jewish prejudice remains. The Anti-Defamation League of B'nai B'rith reported that there were 1,867 hate incidents against Jewish people in 1993. This was an 8 percent increase over the previous year and the second highest number of occurrences in 15 years. Included were such anti-Semitic incidents as the fire bombing of a synagogue, the desecration of Jewish cemeteries, threats of violence, and the distribution of anti-Semitic leaflets.

PATTERNS OF INTERACTION

When populations that differ by race or ethnicity come together, a number of different reactions may occur, some peaceful, others not.[10] **Annihilation** is the elimination of one group by another. It may be intentional, as in the case of the Nazis' attempted extermination of a whole ethnic group—a practice called **genocide**—or it can be unintentional, as in the case of American Indians fallen prey to illnesses brought in by European settlers. **Expulsion** refers to the removal of a group from the territory in which it resides. The Japanese on the West Coast during World War II were removed from their homes and placed in detention camps, and American Indian tribes were forced from their homes onto reservations. **Segregation** refers to the setting apart of one group from another. It is not as extreme as expulsion, but the separation is nevertheless obvious. The segregation may be physical, such as a move to a separate area of the city; or it may be social, as when people living together are limited in how and with whom they may interact. Segregation from the minority's viewpoint may be either involuntary, as with blacks in America, or voluntary. Avoidance (voluntary self-segregation) occurs when a culture such as the Amish in America wants to maintain its identity and uniqueness against the influence of the dominant culture and therefore voluntarily separates itself.

Assimilation and amalgamation are other patterns of interaction that occur between unlike groups. **Assimilation** refers to the mixing and merging of unlike cultures so that two groups come to have a common culture. When two cultures meet, each culture might adopt some of the other's traits so that the result is a true melting pot—the emergence of a new culture different from either of the old cultures. More often, however, assimilation means that the incoming or minority culture adapts to the dominant culture. **Amalgamation** refers to biological (rather than cultural) mixing. Amalgamation seems invariably to accompany the meeting of racial and ethnic groups, despite the best efforts of some to legislate against it. A **marginal person** is defined as one who lives between two antagonistic cultures, a product of each but a member of neither. Studies indicate that the consequences of marginality can be severe—personality disturbances or feelings of inferiority. Marginality is often defined as a product of amalgamation, but it may also result from cultural mixing (the southern rural migrant to the urban North, for example).

Cultural pluralism describes a pattern of interaction in which unlike cultures maintain their own identity and yet interact with each other relatively peacefully. Switzerland is usually cited as the best example of pluralism in that peoples of several nationalities and religions and even three or four different languages are able to get along peacefully. According to Brewton Berry, few countries have been able to successfully accomplish a system of pluralism. Some minorities in the United States have at various times voiced a preference for cultural pluralism, but the dominant culture seems to look either toward integration or toward discrimination and segregation.

MINORITY-GROUP REACTIONS

The patterns of interaction described here are usually dictated by the dominant group. What then of minority-group reactions to domination? Gordon Allport has outlined some types of reactions in his book, *The Nature of Prejudice*.[11] The simplest response is to *deny* that one is a member of that minority group: Members of one race pass as members of another; people change their names to get rid of their ethnic identification. Similarly, **acculturation** refers to attempts by members of minority groups to assimilate, to blend in, and to take on as many characteristics of the dominant culture as possible. Hiding resentment behind *withdrawal* and passivity—"the mask of contentment"—is another reaction. **Secession** refers to the formal withdrawal of a group of people from a political, religious, or national group. The South seceded prior to the Civil War, and the citizens of Quebec have been considering secession from the rest of Canada. *Self-hatred* is another reaction, with the minority group identifying with the dominant group. Studies of Nazi concentration camps showed that after years of suffering, inmates began identifying with the guards. The prisoners wore bits of the guards' clothing, they became anti-Semitic, and in general they adopted the mentality of the oppressor. Likewise, if blacks value light skin more than dark, they may be accepting whites' evaluation of skin pigmentation. Self-hatred may lead to aggression against one's own group, as when a segment of the minority that thinks of itself as being of a higher class takes out its frustrations on a lower-ranked segment. Minorities sometimes show greater *prejudice* against other minorities and in other cases the opposite reaction—greater *sympathy.*

A *strengthening of in-group ties* is another reaction by minorities to domination. **Nativism** is a type of rejection of the dominant culture in which people attempt to improve their own existence by eliminating all foreign people, objects, and customs. This has been seen in the United States with slogans such as "Black Power" and "Don't buy where you can't work," and by the focus on the development of ethnic identity through emphasizing the minority's cultural and historical roots. Minorities may also react to domination by using *aggression.* The aggression may be individual and reflected in crime patterns, or it may be collective and culminate in urban disturbances or riots. In milder form, the aggression may appear in the literature and humor of the minority group as it makes fun of or shows its bitterness toward the majority group. In fact, humor is a common way of expressing feelings about other groups. In the following examples, who is putting down whom?

A black woman from Harlem wins the Irish sweepstakes and decides to buy a fur coat. She tries on a coat at Saks Fifth Avenue and it comes down to her ankles.

As she looks at herself in the mirror, she turns to the saleslady and says, "Do you think this makes me look too Jewish?"

A priest and a rabbi are each driving down a New York street. The rabbi stops at a light and the priest runs into him. An Irish cop comes up to the priest and asks, "Ah, Father, how fast might the rabbi have been going when he backed into you?"

What do you have when you cross a WASP and a chimpanzee? A three-foot-tall, blond company president.

FEMALES AND MALES

How does one decide if a person is upper-middle-class or lower-middle-class? Is an American Indian white? It's difficult to say. However, the question of female and male is no problem at all. There are arguments over how many social classes exist, to say nothing of the problem of determining the social class of a particular individual; the answer depends on the variables and method of analysis used. The boundaries of race and ethnicity are blurred; in many cases, the best answer seems to be that people are what they define themselves to be. Gender categories are more easily defined, at least biologically. This is one distinction that people have felt confident about for ages. If occasional doubts do pop up, a scientific test can be applied. One of the more intriguing aspects of the Olympic Games is the gender test used to make sure that males are not masquerading as females.

The paradox in this situation is that although people confidently feel and always have felt that they have a clear grasp of the biological fact, until recently few have considered the *social* consequences of gender differences. Karl Marx wrote about the consequences of social-class differences over 100 years ago, and he wasn't the first. The consequences of racial differences have been a central issue in the United States for decades. But the consequences of the minority status produced by gender differences have only recently—perhaps in the last 20 years—become a topic drawing widespread and serious attention.

Socialization and Gender Roles

Thirty pairs of parents were questioned within twenty-four hours after the birth of their first child and were asked to "describe your baby as you would to a close relative." Hospital information on the babies showed that the fifteen boy babies and the fifteen girl babies did not differ on such objective data as birth length, weight, irritability, etc. But the parents said that girl babies were softer, littler, more beautiful, prettier, more finely featured, cuter, and more inattentive than the boy babies. The fathers tended to label, or stereotype, the babies in this fashion more than the mothers. The authors of the study suggest that gender typing and gender-role socialization have already begun at birth.[12]

How could this be? One day after birth and already babies are showing definite gender differences in terms of physical appearance and temperament.

The answer, of course, is that *it isn't so*. People assume that males and females are born with different abilities and temperaments. People assume it to be so and then behave as if it *were* so. We tend to act toward children one way if they are male, another way if they are female. We expect them to be a certain way, and they turn out that way. The self-fulfilling prophecy rides again.

To place the assumption about so-called inherited gender characteristics in correct perspective, recall Margaret Mead's description of gender and temperament among the Arapesh, Mundugumor, and Tchambuli. The behavior of the sexes in those societies varied a great deal. Compared with gender roles in America, roles were modified and reversed. The explanation is not that these societies are strange and peculiar, but more simply that the gender roles learned in those societies are different from those learned here.

Like other roles, gender roles are learned through the socialization process. It is possible that in our society the teaching of gender roles starts even earlier than the teaching of other roles. As we saw earlier, parents immediately start acting differently toward their children based on their gender. Look at their toys: Boys get tractors, trucks, tools, guns, and athletic equipment; girls get dolls, cooking sets, play perfume and cosmetic kits, and pretty clothes. Friends and grandparents, scandalized to hear that the mother concerned about gender typing bought her son a doll, were only partly mollified to find that at least it was a boy doll.

The schools continue the pattern. Textbooks from Dick and Jane on up portray boys in active, aggressive, so-called masculine roles and girls in passive, tender, so-called feminine roles. A study of teachers in nursery schools found that they spent more time with the boys in the class than with the girls. Boys were encouraged to work harder on academic subjects. They were given more rewards and more directions in how to do things. Boys were given instructions, then encouraged to complete the task themselves. If the girls did not quickly get the idea, the teacher would often intervene and do the task for them. There was one exception: The teachers did pay more attention to the girls on feminine gender-typed activities such as cooking. Even here girls got praise and assistance, whereas boys received detailed instructions. The boys were given more attention, and the environment was much more of a learning experience for them than it was for the girls. Studies have found the same thing: more attention to males than to females throughout the grades and a "let me do it for you" attitude toward females, even at the college level.[13]

Another group of researchers working in the Boston area found that gender-role differences were well developed in the majority of children by the age of five. The children knew which personality traits were "masculine" and which were "feminine." They knew which jobs were for men and which were for women. The experimenters developed a curriculum that attempted to make the children more flexible in their assumptions about the sexes. The outcome of the program was mixed. To the researchers' surprise, many of the fifth- and ninth-grade boys with whom they worked became *more* stereotyped in their views of women and more rigid and outspoken about what they thought to be the woman's place. The effects on the girls in the program were more positive, showing attitude change away from typical stereotypes and increased self-esteem.[14] The researchers were hopeful about the school program they tested, but that attitude is not enough by itself. The socialization agents include family, peers, teachers, literature, entertainment, mass media, and the whole social tradition that socializes people to believe in the inherent psychological, temperamental, and attitudinal differences between the sexes. It should not surprise us at all when boys grow up to be physicians and mechanics and girls grow up to keep house, have babies, and take care of their families. They have been socialized into these roles from the very beginning.

Consequences of Gender-Role Typing

The results of gender-role typing are many and varied, and not all the benefits are for males. The male is restricted in how he may show emotion: He is strong and silent, he does not show weakness, and he keeps his feelings under careful rein, at least outwardly. The female has far greater freedom to express emotion. The male is subject to much more stress and pressure to achieve and be successful. This is probably part of the reason why males have a shorter life expectancy, more heart disease, and higher rates of suicide and hospitalization for mental illness and related pathologies. Women have been much less involved in crime and deviant behavior than have men, and this too is related to gender-role differences. Some men would like to change roles and to be househusbands, staying home and cooking, working in the garden, and taking care of the kids. What are their chances in a society that sets up gender roles like ours does?

The major consequence of gender-role typing, however, is that women are not able to participate fully in American society, a society in which they make up 51 percent of the members. People are becoming more sensitive to the unfairness of this situation, and women's frustration is increasing.

The second research question at the beginning of this chapter concerned interruptions in group discussions. Are men or women more likely to do the interrupting? A study conducted in the late 1980s discovered a sharp gender difference. The main observation of the research is that men discriminate in their attempts at interruption, whereas women do not. Men interrupt women far more often than they do other men. Women interrupt men and women equally often. Sometimes interruptions are positive or supportive, others are negative. The paper notes that women were three times as likely as men to yield the floor in the face of a negative interruption. The composition of the group also had some effects. For example, males in all-male groups tend to make supportive interruptions of the other males. As the number of women in the group increases, however, male interruptions are much less likely to be supportive. The authors find, perhaps not surprisingly, that everyday conversations both reflect and maintain a society's existing social inequalities.[15]

Ample evidence exists of the effects of gender typing. More women than men graduate from high school (38 percent to 34 percent in 1992), but more men than women receive advanced degrees (9 percent to 5.6 percent). Traditionally, a higher degree was seen as more important for men than for women.

The percentages of male and female workers with income and their average salaries are shown in Table 6.3. The proportion of men working has not changed much in 20 years, whereas the proportion of women working has gone up substantially. But now check their incomes; how fast is that gap closing?

Women represent half of the work force but generally hold lower-skilled and lower-paying jobs. Women are always well represented at the sales and clerical levels but seldom anywhere else. Even when a woman does the same work as a man does, she gets paid less for it. In 1992, on the average, women were paid 71 cents to every dollar paid to men. Women represent more than 50 percent of the nation's registered voters, but in 1989 they held only 17 percent of the state-legislative positions. Although this is a distinct improvement over the 10 percent they held in 1980, it is still well short of parity. In 1993 there were 47 women in the House of Representatives (out of 435), 6 women in the United States Senate (out of 100), and 2 women on the U.S. Supreme Court (out of 9). In 1992, women made up about 21 percent of the nation's lawyers and judges, 20 percent of the physicians, and 9 percent of the dentists.

Table 6.3 Median Income of People 14 Years and Over, by Gender (in 1991 Dollars)

| Years | Males | | Females | |
	Percentage with Income	Median Income	Percentage with Income	Median Income
1970	92	$21,996	66	$ 7,327
1980	95	$20,736	89	$ 8,142
1991	95	$20,469	91	$10,476

United States Bureau of the Census, *Statistical Abstracts of the United States, 1993* (113th ed.), Washington, D.C.

Women constitute 85 percent of the elementary-school teachers but only about 57 percent of the school administrators. In higher education, bachelor's and master's degrees were evenly split between men and women in 1990, but women received 37 percent and men 63 percent of doctorates awarded.

Over the past 25 years, major changes have occurred regarding women's roles and accomplishments. Polls show that many support the aims and efforts of the movement to change women's status in society. More women are elected to public positions. Women's enrollment in business, medical, and law schools has increased substantially. Churches have ordained women and are creating opportunities for women to take leadership roles. Many police departments across the country now have women in jobs equal to those of male police officers. Women's participation in organized sports is increasingly accepted. And even in crime, where men excel, women seem to be trying to catch up. Arrests of males were up 17 percent between 1983 and 1992, but the arrests of females over the same period were up 38 percent.

A 1992 Census Bureau study reported that in the late 1980s women's wages increased sharply in relation to men's. Women's wages, which had been stuck at about 60 percent of men's for some 20 years, increased to 71 percent. A major reason for this has been increasing entry into higher-paying jobs—law, computers, engineering. According to the report and to others evaluating it, the remaining difference in wages is caused by women having less education, work experience, and seniority, by women working for smaller firms, and by continuing gender discrimination in pay, hiring, and promotion policies.

Do these improvements mean that equality is just around the corner? Probably not. Look at the facts: 79 percent of the nation's lawyers and 80 percent of the physicians are males, as are 91 percent of the dentists, and so on. A 1982 study looked at male and female MBA's from Stanford four years after graduation: Women's income was averaging $7,000 a year less than men's. A few women climb the ladder; most do not. Often, the achievements of the successful are widely publicized, perhaps unintentionally giving the impression of great gains. Even those women who are successful inevitably have horror stories to tell of the subtle pressure and overt discrimination used against them. A woman is still paid less than a man for the same work, regardless of level.

The greatest success of the feminist movement has been in raising people's awareness of the existence of gender-role stereotypes and gender discrimination. Change is occurring, and the progress in reducing education and job discrimination is consistent if not dramatic. Changing traditional patterns of socialization for the young and resocializing adults are difficult at best.

THE ELDERLY

Hope Bagger at the age of 70 was called for jury duty but then was told to go home because she was too old. She and many others with similar experiences have become active members of the Gray Panthers, an organization seeking to remedy the way society treats old people. In some parts of the world, old age is seen as a virtue—life in its very highest form. The elderly are revered and are well cared for by society as a whole. In this country, the elderly find that they are forced to retire, they have trouble getting other jobs, and they are exploited if they do get work. As one member of the Gray Panthers put it, the young seem to feel that old people descended from outer space. They forget that the elderly have a past and a future.

A baby born in 1900 had a life expectancy of 48 years; a child born today has a life expectancy of more than 70 years. There were 31.8 million people 65 and older in the United States in 1991, and 13 million over age 75. These age groups are the fastest-growing population groups in the nation. By the year 2030, one fifth of all Americans will be over 65. Many of the elderly are poor and out of work, over 50 percent are widowed, the suicide rate of older males is high, and age often compounds the effects of other minority statuses.

Aging affects people both physically and socially. In a general sense, physical aging begins in people around the age of 30 and continues throughout the life span. Certain body functions, muscle tone and strength, and the senses all reach their peaks around 30 and decline thereafter. Although many people tend to view it as such, aging is not a disease. People are more likely to die as they get older because the body becomes more vulnerable to outside influences, less adaptable, and more affected by stress and crisis.[16]

Social aging is more treacherous, yet potentially easier to deal with, than physical aging. Social aging refers to the attitudes we have about older people. We are indoctrinated with the idea that the old are different and inferior to the rest of us. We stereotype the aged, and they are victims of prejudice and discrimination. In fact, American society treats the elderly with the same sort of stereotyping, the same institutionalized discrimination, as that based on gender and race. The old are thought to be unintelligent, asexual, unemployable, mentally ill, and difficult to get along with. The fact that many if not most older people are *not* this way is beside the point. We continue to make them retire before they want to, to refuse to hire them, and to pay them less when they do get a job. And the self-fulfilling prophecy suggests to us that if people are viewed and treated a certain way long enough, they tend to become that way.

What do the elderly want? Our stereotypes tell us that they want a life of leisure (sitting out the last 20 years on the porch swing or in front of the tube); they want to grow old gracefully; they want a time of disengagement (to be able to relax now because the important things are done). On the contrary, to remain in good health old people need what everyone needs: work to do, money to live on, a place to live in, and other people to care for and to care for them. Society should recognize that the elderly are an important resource.[17]

As with many other victims of minority status, the elderly are becoming politically active. It probably starts with consciousness raising and moves to more militant stands on issues through collective action. Groups like the Gray Panthers, the National Retired Teachers Association, the National Council of Senior Citizens, and the American Association of Retired Persons (AARP) have formed a "gray lobby" of enormous strength in Washington. The AARP, founded in 1958 to provide insurance for retirees, now has some 27 million members and has become the nation's largest special-interest group. One of nine Americans belongs, and its members are flexing their political muscles.

It is important, then, that the elderly not be viewed as social obsoletes. They *do* have a past and a future. And remember that unlike many minority statuses, this is one that happens to *all* of us—if we're fortunate.

THEORY AND RESEARCH: A REVIEW

Conflict theorists would have no difficulty explaining the sort of social differentiation we talked about in this chapter. Minority-status people are dominated and exploited by others because of the continuing struggle for scarce resources such as jobs, food, land, wealth, and power. Those who hold privileged positions attempt to ensure that they are safe by limiting access to these positions to others by whatever means possible. Racial, ethnic, gender, and age differences are easily identified and therefore become perfect "excluders." Color of skin, inability to speak the language well, or lack of citizenship become ready criteria for limiting access to jobs, education, or other valuable benefits. As in the discussion of stratification in the previous chapter, *functional analysis* would suggest that complex societies have numerous roles that must be performed. Some of these are more unpleasant than others, but all must be filled for society to run smoothly. Therefore, the stratification or ranking of society's members represents a good functional adaptation. By making distinctions between female and male, old and young, and recent arrivals and old timers, all slots are filled and all needed tasks from farm laborer and garbage collector to homemaker and company president are performed.

The data used in this chapter are descriptive and, for the most part, are census data obtained by *questionnaire or interview surveys*. Several studies cited in this chapter (Hartley's studies, Coleman's studies, the study of newborn babies, and the study of Stanford grads) were also *reactive surveys*.

SUMMARY

This is the second of two chapters on social differentiation and social inequality. Here we examined the characteristics of race, ethnic, gender, and age categories. We found in each case that elements within the categories are ranked and that inequality follows. One common component of these categories is that of *minority status*. The fact that categories are ranked and the consequences of this ranking were the central topics of this chapter. *Prejudice, discrimination,* and *stereotyping* are commonly used to reaffirm the status ranking we believe in. They help us keep others "in their place." We examined the types of reactions that occur when populations of different races or ethnic backgrounds meet, as well as the reactions of minority groups themselves to discrimination.

We examined a number of statistical examples of differential treatment—differences in education, jobs, income, opportunities in life, crime and deviant behavior, sickness, and others. Less easily displayed statistically but more important as a human concern is the effect of continued differential treatment on *how one sees oneself.* Some concepts studied earlier—socialization, role, self-concept, and self-fulfilling prophecy—help us understand the critical significance of minority status for the individual's development.

The first two readings that follow deal with the perceptions and judgments people make about others without knowing them—the foundation upon which stereotyping, prejudice, and discrimination are built. The first article is a first-person account by a black man, who describes his urban adventures in being mistaken for someone he's not. The second article describes "the jeweler's dilemma"—in a dangerous world, whom do you let into your store? The third reading examines gender discrimination in a work situation, the work situation being a bar near a college campus. The final reading is an excerpt from Dee Brown's *Bury My Heart at Wounded Knee,* which describes the consequences for the American Indians of the western expansion of the United States between 1860 and 1890.

TERMS FOR STUDY

acculturation (195)

amalgamation (194)

annihilation (194)

assimilation (194)

authoritarian personality (185)

cultural pluralism (194)

discrimination (184)

ethnicity (183)

expulsion (194)

genocide (194)

marginal person (194)

minority status (181)

nativism (195)

prejudice (184)

race (182)

racism (184)

secession (195)

segregation (194)

stereotyping (185)

Reading 6.1

JUST WALK ON BY

Brent Staples

*Stereotyping, prejudice, and discrimination are based on attitudes and judgments
we make about others without knowing them. We ignore the individual and generalize and categorize, often falsely. In this article, a black man describes his experiences of being mistaken for someone he's not, continually, and beginning "to
know the unwieldly inheritance I'd come into—the ability to alter public space
in ugly ways."*
Brent Staples is Assistant Metropolitan Editor of the New York Times.

My first victim was a woman—white, well dressed, probably in her early twenties. I
came upon her late one evening on a deserted street in Hyde Park, a relatively affluent
neighborhood in an otherwise mean, impoverished section of Chicago. As I swung onto
the avenue behind her, there seemed to be a discreet, uninflammatory distance between us. Not so. She cast back a worried glance. To her, the youngish black man—a
broad six feet two inches with a beard and billowing hair, both hands shoved into the
pockets of a bulky military jacket—seemed menacingly close. After a few more quick
glimpses, she picked up her pace and was soon running in earnest. Within seconds
she disappeared into a cross street.

That was more than a decade ago. I was twenty-two years old, a graduate student
newly arrived at the University of Chicago. It was in the echo of that terrified woman's
footfalls that I first began to know the unwieldy inheritance I'd come into—the ability to
alter public space in ugly ways. It was clear that she thought herself the quarry of a
mugger, a rapist, or worse. Suffering a bout of insomnia, however, I was stalking sleep,
not defenseless wayfarers. As a softy who is scarcely able to take a knife to a raw
chicken—let alone hold one to a person's throat—I was surprised, embarrassed, and
dismayed all at once. Her flight made me feel like an accomplice in tyranny. It also
made it clear that I was indistinguishable from the muggers who occasionally seeped
into the area from the surrounding ghetto. That first encounter, and those that followed,
signified that a vast, unnerving gulf lay between nighttime pedestrians—particularly
women—and me. And I soon gathered that being perceived as dangerous is a hazard
in itself. I only needed to turn a corner into a dicey situation, or crowd some frightened,
armed person in a foyer somewhere, or make an errant move after being pulled over
by a policeman. Where fear and weapons meet—and they often do in urban America—
there is always the possibility of death.

In that first year, my first away from my hometown, I was to become thoroughly familiar with the language of fear. At dark, shadowy intersections, I could cross in front of a
car stopped at a traffic light and elicit the *thunk, thunk, thunk, thunk* of the driver—
black, white, male, or female—hammering down the door locks. On less traveled
streets after dark, I grew accustomed to but never comfortable with people crossing to

the other side of the street rather than pass me. Then there were the standard unpleasantries with policemen, doormen, bouncers, cabdrivers, and others whose business it is to screen out troublesome individuals *before* there is any nastiness.

I moved to New York nearly two years ago and I have remained an avid night walker. In central Manhattan, the near-constant crowd cover minimizes tense one-on-one street encounters. Elsewhere—in SoHo, for example, where sidewalks are narrow and tightly spaced buildings shut out the sky—things can get very taut indeed.

After dark, on the warrenlike streets of Brooklyn where I live, I often see women who fear the worst from me. They seem to have set their faces on neutral, and with their purse straps strung across their chests bandolier-style, they forge ahead as though bracing themselves against being tackled. I understand, of course, that the danger they perceive is not a hallucination. Women are particularly vulnerable to street violence, and young black males are drastically overrepresented among the perpetrators of that violence. Yet these truths are no solace against the kind of alienation that comes of being ever the suspect, a fearsome entity with whom pedestrians avoid making eye contact.

It is not altogether clear to me how I reached the ripe old age of twenty-two without being conscious of the lethality nighttime pedestrians attributed to me. Perhaps it was because in Chester, Pennsylvania, the small, angry industrial town where I came of age in the 1960s, I was scarcely noticeable against a backdrop of gang warfare, street knifings, and murders. I grew up one of the good boys, had perhaps a half-dozen fistfights. In retrospect, my shyness of combat has clear sources.

As a boy, I saw countless tough guys locked away; I have since buried several, too. They were babies, really—a teenage cousin, a brother of twenty-two, a childhood friend in his mid-twenties—all gone down in episodes of bravado played out in the streets. I came to doubt the virtues of intimidation early on. I chose, perhaps unconsciously, to remain a shadow—timid, but a survivor.

The fearsomeness mistakenly attributed to me in public places often has a perilous flavor. The most frightening of these confusions occurred in the late 1970s and early 1980s, when I worked as a journalist in Chicago. One day, rushing into the office of a magazine I was writing for with a deadline story in hand, I was mistaken for a burglar. The office manager called security and, with an ad hoc posse, pursued me through the labyrinthine halls, nearly to my editor's door. I had no way of proving who I was. I could only move briskly toward the company of someone who knew me.

Another time I was on assignment for a local paper and killing time before an interview. I entered a jewelry store on the city's affluent Near North Side. The proprietor excused herself and returned with an enormous red Doberman pinscher straining at the end of a leash. She stood, the dog extended toward me, silent to my questions, her eyes bulging nearly out of her head. I took a cursory look around, nodded, and bade her good night.

Relatively speaking, however, I never fared as badly as another black male journalist. He went to nearby Waukegan, Illinois, a couple of summers ago to work on a story about a murderer who was born there. Mistaking the reporter for the killer, police officers hauled him from his car at gunpoint and but for his press credentials would probably have tried to book him. Such episodes are not uncommon. Black men trade tales like this all the time.

Over the years, I learned to smother the rage I felt at so often being taken for a criminal. Not to do so would surely have led to madness. I now take precautions to make myself less threatening. I move about with care, particularly late in the evening. I give a wide berth to nervous people on subway platforms during the wee hours, particularly when I have exchanged business clothes for jeans. If I happen to be entering a building behind some people who appear skittish, I may walk by, letting them clear the lobby before I return, so as not to seem to be following them. I have been calm and extremely congenial on those rare occasions when I've been pulled over by the police.

And on late-evening constitutionals I employ what has proved to be an excellent tension-reducing measure: I whistle melodies from Beethoven and Vivaldi and the more popular classical composers. Even steely New Yorkers hunching toward nighttime destinations seem to relax, and occasionally they even join in the tune. Virtually everybody seems to sense that a mugger wouldn't be warbling bright, sunny selections from Vivaldi's *Four Seasons*. It is my equivalent of the cowbell that hikers wear when they know they are in bear country.

QUESTIONS 6.1

1. Select at least five of the concepts discussed in this chapter and show how they are illustrated in this article.

2. Write a short essay on the effects of prejudice and discrimination on the individual, using concepts like socialization, self, looking-glass self, and significant and generalized other.

3. What sorts of patterns of interaction are illustrated in this article?

Reading 6.2

THE JEWELER'S DILEMMA

A controversy broke out in Washington, D.C., in 1986 over the practice in certain jewelry stores of admitting customers only through a buzzer system. Some store owners use this system to exclude young black males on the grounds that these people are most likely to commit robbery. Some people argued that this was racism; others felt that the practice was justified. The New Republic *magazine asked some of its readers to comment on this situation.*

The Intelligent Bayesian

Men are not gods. Therefore men face challenges gods would not have to endure—ignorance and uncertainty. To make decisions, we need to have information about the world around us. The information we gather is not only imperfect, it is costly as well. So we learn to economize by guessing, prejudging, and using stereotypes.

Imagine you are challenged to a basketball game and must select five out of 20 people who appear to be equal in *every* respect except race and sex. There are five black and five white females, five black and five white males. You have no information about their basketball proficiency. There is a million-dollar prize for the contest. How would you choose a team? If you thought basketball skills were randomly distributed by race and sex, you would randomly select. Most people would perceive a strong associative relationship between basketball skills on the one hand, and race and sex on the other. Most would confine their choice to males, and their choice would be dominated by black males.

Can we say such a person is a sexist/racist? An alternative answer is that he is behaving like an intelligent Bayesian (Sir Thomas Bayes, the father of statistics). Inexpensively obtained information about race and sex is a proxy for information that costs more to obtain, namely, basketball proficiency.

There is a large class of human behavior that generally falls into the same testing procedure. Doctors can predict the probability of hypertension by knowing race, and osteoporosis by knowing sex. A white jeweler who does not open his door to young black males cannot be labeled a racist any more than a black taxi driver who refuses to pick up young black males at night. Black females and white females and white males commit holdups, but in this world of imperfect information cabdrivers and jewelers play the odds. To ask them to behave differently is to disarm them. The other side of the argument is that they are losing money if they err—i.e., turning away an honest customer.

People can be stupid Bayesians. In 1971 I owned a house in Washington's exclusive Chevy Chase suburb. My Saturday chore was to pick up motorists' litter along the side of my house. One time I was approached by a gentleman who stopped his car and said, "Sir, when you're finished working at this house, could you come over to my place and work for me?" I politely told the gentleman I would not have time, since the rest of the day would be spent working on my Ph.D. dissertation. The gentleman apologized profusely. (My wife once encountered another incompetent Bayesian. Awaiting a bus during a downpour, she was offered a ride to Chevy Chase Circle by a black woman who turned out to be a domestic servant. "Don't you just hate working for these white people out here?" her benefactor observed.) Mistaken identity is always at least a bit unpleasant, but it is not the same as racism.

WALTER E. WILLIAMS
Walter E. Williams is John M. Olin Distinguished Professor of Economics
at George Mason University.

The Weight of Suspicion

Who would object to the refusal of a storekeeper, particularly of a store so obviously a target for crime as a jewelry store, to admit a stranger who looked "suspicious"? What is the purpose of a locked door and a buzzer system unless to keep out strangers who might wish to do harm? And since they are by definition strangers, how does anyone expect to distinguish between the innocuous and the malevolent strangers except by finding the latter "suspicious"?

I would find anyone suspicious who appeared different in certain ways from most of those who come into my shop. Teenagers, unless they were conventionally dressed, would be suspicious. People whose carriage and movements gave the impression of nervous tension would be suspicious. People whose voice on the intercom sounded tense, artificial, unclear would be suspicious. My decision to admit would be based on an intuitive calculation of the number of categories of suspicion and the intensity of my feeling, balanced by the knowledge that I must take adverse chances or encounter the greater risk of losing my store. And, of course, I add two decisive categories of suspicion, assuming one or more of the categories previously mentioned were positively present: sex and race.

To pretend that a roughly dressed, nervous, slurring, tense young black male has a democratic right to enter a jewelry store is to distort a rational process with an irrelevant abstraction. Does it salve my conscience to say that I exclude him because he's suspicious, not because I ascribe an innate criminal potentiality to males with black skins? Perhaps. My parents used to whisper the phrase "colored people" lest our Negro maid be embarrassed.

I too like to feel virtuously unprejudiced, and I have convinced myself that only so long as young black males commit an inordinate number of robberies will I continue to find decisive some suspicion-raising attributes I would overlook in others.

ROGER STARR
Roger Starr is an editorial writer at the New York Times.

The Lazy Man's Substitute

Neither black nor white store owners are in business to display the virtues of admitting people of all colors, creeds, and fashions to their stores. They are in busines to make money. I would want to take precautions to prevent robbery; I would look closely at people entering the store. The race of a potential customer would be one factor among many to be considered as I girded myself against thieves.

But in Washington and almost all other major cities, blacks do patronize jewelry stores. A jeweler in Beverly Hills who closed his or her door to heavily bejeweled Mr. T. would be foolishly closing his cash register. Unless I am a racist, race and age cannot be the sole deciding factors in calculating whom I will or will not let into my store. And I certainly would not close my door to, say, all young black men—not even to those who are casually dressed and behaving nervously. I would act cautiously in dealing with them, as I would with an antic, strangely dressed white man.

As a cabdriver I would apply the same considerations. Discrimination can be used judiciously. I would certainly exclude one class of people: those who struck me as dangerous. Nervous-looking people with bulges under their jackets would not be picked up; nor would those who looked obviously drunk or stoned. It all comes down to a subjective judgment of what dangerous people look like. This does not necessarily entail a racial judgment. Cabdrivers who don't pick up young black men as a rule are making a poorly informed decision. Racism is a lazy man's substitute for using good judgment.

. . . In this situation and all others, common sense is my constant guard. Common sense becomes racism when skin color becomes a formula for figuring out who is a danger to me.

JUAN WILLIAMS
Juan Williams is a reporter for the *Washington Post.*

Black Man's Burden

While I am putting myself in the shoes of the jewelry store owner, I will also put myself in the skin of a black person. This makes the dilemma even more interesting.

In downtown Washington, where my jewelry store is located, I know that it is possible, but not likely, that a white man is going to rob me. To the extent that I worry about robbers, I mostly worry about blacks. Thus one effective method of keeping robbers out of my store will be to scrutinize blacks more carefully.

I don't much like to do this because I know how easily such occasional and sensible "discrimination" can turn into widespread and simple bigotry. For example, I read in the *Washington Post* recently that when I go out to look for a house or an apartment in the Washington area, I have a 50 percent chance of being steered away from white neighborhoods or not being shown the nicest apartment or not being told about all the amenities. Now, most of these real estate agents in Washington are white and most of them very courteous. You just never know when one of them may rip you off.

As a businessman, I like to be consistent in my dealings with the public. If I don't let a black guy in the door because he's black and he's wearing a baseball cap, maybe I should refuse to deal with a white real estate agent the next time I am apartment hunting. Or maybe I should refuse to be shown around by white real estate agents with Southern accents who voted for Ronald Reagan. But what would white people think if they knew that successful black men like myself were refusing to do business with them just because of their skin color?

The last thing black people need today is a lot of white people getting pissed off at them. So when facing a situation in which I might get ripped off, I think it's prudent to take a chance. The humiliation of getting ripped off by some speedy black kid who snatches one of my gold necklaces is comparable to the humiliation of getting discriminated against by some smiling white real estate agent. But, of course, the odds that black kid outside is going to rob me are nowhere near 50 percent. Someday I hope I won't have to wonder about white real estate agents. So I figure: unless he's really a dangerous-looking kid, let him in. A lot of white people won't understand that it's them, more than the kid, I'm giving a break. But maybe someday.

JEFFERSON MORLEY

Comments: *To further complicate the situation, let's add some other information: (1) As the editors of* The New Republic *point out, what the jewelers are doing is illegal. If race is a factor in deciding whom you admit to your store, you're breaking the law. (2) Juan Williams (one of the letter writers) reports that the assumption that blacks commit more robberies is mistaken, "a convenient racist lie." He reports that young blacks' robbery rate is no higher than whites of the same age. (3) In 83 percent of robberies by blacks, their victims are black.*

QUESTIONS 6.2

1. "The behavior of the jewelers in denying admittance to certain people is blatant racism." Support or refute.

2. "Playing the odds in situations like this is only good and prudent business." Support or refute.

3. What does Walter Williams mean by "being an intelligent Bayesian"? Give examples in other types of social relationships.

4. How is the coach who selects five black basketball players to start the game while leaving several white players on the bench behaving differently from the jeweler who won't admit the black teenager to his store? Or is it the same? Discuss in terms of the meanings of stereotypes, prejudice, and discrimination.

Reading 6.3

THE COCKTAIL WAITRESS

James Spradley and Brenda Mann

As part of gender-role socialization, women and men learn a division of labor; men do some types of work, women do others. We would like to believe that there are functional reasons for such a division of labor, but perhaps it's more likely that the separation of roles is another, often unconscious, way of "keeping people in their place."

In this excerpt from their book, The Cocktail Waitress: Woman's Work in a Man's World, *anthropologists Spradley and Mann describe a situation in which the division between women's work and men's work is very clearly established. Their research was done in a college bar located in a large midwestern city in which Mann worked as a cocktail waitress.*

. . . Even in the smallest society it is impossible for everyone to work at all the things that need to be done. For this reason every culture contains rules for allocating jobs, every society has a division of labor. In a Bushmen band in South Africa, for example, men track down wild animals to provide meat for the people, while women dig roots and gather berries to add to the common food supply. Among the Kurelu of New

From *The Cocktail Waitress: Woman's Work in a Man's World* by James Spradley and Brenda Mann. Copyright © 1975 by Newbery Award Records Inc.

Guinea, men protect the borders of the tribal territories from attack and participate in offensive warfare. The women, on the other hand, have their assigned tasks such as gathering salt by soaking banana leaves in a salt spring, drying the leaves, and then burning them to retrieve the salt. In some societies, young girls are assigned the task of caring for small children, boys take the herds of goats or cattle for pasture and water, old men stay home to protect women and children, and young men do the heavy work of house-building. Every culture, then, has a division of labor, and we were not surprised to find this a feature of bar culture. Indeed, if the fifteen or twenty people who work at Brady's Bar all came to work on the same night and tried to carry out the same activities at the same time, chaos would reign. If all the employees took orders from customers, if all crowded together behind the bar at the same time to mix drinks, and if, each evening, everyone tried to do everything, the confusion and disorder would eventually destroy the bar. Like a small tribal band, the people who work at Brady's share a set of rules for allocating jobs and dividing up necessary tasks.

The men who work in the bar mix drinks, serve the customers seated around the long horseshoe bar, control the money, and manage business transactions with the outside world. The women, on the other hand, focus their activities on serving the customers who sit at tables.

One of our first goals was to discover the nature and meaning of these female role assignments. Our primary question was, "What does a person have to know in order to do what a cocktail waitress does?" The answer initially appeared rather simple—she has to find out what customers want to drink, tell the bartender, carry drinks back to the table, collect the payments, and later clear away empty glasses.

$$\cdots$$

It is 12:40 A.M. and Sandy knows that in twenty minutes the bar will close. She would like to tell the six guys at the corner table that the time has come to order their last drinks for the evening because it is near closing. Last call generally occurs around 12:45 and if she doesn't give last call soon, it will be impossible to get them to leave before 1:30 A.M., since they will take that long to finish their drinks. But the appointed time comes and goes. Although it would be a simple matter for her to tell the table, "Last call, would you like anything else to drink?" and in fact, she wanted to announce it fifteen minutes ago, she cannot take this simple step on her own initiative. At 12:55, Jim, the bartender working the lower section of the bar tells her "Give last call." Finally, she can make this announcement to the customers seated in her section. Because he delayed his permission, however, it will be closing time when Sandy delivers the last round of beers to the six guys from St. John's, and much later before she can retrieve the empty bottles. The cultural rules at Brady's Bar require that a male must tell the waitress when it is time for last call, and they require that a female must serve the last round to the tables. Over and over again we discovered such arbitrary rules interlocking male and female performances.

$$\cdots$$

"Could I please have a vodka tonic, Ted?" Denise asked, giving him her sweetest smile. She knew what the answer would be but it never hurt to try. Sometimes you can catch the bartender in a benevolent mood.

"Look, you know you can't have a drink. You start work in a little while and you might fall on that pretty face of yours. Do you want a Coke?"

"Okay. But how about some quarters for the juke box?" Ted handed her four quarters, she picked out her favorites and then went back to eating her sandwich and sipping her Coke. She sat there, listening to her music, and watching Ted clean up the bar—restocking the beer, and washing the dozens of dirty glasses that had accumulated during the lunch hour rush.

Two employees sitting at the bar, hanging around before work; one has the right to drink anything, the other does not. The reason for this restriction? Being a female. This particular rule may be unique to Brady's Bar, but we found this simple pattern of interaction repeated itself again and again in other situations. These specific cases merely reflect an important rule in our culture that governs much of male-female interaction, what we call the *handicap rule*. At every turn, we discovered evidence that the cultural rules at Brady's Bar place certain handicaps on all those players who were born female.

Let's go back to the social encounters between male and female workers at Brady's Bar. Here, the arbitrary handicap rules imposed on females are often hidden or at least justified as instinctual, God-given traits: "Women can't hold their liquor," "Women are too emotional." As in the wider culture, females are repeatedly required to play with an arbitrarily imposed handicap. It is as if all the players in the game made a tacit agreement that women must play by different rules than men. Even a suggestion to make these rules the same arouses male anger, and all the waitresses soon find this out as the following example makes clear. One night Denise was hanging around before work, and Ted asked her to watch the bar for a few minutes. Dave, one of the bartenders for the evening, came in and sat down. "Hey, bartender!" he called to Denise as he playfully banged his fist down on the bar. "I'll have a bourbon sour." Feeling especially powerful behind the bar, Denise replied, "Oh, I'm sorry, Dave, but you're working tonight. No drink for you. How about a nice big Coke?" The tone of her voice indicated to Dave that she was teasing, and Dave's response was also outwardly light: "Be a good bartender now and give me my drink!" And, of course, she complied. But Denise could tell she had said the wrong thing. She had unmasked the handicap rule by her actions, momentarily exposing its arbitrariness. Later that night, after the bar closed and all the employees gathered at the bar for a drink, Dave served the others, ignoring Denise's request for a gin and tonic. Denise was then forced into the position of having to beg Dave before he would serve her a drink—a kind of ritual reminder that she is not to question the implicit handicap rule.

· · ·

"Hey, Mike. Stubbs. How are you?" Mark stood up and playfully punched Mike on the shoulder and he returned the greeting. Both of the guys nodded and smiled at Denise. Mark turned to Denise and said, "Hey, Denise. Be a good girl and get behind the bar for me." The way he said it, she felt he was both giving her permission and giving her an order, but this was an opportunity that no waitress passed up. She crawled under the gate that closes at the end of the bar and turned to Mike and Stubbs for their order. "Schlitz for both of them, " said Mark. Denise opened one cooler after another until she found the one containing the Schlitz, found the opener attached to the side of the cooler, opened them and placed them on the bar in front of Mike and Stubbs. She looked around for the beer glasses, and by the time she had placed a napkin with a glass on it in front of each of them, they were halfway finished with their bottles. The minutes dragged slowly by as the men talked, but Denise busied herself exploring behind the bar; opening coolers to find out where the different kinds of beer were, checking the ice supply, examining the different bottles of liquor. Another customer came and looked surprised to see her there but ordered a bourbon sour as if nothing was out of order.

"Bartender. Oh, bartender. Can I have another beer?" That was Stubbs, calling attention to this reversal of roles. Mike joined in, "Me too. Isn't that a funny looking bartender? Mark sure has changed. That's a nice dress you have on, Markie." Denise smiles and concentrates on opening two more beers. She places them in front of the two guys, "Here you are." She took one of the dollar bills from the counter and rang it up on the cash register. Same old jokes every time a girl gets behind the bar. She had seen it happen to others and knew what to expect. She always felt somewhat uncomfortable back there but it was fun, the opportunity to mix drinks. Like the other girls, she watches the bartenders make drinks all night, every night, and so she knows almost as well as they do how to make a Vodka Gimlet, a Tequila Sunrise, an Old Fashioned, or anything else. While it never lasted very long, it was fun to work behind the bar, a special treat. All of the girls respond eagerly when Mark or one of the other bartenders sets aside the rules governing female behavior that normally keep them out of that area.

What we have witnessed here is a kind of role change, what we call the *cross-over phenomenon*. Situations in which men and women temporarily reverse their roles are not uncommon in our society. Whenever male and female activities are linked together, opportunities for individuals to cross over present themselves. Mr. Jones, for example, normally drives when his wife is in the car, but he lets her drive this morning so he can read some papers for an early meeting at the office. On other mornings he pulls up to his office building, and she slides over to take his place behind the wheel and drive home. Mrs. Jones normally bathes their two children after dinner each night, but on Tuesdays her husband takes over and performs this task so she can attend the University. The cross-over phenomenon probably occurs in all cultures of the world. The cultural rules for crossing over and doing tasks assigned to the opposite sex could easily be *symmetrical*. That is, on occasion men do women's work and vice versa. In the process there is an open appreciation for this exchange of responsibilities that are usually divided. But the rules that regulate the cross-over phenomenon in Brady's Bar are not the same for each sex. They are *asymmetrical*, functioning in such a way as to put women at a disadvantage in the game of social interaction.

Although Denise and Sandy and the other waitresses get behind the bar occasionally, and they are always eager to do so, bartenders *never* cross over to perform some of the female tasks in the bar. If Rob and Denise were sitting at the bar and a group of customers came in and sat at a table, Denise could never turn and say, "Rob, will you take care of those customers for me?" Even if she were busy taking his place behind the bar and customers entered, no bartender would voluntarily get up to wait on the tables for the cocktail waitress. Denise would simply come from behind the bar, wait on the table, go back behind the bar and fix the drinks herself, and then serve them. This is expected of her. Furthermore, none of the waitresses would think of asking the men to clean the tables or to wait on their customers. They have learned the rules well. Both men and women seem to act as if the tasks assigned to women at Brady's might have a polluting effect on men, contaminating their ritual purity. The reverse is not true. A man loses if he does women's work and participates in the cross-over phenomenon, and so he avoids it or refuses to switch. A woman gains and is usually eager to cross over.

An interesting dimension of this cross-over phenomenon relates to the way it requires the female to express gratitude. When a man crosses over to assist a woman, she should thank him, expressing gratitude to him for his assistance. If one of the male employees ever stepped in to assist a busy waitress, it would be seen as an act of chivalry and the girl in question would openly express her gratitude. But when a woman crosses over to assist a man, engaging in some typical male activity, such as tending bar, *she must still express gratitude* for he has allowed her to partake of a more valuable social world. It is doubtful that a husband would thank his wife for driving the car when they are traveling together. After all, in our cultural idiom he has "let her drive." His wife, on the other hand, would express gratefulness if he cooked dinner or did the dishes for her. When a waitress at Brady's is allowed to tend bar she feels that a privilege has been extended to her. Not only does she operate with a handicap but she almost always sees these particular rules as legitimate.

Denise has just waited on a table and after she serves them she stops to chat with a regular customer who is sitting next to the table. While she's talking, Dave calls her from across the bar so she quickly excuses herself and goes to see what he wants. "I need some stuff from the cooler. Get me some Heineken lights and a bottle of juice. Okay? That's a sweet girl." He turns back to his discussion with a friend sitting across the bar while Denise heads back through the kitchen to the cooler. "Why can't Dave get this stuff himself anyway?" she grumbles to herself. "He's always sending us back to carry this stuff. It should be his job, not the waitresses'." Denise thought about the time she had brought this up one night after work and had gotten nowhere. "Bartenders are never idle," she was told, "When they aren't mixing drinks and serving the bar, they are P. R.'ing and that is good for the bar." "So," thought Denise, "They call standing around talking to friends, P. R.'ing, but if we do it, they call it loafing." Although she may be able to visit a little later in the evening, even then, because of this handicap rule, she will have to do it so as to avoid the accusation that she is merely wasting time, loafing around.

. . . Waitresses feel that the division of labor is far from equal. George performs the major task of mixing drinks and ringing up the total for the order on the cash register. And Stephanie must do all she can to make this easy for George: translating the customer's order into language the bartender can immediately recognize; adding straws and fruit to the drinks; stirring the drinks as the mix is added to the alcohol; totaling the order for him; handing him the correct amount of money for the order; and finally, placing the drinks on the tray. While it would be possible for George to help Stephanie by performing some of these tasks himself, he does not. New girls often wait for the bartender to do things like place drinks on the tray or add up the prices, but they rapidly learn that it is their duty to make things as easy as they can for the bartender. "When I first started," says Joyce, "I gave the bartender the order and just stood there." But Joyce soon noticed the other girls helping by adding the nuts and straws to the drinks. A couple of times, Dave asked Joyce to do those things for him, and it wasn't long before Joyce just started doing all those things automatically.

The division of labor between bartenders and waitresses is clearly based on sex. Work is divided up in an arbitrary manner, men sometimes doing in their part of the bar what women are doing in theirs. But although arbitrary, this division is not casual nor easily changed. A female bartender or a male waiter are innovations which would be strongly opposed. The *handicap rule* helps keep women in their place and reinforces the centrality of males in social life. As we saw, even the suspension of the handicap rule in one or another situation, as when a benevolent bartender occasionally offers a waitress a drink before work, is a male prerogative, one that elicits expressions of appreciation from females. Again, *in the cross-over phenomenon,* we saw the subtle ways inequality can express itself—women seeking to do men's tasks and men avoiding female tasks, even though some of the activities are seemingly identical. The division of labor at Brady's Bar, and in society generally, has a much deeper significance than efficient service to paying customers. *Routine tasks are transformed into symbols of sexual identity.* The "job" becomes a "ritual" to publicly announce the significant differences our culture attaches to sexual gender. It says to everyone in the bar: "There is a woman, there is a man."

QUESTIONS 6.3

1. Using this selection, illustrate examples of discrimination based on gender. A double standard exists—show how.

2. What is the "handicap rule"? Give examples of this in other aspects of life.

3. What is the "cross-over phenomenon"? Give examples of this in other aspects of life.

4. This study involved participant observation (see Chapter One). Illustrate potential strengths and weaknesses of this approach using this selection as an example.

Reading 6.4

WOUNDED KNEE

Dee Brown

In his book, Bury My Heart at Wounded Knee, *Dee Brown writes about the critical period between 1860 and 1890, the 30 years during which the culture and civilization of the American Indian was destroyed. Brown tells the story in an unusual way—he uses the viewpoint and the words of the victims, the Indians. The story is a haunting and depressing one and includes descriptions of the long walk of the Navahos, the war to save the buffalo, the war for the Black Hills, the flight of the Nez Perce, and the story of Geronimo, the last of the Apache chiefs. The following excerpt from Brown's book describes the end of it, the "battle" at Wounded Knee Creek in South Dakota in late December 1890. The Indians' lands were gone then, their chiefs were dead or imprisoned, their own numbers decimated by 30 years of battles, by disease, by living on crowded reservations in unfamiliar places, and by heartbreak. In early December 1890 the great chief Sitting Bull was assassinated. Then . . .*

As soon as Big Foot learned that Sitting Bull had been killed, he started his people toward Pine Ridge, hoping that Red Cloud could protect them from the soldiers. En route, he fell ill of pneumonia, and when hemorrhaging began, he had to travel in a wagon. On December 28, as they neared Porcupine Creek, the Minneconjous sighted four troops of cavalry approaching. Big Foot immediately ordered a white flag run up over his wagon. About two o'clock in the afternoon he raised up from his blankets to greet Major Samuel Whitside, Seventh U. S. Cavalry. Big Foot's blankets were stained with blood from his lungs, and as he talked in a hoarse whisper with Whitside, red drops fell from his nose and froze in the bitter cold.

Whitside told Big Foot that he had orders to take him to a cavalry camp on Wounded Knee Creek. The Minneconjou chief replied that he was going in that direction; he was taking his people to Pine Ridge for safety.

Turning to his half-breed scout, John Shangreau, Major Whitside ordered him to begin disarming Big Foot's band.

"Look here, Major," Shangreau replied, "if you do that there is liable to be a fight here; and if there is, you will kill all those women and children and the men will get away from you."

Whitside insisted that his orders were to capture Big Foot's Indians and disarm and dismount them.

"We better take them to camp and then take their horses from them and their guns," Shangreau declared.

"All right," Whitside agreed. "You tell Big Foot to move down to camp at Wounded Knee."

The major glanced at the ailing chief, then gave an order for his Army ambulance to be brought forward. The ambulance would be warmer and would give Big Foot an easier ride than the jolting springless wagon. After the chief was transferred to the ambulance, Whitside formed a column for the march to Wounded Knee Creek. Two troops of cavalry took the lead, the ambulance and wagons following, the Indians herded into a compact group behind them, with the other two cavalry troops and a battery of two Hotchkiss guns bringing up the rear.

Twilight was falling when the column crawled over the last rise in the land and began descending the slope toward Chankpe Opi Wakpala, the creek called Wounded Knee. The wintry dusk and the tiny crystals of ice dancing in the dying light added a supernatural quality to the somber landscape. Somewhere along this frozen stream the heart of Crazy Horse lay in a secret place, and the Ghost Dancers believed that his disembodied spirit was waiting impatiently for the new earth that would surely come with the first green grass of spring.

At the cavalry tent camp on Wounded Knee Creek, the Indians were halted and children counted. There were 120 men and 230 women and children. Because of the gathering darkness, Major Whitside decided to wait until morning before disarming his prisoners. He assigned them a camping area immediately to the south of the military camp, issued them rations, and as there was a shortage of tepee covers, he furnished them several tents. Whitside ordered a stove placed in Big Foot's tent and sent a regimental surgeon to administer to the sick chief. To make certain that none of his prisoners escaped, the major stationed two troops of cavalry as sentinels around the Sioux tepees, and then posted his two Hotchkiss guns on top of a rise overlooking the camp. The barrels of these rifled guns, which could hurl explosive charges for more than two miles, were positioned to rake the length of the Indian lodges.

Later in the darkness of that December night the remainder of the Seventh Regiment marched in from the east and quietly bivouacked north of Major Whitside's troops. Colonel James W. Forsyth, commanding Custer's former regiment, now took charge of operations. He informed Whitside that he had received orders to take Big Foot's band to the Union Pacific Railroad for shipment to a military prison in Omaha.

After placing two more Hotchkiss guns on the slope beside the others, Forsyth and his officers settled down for the evening with a keg of whiskey to celebrate the capture of Big Foot.

The chief lay in his tent, too ill to sleep, barely able to breathe. Even with their protective Ghost Shirts and their belief in the prophecies of the new Messiah, his people were fearful of the pony soldiers camped all around them. Fourteen years before, on the Little Bighorn, some of these warriors had helped defeat some of these soldier chiefs—Moylan, Varnum, Wallace, Godfrey, Edgerly—and the Indians wondered if revenge could still be in their hearts.

"The following morning there was a bugle call," said Wasumaza, one of Big Foot's warriors who years afterward was to change his name to Dewey Beard. "Then I saw the soldiers mounting their horses and surrounding us. It was announced that all men should come to the center for a talk and that after the talk they were to move on to Pine Ridge agency. Big Foot was brought out of his tepee and sat in front of his tent and the older men were gathered around him and sitting right near him in the center."

After issuing hardtack for breakfast rations, Colonel Forsyth informed the Indians that they were now to be disarmed. "They called for guns and arms," White Lance said, "so all of us gave the guns and they were stacked up in the center." The soldier chiefs were not satisfied with the number of weapons surrendered, and so they sent details of troopers to search the tepees. "They would go right into the tents and come out with bundles and tear them open," Dog Chief said. "They brought our axes, knives, and tent stakes and piled them near the guns."

Still not satisfied, the soldier chiefs ordered the warriors to remove their blankets and submit to searches for weapons. The Indians' faces showed their anger, but only the medicine man, Yellow Bird, made any overt protest. He danced a few Ghost Dance steps, and chanted one of the holy songs, assuring the warriors that the soldiers' bullets could not penetrate their sacred garments. "The bullets will not go toward you," he chanted in Sioux. "The prairie is large and the bullets will not go toward you."

The troopers found only two rifles, one of them a new Winchester belonging to a young Minneconjou named Black Coyote. Black Coyote raised the Winchester above his head, shouting that he paid much money for the rifle and that it belonged to him. Some years afterward Dewey Beard recalled that Black Coyote was deaf. "If they had left him alone he was going to put his gun down where he should. They grabbed him and spinned him in the east direction. He was still unconcerned even then. He hadn't his gun pointed at anyone. His intention was to put that gun down. They came on and grabbed the gun that he was going to put down. Right after they spun him around there was the report of a gun, was quite loud. I couldn't say that anyone was shot, but following that was a crash."

"It sounded much like the sound of tearing canvas, that was the crash," Rough Feather said. Afraid-of-the-Enemy described it as a "lightning crash."

Turning Hawk said that Black Coyote "was a crazy man, a young man of very bad influence and in fact a nobody." He said that Black Coyote fired his gun and that "immediately the soldiers returned fire and indiscriminate killing followed."

In the first seconds of violence, the firing of carbines was deafening, filling the air with powder smoke. Among the dying who lay sprawled on the frozen ground was Big Foot. Then there was a brief lull in the rattle of arms, with small groups of Indians and soldiers grappling at close quarters, using knives, clubs, pistols. As few of the Indians had arms, they soon had to flee, and then the big Hotchkiss guns on the hill opened upon them, firing almost a shell a second, raking the Indian camp, shredding the tepees with flying shrapnel, killing men, women, and children.

"We tried to run," Louise Weasel Bear said, "but they shot us like we were buffalo. I know there are some good white people, but the soldiers must be mean to shoot children and women. Indian soldiers would not do that to white children."

"I was running away from the place and followed those who were running away," said Hakiktawin, another of the young women. "My grandfather and grandmother and brother were killed as we crossed the ravine, and then I was shot on the right hip clear through and on my right wrist where I did not go any further as I was not able to walk, and after the soldier picked me up where a little girl came to me and crawled into the blanket."

When the madness ended, Big Foot and more than half of his people were dead or seriously wounded; 153 were known dead, but many of the wounded crawled away to die afterward. One estimate placed the final total of dead at very nearly 300 of the original 350 men, women, and children. The soldiers lost twenty-five dead and thirty-nine wounded, most of them struck by their own bullets or shrapnel.

After the wounded cavalrymen were started for the agency at Pine Ridge, a detail of soldiers went over the Wounded Knee battlefield, gathering up Indians who were still alive and loading them into wagons. As it was apparent by the end of the day that a blizzard was approaching, the dead Indians were left lying where they had fallen. (After the blizzard, when a burial party returned to Wounded Knee, they found the bodies, including Big Foot's, frozen into grotesque shapes.)

The wagonloads of wounded Sioux (four men and forty-seven women and children) reached Pine Ridge after dark. Because all available barracks were filled with soldiers, they were left lying in the open wagons in the bitter cold while an inept Army officer searched for shelter. Finally the Episcopal mission was opened, the benches taken out, and hay scattered over the rough flooring.

It was the fourth day after Christmas in the Year of Our Lord 1890. When the first torn and bleeding bodies were carried into the candlelit church, those who were conscious could see Christmas greenery hanging from the open rafters. Across the chancel front above the pulpit was strung a crudely lettered banner: PEACE ON EARTH, GOOD WILL TO MEN.

QUESTIONS 6.4

1. Describe the treatment of the Indians in terms of the patterns of interaction described in the text. What is the pattern today?

2. Discuss the concept of Manifest Destiny: What might have caused it to appear? Were there other victims in ad-

dition to the Indians? What are the consequences of this doctrine for its adherents as well as for its victims? Is this doctrine still with us?

3. Research the Indian situation in your area; what tribe lived there and what happened to them?

NOTES

1. This distinction is suggested by Brewton Berry in *Race and Ethnic Relations*, 3d ed. (Boston: Houghton Mifflin, 1965), ch. 2.

2. See "Racial Identity Among Caribbean Hispanics: The Effect of Double Minority Status on Residential Segregation," by Nancy Denton and Douglas Massey, *American Sociological Review* 54 (October 1989), pp. 790–808.

3. The definition of racism is from Martin Marger, *Race and Ethnic Relations*, 2d ed. (Belmont, Calif.: Wadsworth, 1991), p. 27. For more on the mysterious Danireans and Wallonians, see E. L. Hartley, *Problems in Prejudice* (New York: Kings Crown Press, 1946).

4. Robert K. Merton, "Discrimination and the American Creed," in *Discrimination and National Welfare*, edited by R. M. MacIver (New York: Harper & Brothers, 1949), pp. 99–126. Also see Berry's summary in *Race and Ethnic Relations*, 3d ed., pp. 300–302 (see Note 1).

5. T. W. Adorno et al., *The Authoritarian Personality* (New York: Harper & Row, 1950).

6. The research discussed in this paragraph was from papers presented at meetings of the American Psychological Association, 1987, as reported in *Time*, Sept. 14, 1987, p. 74.

7. "The Meaning of the New Coleman Report," by Diane Ravitch, *Phi Delta Kappan*, June 1981, pp. 718–720.

8. General Carleton's views on Manifest Destiny and the Indians are quoted by Dee Brown in *Bury My Heart at Wounded Knee* (New York: Holt, Rinehart & Winston, Bantam Books, 1971), p. 31.

9. Vine Deloria, Jr., *Custer Died for Your Sins* (New York: Macmillan, 1969), pp. 28–29.

10. These paragraphs on patterns of interaction are drawn from Berry, *Race and Ethnic Relations*, 3d ed., chs. 7–12 (see Note 1).

11. See Gordon Allport, *The Nature of Prejudice* (Garden City, N.Y.: Doubleday, 1958), ch. 9.

12. Jeffrey Rubin, Frank Provenzano, and Zella Luria, "The Eye of the Beholder: Parents' Views on Sex of New Borns," *American Journal of Orthopsychiatry* 44 (July 1974), pp. 512–519.

13. Lisa Serbin and Daniel O'Leary, "How Nursery Schools Teach Girls to Shut Up," *Psychology Today* 9 (December 1975), p. 57, and Myra and David Sadker, "Sexism in School," *Psychology Today* 19 (March 1985), pp. 54–57.

14. This research is summarized by Carol Tavris in "It's Tough to Nip Sexism in the Bud," *Psychology Today* 9 (December 1975), p. 58. Tavris states that the full report of this research by Marcia Guttentag is available in *Undoing Sex Stereotypes* by Marcia Guttentag and Helen Bray (New York: McGraw-Hill, 1976).

15. See "Interruptions in Group Discussions: The Effects of Gender and Group Composition," by Lynn Smith-Lovin and Charles Brody, *American Sociological Review* 54 (June 1989), pp. 424–435.

16. Anita Harbert and Leon Ginsberg, *Human Services for Older Adults: Concepts & Skills* (Belmont, Calif.: Wadsworth, 1979), chs. 1 and 2.

17. Alexander Comfort, "Age Prejudice in America," *Social Policy*, November/December 1976, pp. 3–8.

CHAPTER SEVEN

Institutions: Family

TWO RESEARCH QUESTIONS

1. Do children repeat their parents' marital experiences (age at marriage, likelihood of being pregnant when married, chances of divorce, and chances of remarriage), or are children's patterns independent of their parents'?

2. What is the effect of having children on marital stability? Is the presence of children conducive to marital happiness? Which is more beneficial to marital happiness—young children or older (teenage) children?

So FAR WE have peered through several windows to discover the social organization of society. One view of social organization is through study of the nature and structure of groups. Analysis of the characteristics of categories and social differentiation provides another view. In this chapter, the study of social institutions will provide us with a third perspective.

One way of understanding social institutions as a concept is to describe the emergence of an institution. All societies have a set of constant and important central needs or problems that must be dealt with. A society, if it is to survive, must deal with such issues as reproduction of the species, socialization of the young, distribution of goods and services, care for the sick, and so on. Norms and roles emerge to organize and regulate behavior as society deems appropriate. And so a system of social relationships—a pattern of organization—develops. **Social institutions,** then, refer to this organized system of social relationships, common to all societies, that emerges to deal with these basic problems.[1]

For example, society's basic task of reproducing the species leads to the development of particular patterns of behavior. A society supports some of these behaviors (say, heterosexual contact) and discourages others (indiscriminate mating outside of marriage). This support reflects the development of norms: consistent views about the way things should and should not be done. The development of these norms and roles whereby the experimental becomes expected and the spontaneous becomes formalized is the process of **institutionalization.** As this process proceeds and the system of social relationships becomes more and more organized, a so-called institutional area emerges—in this case, the institution is the *family.*

The study of social organization makes a good deal more sense if we integrate several of the concepts we have studied. For example, if you merge the concepts of social institutions and groups, as shown in Table 7.1, you will discover that specific basic problems lead to the development of corresponding institutional areas. Next, notice that groups perform the functions that are important to the institutional area. Two important factors—the individual's inter-

Table 7.1 Institutions in the United States

Basic Problem or Issue Faced by Society	Institutional Area	Institutional Organization or Large Organizations	Institutional Units or Small Groups
Reproduction of the species	Family	Planned Parenthood; an organization of marriage counselors	My family
Socialization and training of the young	Education	University of California; Boston school system	A kindergarten class; a graduate seminar
Explanation for the unknown or uncertain	Religion	Catholic Church; the Council of Churches	Wednesday evening Bible-study group; a Sunday-school group
Distribution of goods and services	Economics	General Motors; IBM	Assembly-line workers installing brakes on Fords; the steno pool handling the Phillips account
Development of rules and regulations to govern people's behavior	Government	United States Senate; state assembly	Township subcommittee on dogcatching; committees for the preservation of historic landmarks
Use of free time; time away from work	Leisure	Professional Golfers Association; NCAA; Sacramento Recreation Association	A professional basketball team; a bowling team
Care and treatment of the sick	Health care	Massachusetts General Hospital; Kaiser health plan	Physicians and nurses on a cardiac ward; technicians in a medical lab
Collection and organization of empirical knowledge	Science	National Science Foundation; Salk Institute	Astronomers in an observatory studying the universe; a group of biologists experimenting with virus strains

ests and the institutional norms—come together in groups. These groups are of various sizes and complexities: from large, formal organizations to small, personal primary groups.

Let's follow an example across Table 7.1. People in most societies are concerned with life after death and seek an explanation for what is unknown or uncertain. Great power or faith may be invested in an idol, a god, a person, or an idea. At first experimentally and spontaneously, then more consistently, and finally in a formalized fashion, norms and roles emerge defining behaviors and beliefs felt to be appropriate in that society. This could be called the institutional area of religion. In America the institution of religion is organized around norms favoring Christianity. Large organizations such as the Catholic, Presbyterian, and Baptist churches and small groups such as the Wednesday evening Bible-study groups or Sunday-school groups make up the *locus* in which individual interests and institutional norms intersect. In these groups, institutional norms are played out. This same pattern is repeated in other institutional areas, as can be seen in the table.

In some institutional areas—particularly the family—smaller groups are all-important and assume the major, almost exclusive, role. In other institutional areas, especially economic and governmental ones, larger groups play a major role. The important fact, however, is that we can combine group and institution to understand social organization.

As you look at Table 7.1, it becomes apparent that institutional areas are not necessarily separate and isolated from each other. In fact, there is a good deal of overlap between them. Socialization of the young takes place in the family, and in educational and religious institutions as well. Groups from many institutional areas (science, health care, religion, education, leisure, government) fill economic functions: Scientists, physicians, clergy, teachers, professional basketball players, and lawyers all earn a living. So these institutional areas are overlapping circles, each intersecting with others in sometimes complicated ways.

Because the institutional areas are interrelated, we should expect that change in one institutional area will be related to change in another. Child-labor laws, reform movements, and the organization of labor have led to shorter work days and work weeks and therefore much more free time, which in turn has given us the institutional area of leisure. Increased employment of women (economic institutional area) has affected the family. The increase of women's rights in general is having a tremendous impact as women are treated as equals in all institutional areas. An economic decline in the 1970s and early 1980s had consequences for higher education: College students became more concerned about getting jobs and moved to business administration and other so-called job-related majors and away from the liberal arts and social sciences. Politicians running for office must contend with the effect of economic factors on voting behavior. When the Soviet Union launched its first Sputnik in 1957, many people in the United States became concerned that we had "fallen behind." One response was a renewed emphasis on science in schools: change in one area (education) to bring change in another (science). So it is clear that although we have introduced the institutional areas as separate entities, they are closely related and intersect at many points. As we said earlier, the social organization of society is best seen as a social fabric—a network of norms, roles, and values through which people interact individually and in groups, and this point is evident again as we examine social institutions.

Although the same institutional areas are found in many societies, they may not *look* the same in every society. The mountain-dwelling Arapesh of New Guinea and the campus-dwelling college students of America both operate within the institutional area of the family, for example, which involves elaborate procedures for courtship, marriage, and the raising of children; and yet the specific practices of the two societies are vastly different. Not only will the content be different, but the same institutional area in different societies may, at least to an extent, focus on different cultural concerns and serve different functions.

Institutions change, as do all aspects of human behavior. Change in institutions probably occurs more slowly than does the change in behavior and values of smaller elements like groups and individuals. Institutional change also varies from one society to another. To make a comparison of extremes, modern industrialized countries, for the most part, change more rapidly than do primitive preliterate societies. Even within one society, one institutional area may change slowly, another rapidly. We will discuss this topic of social change in more detail in Chapter Twelve.

In summary, social organization can be examined at several levels. We can understand people's behavior by focusing on the dynamics and characteristics of group behavior and on the particular types of groups to which people belong. We can also understand behavior by focusing on the attitudes, values, norms, procedures, and symbols that are part of certain institutional areas. Finally, the analysis is improved further by combining concepts and observing that groups operate in the context of the norms, roles, and values of an institutional area. To get a better idea of how the concept of institution is applied to specific areas of human behavior, the remainder of this chapter and the next three chapters will examine several institutional areas in more detail.

FAMILY

Sociologists have always looked with great interest on the institution of the family, mainly because we tend to see the family as the most basic of all institutions, and perhaps also because family practices vary tremendously from one culture to another. These variations never cease to fascinate, and ethnocentrism plays a part in this fascination. Many people are convinced that their own way is the right and best way, and they are amazed to find other people behaving differently. This ethnocentrism is unfortunate. Perhaps these issues can be placed in better perspective if we examine cross-culturally some similarities and differences in family structure.

Structural Variabilities: How Do Family Structures Differ?

Anthropologists, who study primitive societies all over the world, find that the structure of the family is almost infinitely variable. Imagine almost anything, and a society somewhere probably practices it (of course, the people there would be aghast if they knew the strange things *we* were doing).

It usually begins with some form of mating ritual: A couple may go on a first date; or a woman may inherit a husband; or a man may buy a spouse or two, or fall in love, or capture a mate. On the east coast of Greenland, the man goes to the woman's tent, grabs her by the hair, and drags her off to his tent.

In some cultures, parents are responsible for arranging marriages for their children. Chinese imperial law from around 1700 stated that a family elder who failed to find a husband for any young woman belonging to his household was condemning her to an unfulfilled life and was liable to receive publicly 80 blows with the bamboo.[2] In India, if a young woman's parents had been unable to find a husband for her as late as three years after she had reached puberty, then she could arrange a marriage for herself, but in so doing she would bring great disgrace on her family. Normally the marriage partner was chosen by the parents together. In China, however, the final decision on a prospective husband was typically made by the young woman's father; the mother, on the other hand, would select her son's future wife. Correct order had to be observed in marrying off children: Eldest children must be found mates first. This led to problems if the eldest child was unattractive or if younger children were especially attractive and precocious. This in turn forced desperate parents to trickery, switching one daughter for another at the last minute. Recall from the Bible that Jacob thought he had married Rachel but found out the next morning that the bride whom he had spent the night with was her elder sister Leah.

Beyond these prescribed, ritualistic mate-selection practices, there is also "love" as the basis for choosing a partner. This works in various ways: **Exchange theory** proposes the idea of a marketplace in mate selection. Mate seekers want to maximize their chances for a happy marriage and look for a good deal, a bargain if possible, in which their own assets and liabilities are compared with those of the potential partner. They continue the relationship on the assumption that they will get more out of it than it will cost. Anyone different from and especially below their own level of exchange is likely to be out of luck. **Complementary needs theory** suggests that people select partners who make up for, balance, or supply characteristics that they themselves don't have. Thus the talker looks for a listener, the eater for a cook, or the Mercedes-Benz fancier for a millionaire.

Role and value theories of mate selection suggest that people who share common values and common definitions about roles are more likely to select each other. If a woman is a big-city, sports-minded Episcopalian and a man is a small-town, sedentary Baptist (different values), or if she wants to stay home and cook and he pushes her to study medicine (differing role expectations), they will probably look elsewhere. **Process theory** suggests that mate selection involves a number of social and psychological processes. The field of potential mates is progressively narrowed down by physical attraction, religious and racial differences, role and value similarities, psychological influences, and so on. Bernard Murstein, an exchange theorist, believes that most of us may settle for a partner rather than choose one; only people with many assets and few liabilities actually choose each other—the rest of us just settle.[3] How depressing . . .

At any rate, whomever you date, buy, inherit, are forced to accept, capture, or settle for is related, in part, to whether your society practices **endogamy** (marriage within a certain group) or **exogamy** (marriage outside that certain group). In the United States, as in most societies, both are practiced. We forbid marriage within the immediate family, and it is therefore necessary to marry an outsider: exogamy. At the same time interracial marriages are often discouraged, and some religious groups encourage their members to marry people of the same faith: endogamy. The number of spouses you may have at one time varies depending on the culture in which you live. **Monogamy** means having one spouse at a time. It is estimated that most people in the world practice monogamy. The United States is technically monogamous, but with our high rates of divorce and family instability, it might be more accurate to say that we practice a type of serial or musical-chairs monogamy.

Polygamy is marriage to more than one spouse and includes three types: polygyny, polyandry, and group marriage. **Polygyny** refers to the practice of one man having several wives at a time. G. P. Murdock studied 250 societies and found that polygyny was the preferred form of marriage in 77 percent of them, although, as mentioned above, most people in the world practice monogamy.[4] Even in polygynous societies, most people practice monogamy. Although both sexes favor polygyny in such societies, the wealthy are more able to afford it and are more likely to practice it. **Polyandry** refers to the very rare practice of one woman having several husbands and, when it occurs, is usually found in poor societies with a shortage of women. Occasionally such societies kill female babies to keep the population down. When the woman becomes pregnant, deciding which husband is the father is a problem. This is usually solved by some sort of ceremony, possibly by shooting arrows at a distant target and designating the owner of the closest arrow to be the father. **Group marriage** refers to the marriage of two or more men to two or more women at the same time. The successful, long-term practice of group marriage is extremely rare and, when it does occur, is usually limited to closed groups or communes.

Family practices vary in many other ways in societies across the world. Where do newlyweds live after marriage? They may be expected to live either with the husband's family or with the wife's family, or to alternate between the husband's and wife's families. In some societies (such as the United States) the couple is expected to live apart from both families. How is ancestry traced? In some societies family lineage is traced through males, in others through females, and in still others through both sides equally. As to who has the formal power and legal authority in the family and who makes the decisions, in **patriarchal societies** it is the male and in **matriarchal societies** the female. It is interesting to note that many of the various practices we have mentioned arise at least partially in reaction to the incest taboo. Almost all societies forbid or at least limit sexual contact between family members with restrictions that are called **incest taboos.** The particular lines of the taboo vary from one society to another. To help regulate the situation, various rules— residence, descent, and so on—were developed. It should be clear by now that

COURTSHIP AMONG THE DOBU

Marriage is set in motion by a hostile act of the mother-in-law. She blocks with her own person the door of her house within which the youth is sleeping with her daughter, and he is trapped for the public ceremony of betrothal. Before this, since the time of puberty, the boy has slept each night in the houses of unmarried girls. By custom his own house is closed to him. He avoids entanglements for several years by spreading his favours widely and leaving the house well before daylight. When he is trapped at last, it is usually because he has tired of his roaming and has settled upon a more constant companion. He ceases to be so careful about early rising. Nevertheless he is never thought of as being ready to undertake the indignities of marriage, and the event is forced upon him by the old witch in the doorway, his future mother-in-law. When the villagers, the maternal kin of the girl, see the old woman immobile in her doorway, they gather, and under the stare of the public the two descend and sit on a mat upon the ground. The villagers stare at them for half an hour and gradually disperse, nothing more; the couple are formally betrothed.

From this time forward the young man has to reckon with the village of his wife. Its first demand is upon his labour. Immediately his mother-in-law gives him a digging-stick with the command, "Now work." He must make a garden under the surveillance of his parents-in-law. When they cook and eat, he must continue work, since he cannot eat in their presence. He is bound to a double task, for when he has finished work on his father-in-law's yams, he has still to cultivate his own garden on his own family land. His father-in-law gets ample satisfaction of his will to power and hugely enjoys his power over his son-in-law. For a year or more the situation continues. The boy is not the only one who is caught in this affair, for his relatives also are loaded with obligations. So heavy are the burdens upon his brothers in providing the necessary garden stuff and the valuables for the marriage gift that nowadays young men at their brother's betrothal escape from the imposition by signing up with the white recruiter for indentured labour.

From Patterns of Culture *by Ruth Benedict, copyright renewed 1962 by Ruth Valentine, pp. 133–135. Reprinted by permission of the publisher, Houghton Mifflin Company.*

almost any imaginable practice exists in some society somewhere. And all of it seems to work.

A number of other terms have emerged that help describe aspects of family structure. The family one is born into is called the **family of orientation.** The family of which one is a parent is called the **family of procreation.** The **nuclear family** includes the married couple and their children. The **extended family** includes more than two generations in close association or under the same roof, and the **compound family** includes multiple spouses—several wives and/or husbands at the same time.

In summary, we could describe the peculiar American family as a nuclear, monogamous, patriarchal, exogamous structure practicing marriage by romance.

Functional Uniformities: How Are Families Alike?

The institution of the family exists in all societies in some form. Functional analysis explains it this way: A family must serve a purpose, fill basic needs, and perform essential tasks. Those who have studied family institutions have discovered a number of such functions, some obvious and important and some less obvious. An obvious function of the family is to provide for continuation of the species. Societies, to exist, need people; the institution of the family functions to control reproduction. A second function of the family is to control sexual expression. The family institution attempts to deal with powerful sexual needs and desires by defining the who, when, where, why, and how of sexual activity.

A third function of the family is to care for and socialize children. The baby cannot survive without the care that the family provides. Socialization of the young by the family transmits the culture and prepares the child for participation in the adult world. Much of the child's later life is patterned on family models. It is possible that personal pathology and individual unhappiness will follow from inadequate family models or from a bad family atmosphere. It may frequently be the case that the child is the defenseless target of the parents' own frustrations, anxieties, neuroses, and maladjustments.

The first research question at the beginning of this chapter was essentially a socialization question: Do children repeat their parents' marital experiences or do they take off in new directions? Probably some of both, but a recent paper discovered some striking similarities between mothers' and daughters' behaviors. The author used data drawn from interviews conducted with mothers and children in the Detroit area between 1962 and 1985. The researcher found that daughters of mothers who married young and who were pregnant at marriage entered into cohabitational and marital unions sooner and at a higher rate than other children. Daughters were also quite likely to repeat their mother's pattern regarding the aftermath of divorce. Daughters were far more likely to remarry if their mothers did than if their mothers did not. Even parents' marital experiences become part of the socialization process and are clearly important for their children.[5]

The family may also function to provide close affectional and emotional ties for the individual. In some societies, many groups and associations may perform this function, but in mass, industrialized secondary societies such as the United States, the family may be one of the few remaining institutions in which primary relationships are possible. The family also functions to provide placement or status ascription. The family we are born into provides us with a *place* in society. If a person is a Protestant, middle-class, midwestern American Democrat, it is more than likely to be because he or she inherited certain statuses, such as religion, social class, region, nationality, and political preference, from his or her family of orientation.

These are some of the functions of the family, and there are probably others. They are not necessarily completely separate from each other; for example, reproduction of the species and control of sexual expression seem to be related. And further, other institutions may serve some of the same functions the family institution serves. The institution of education certainly deals with socialization of the young and transmission of culture, as does the institution of religion.

THE FAMILY: CONTRAST AND CHANGE

We have made the point that even though the family exists everywhere, there are many variations in the form it takes. Also, the family continually changes, in some societies more rapidly than in others, and a variety of forces produce these changes. A brief look at another family system will help illustrate this point.

The Family in China

In his book Ross Eshleman describes several stages of change in the Chinese family.[6] China is a country with a population four or five times as large as the U.S. population and with a family system far different from ours. Until 1900 or so, the most prevalent type of family in China was the traditional family. The clan—the central characteristic of the traditional family—was made up of all people with a common surname who descended from a common ancestor. A clan could be as large as several thousand people and included gentry (intellectuals, landowners, government officials) and peasants who cultivated the land. The clan performed major functions such as lending money to members, establishing schools, settling disputes, collecting taxes, and maintaining ancestral burial grounds. Ancestors were worshipped (given offerings of goods and food) even after death.

Males were dominant in the Chinese family. A wife had to obey three people: her father before marriage, her husband after marriage, and her son after her husband died. Female infanticide was common, especially among the poor. Marriages were arranged through matchmakers; many women met their husbands for the first time on their wedding day. Women's duties were to have male children and aid in the work. Men could have additional wives, but women could have only one husband and were not allowed to remarry if widowed. Binding of women's feet was practiced; this reduced the foot in length by about three inches and resulted in permanent crippling. Footbinding was practiced more among the upper class than among the peasants because the latter needed freedom of movement to work in the fields. Suicide was seen as a socially acceptable way for women to deal with their problems.

Change began to modify the traditional Chinese family in the 1900s. These changes were influenced by events occurring in the rest of the world, especially in Russia. Karl Marx's collaborator Friedrich Engels felt that in a capitalist society the family becomes the chief source of female oppression. Private property leads to subjugation and slavery of women and elevation of the status of men. One of the consequences of the revolution in Russia was the passage there of new marriage and divorce laws, the granting of abortion on demand, and the granting of legal equality to women.

China did not follow the Russian pattern precisely but did work to reform what were seen as the evils of the traditional family. A Communist Party conference in 1922 called for equal voting rights for women and an end to the traditional maltreatment of women. The Marriage Law of 1950 represented a major attempt to restructure the traditional family. This law enacted a new democratic marriage system based on free choice of partners, monogamy, and equal rights for both sexes. It prohibited bigamy, child betrothal, and infanticide. It gave husband and wife equal status in the home, granted divorce if both wanted it, and allowed the divorced woman to keep property that belonged to her before her marriage.

So, what emerges as the Chinese family of today? As we might expect, characteristics of both the traditional family and the family changed by law blend together. Rural and urban differences are great; thus, the rural families keep more of the traditional practices than do the urban ones. Generally, although women have made progress toward equality, discrimination still exists in access to jobs and adequate pay. Most of the important jobs are held by men; and sons are valued over daughters. Parental arrangement of marriages still occurs, while premarital sex and illegitimate births are infrequent and divorce is not widespread.

In conclusion, changes in the political system in China produced pressures for change in other institutions. Communist ideology and the economic goals of the state pushed for an altered family in which women were more equal and able to participate in production, whereas family or clan-held property was confiscated and clans became communes. The contemporary Chinese family is consequently vastly different from the traditional one, but it has many remnants of the traditional family and is described as strong and stable.

The Family in America

If we look at the American family over the last 100 years or so we can see a number of changes. The family is smaller than it used to be; there are fewer children and fewer adults. Birth rates have been dropping in the United States over a long period of time, and therefore the 1970s had the lowest rates on record. The large-family model of the 19th century has given way to the smaller family of two or fewer children. Some of the changing feelings about appropriate family size have happened fairly recently. Figure 7.1 reflects attitudes of married women regarding the number of expected births, and we can see substantial change between 1967 and 1988. In 1967, few women wanted no children or only one—two, three, or four or more were preferred. In 1988, more women chose none or one and most wanted two; the idea of four or more was distinctly unpopular. The American family is also small in that it is a *nuclear* rather than an *extended* family. The extended family of a century ago had aunts, uncles, parents, grandparents, and children all living under the same roof. Compare that with the nuclear family of today and think of the differences in roles, division of labor, diffusion of responsibilities, child care, expression of affection and emotion, and individual freedom. The shift from the extended to the nuclear family brought tremendous changes in the roles of individual family members. The family of today is segregated: Each seg-

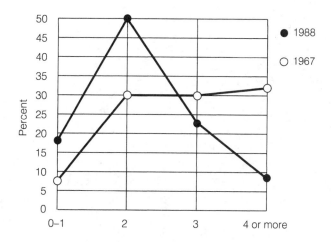

FIGURE 7.1 *Lifetime births expected by married women.*
United States Bureau of the Census, *Statistical Abstracts of the United States, 1990* (110th ed.), Washington, D.C.

ment of that large extended family of 100 years ago now lives by itself as a separate family unit.

Today's family is probably more egalitarian and less patriarchal: People share authority more than they used to, and women are making many of the important decisions. The woman has developed equality in other ways relating to family structure. She is more likely to be working outside the home while married. The possibility of a career has given her alternatives to early marriage or even to marriage at all. If she does marry, she has a greater say in which mate to choose and in the type of role she will play as wife. Likewise, equality has brought major changes in the roles of males: Words like *house-husband* and *paternity leave* are creeping into the vocabulary.

The American family has moved from farm to city. Last century's farm family now lives in or near an urban area. And the family is much more geographically mobile than it used to be. Currently, 17 percent of Americans move from one house to another in a year, and 3 percent move from one state to another.

The family used to be the all-powerful institution, but gradually other institutions—religious, educational, governmental—have become involved in some of the functions that the family formerly handled exclusively. So far this has resulted not in the disappearance of the family, but in increased emphasis on and jealous guarding of the remaining functions that the family fills: "Sex education in the schools?? Certainly not! Keep it in the home where it belongs!" As mentioned before, perhaps one of the major functions of the family in mass society is to provide for the affectional and emotional needs of people. The family is nearly the last refuge of the true primary type of relationship, and if it is accurate to call this a basic social need, then this function of the family becomes exceedingly important as other functions decrease in importance.

The second research question at the beginning of this chapter concerned the relationship between children and marital stability. It is often suggested that the presence of children is one factor that might hold marriages together. Is this a "commonsense" suggestion, and is it really true? A recent study examined this question by looking at interviews with about 5,000 couples who were first married between 1968 and 1985. The researchers found some interesting patterns. The first birth has a major effect in *increasing* the stability of a marriage. (Exception: A child born before the parents were married decreases marital stability.) The presence of additional younger children (less than six years old) also seems conducive to helping a marriage stay together. However, the presence of children between 6 and 12 years of age appears to be a neutral factor—they neither increase nor decrease the chances that their parents' marriage will break up. Finally, older children (teenagers) are a problem for marital stability—their presence significantly *increases* the chances that parents will dissolve their marriage! The researchers suggest that the presence of young children may hold a marriage together because couples who plan to stay together are more likely to plan having children; their first child is a "signal" of their stability. Instability in families with teenagers perhaps reflects marriages in which the parents were unhappy for some time but waited to divorce until the children were older, or perhaps it reflects the "outside-world" problems that teenagers have and the resulting strains these problems place on their parents' marriages. Does having children increase marital stability? As the authors of the study say, "clearly, yes and no."[7]

Several interesting and important changes are taking place in the American family. The contrast between the family of the 1960s and the family of today is sharp.[8] Note the following:

Later Marriage The median age at first marriage (see Figure 7.2), which didn't change much between 1950 and 1970, climbed steadily between 1970 and the present. In 1992, the median age of first marriage for men was 26.5 years, higher than any previous figure going back 100 years. For women it was 24.4, also significantly higher than any previous figure.

Staying Single A larger number of late marriages means that the proportion of adults staying single is increasing. This trend, which started in the 1960s and continued through the 1970s, is especially notable in age groups in which most people have traditionally married, as is shown in Table 7.2. The propor-

FIGURE 7.2 *Median age at first marriage.*

United States Bureau of the Census, "Marital Status and Living Arrangements: March 1992," *Current Population Reports,* Series P-20, No. 468.

Table 7.2 Family Change, 1970 to 1992

	1970	1992
Households made up of married couples with children	40%	26%
Persons per household	3.1	2.6
5 or more persons in the household	21%	10%
Single parent families	13%	30%
Never married women, 20–24	36%	66%
Never married men, 20–24	55%	80%
Never married women, 25–29	11%	33%
Never married men, 25–29	19%	49%
Children under 18 living with two parents	85%	71%
Median age, first marriage, men	23.3	26.5
Median age, first marriage, women	20.8	24.4

United States Bureau of the Census, "Marital Status and Living Arrangements: March 1992," *Current Population Reports,* Series P-20, no. 468.

tion of women 30 to 34 years old who had not yet married tripled between 1970 (6 percent) and 1992 (19 percent). One result of this is an increased number of young adults living with their parents. One third of unmarried people 25 to 29 years old lived at home with their parents in 1992.

Living Alone Twenty-seven million people lived alone in 1992 (as compared with 11 million in 1970), about one quarter of all households. Most of the growth in this category came from divorced or never-married people. More women (60 percent) than men lived alone.

Unmarried Couples The number of households occupied by two unrelated adults of the opposite sex doubled between 1980 (1.6 million) and 1993 (3.5 million). The number of unmarried-couple households is relatively small when compared with the number of married couples; however, the ratio of unmarrieds to marrieds is growing.

Divorce Marriage and divorce rates in the United States are shown in Figure 7.3. Marriage rates have fluctuated over the years with a downward trend since 1980. Divorce rates went up sharply, starting in the middle 1960s with a high in 1980, and have declined slightly since then. One of the many consequences of this higher divorce rate is that in 1992, 71 percent of children under 18 lived with two parents, compared with 85 percent in 1970. The list below gives some other facts about marriage and divorce in the 1990s.

Census Bureau Facts About Marriage and Divorce in the 1990s

- Women in their thirties and early forties have the highest divorce rate.

- First marriages that end in divorce typically last about six years.

- Women who divorce typically wait 2.5 years before remarrying.

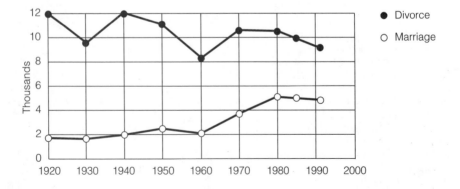

FIGURE 7.3 *Marriage and divorce rates (per 1,000 people) in the United States.*

United States Bureau of the Census, "Marital Status and Living Arrangements: March 1992," *Current Population Reports*, Series P-20, no. 468.

- Three out of ten divorced women who remarry divorce again.

- Four out of ten first marriages end in divorce.

- The proportion of women who never marry went from five to ten percent.

- Only 75 percent of black women will ever marry compared with 91 percent of white women.

The American family of today, then, is many things. The "traditional family," in which the husband works and the wife stays home with the children, represents one quarter to one third of all families. Young people seem to be postponing marriage, perhaps because of unstable economic conditions, perhaps because a new variety in living arrangements enables people to enjoy primary group benefits without the necessity of a traditional marriage. People who do marry do so later. Some population analysts take this as a hopeful sign. Maybe Americans are being more careful about selecting mates. Taking more time and marrying at a later age may lead to more stable marriages in the long run, and perhaps today's high divorce rates will begin to decline. At any rate, if the marriage is not happy, it is likely to be dissolved quickly. There are now more single-parent families due to both divorce and to single people being allowed to adopt children. The nuclear family (two parents plus children) is a major part of American family life, but it continues to shrink in size as the birth rate drops. The retired couple in their sixties and seventies represents another facet of the American family, one whose numbers are growing substantially as our population grows older and as the birth rate decreases.

Perhaps, then, we see two contrasting threads in the American family. One is an experimental, postponing-or-never-marrying, alternative-lifestyle thread; the other is the traditional nuclear family, progressing from young to middle-aged parents with their children to elderly retired couples. Time has changed the nuclear family: It is a smaller unit with more equal relationships between partners. Although the two threads are contrasting, they also intersect; many people may find themselves first being experimental and later following a more traditional path as their own needs and definitions undergo change.

FAMILY PROBLEMS—THE CONFLICT APPROACH

Several times in this book we have examined the contrasting viewpoints provided by functional theorists and conflict theorists. When the family is studied, one generally starts (as we did earlier in this chapter) with the functions that the family provides. The family in all societies does have numerous and important functions, but the conflict-theory approach deserves examination as well. Just as the family fulfills certain functions, it provides a setting for conflict. As we noted earlier, Engels felt that the capitalist family was the chief source of oppression of females and that the "battle between the sexes" is probably waged more often in the family arena than anywhere else. Conflict is a sign that all is not well, and in this section we will examine the evidence of problems and instability in the American family.

Divorce is a good place to start in looking at family problems. There were 1.2 million divorces in 1991. The divorce rate climbed sharply from the middle 1960s to 1980 and is at about that same level currently. Divorce is rising in many countries, but the United States continues to have one of the highest rates in the world. Why? Among the many explanations offered, the following seem most convincing: Divorce is becoming more and more accepted. It is not the stigma it once was, and people are much more inclined to go through it today. Laws have changed; many states have enacted no-fault divorce laws, making divorces much easier to obtain. Another explanation has to do with the emphasis on individuality and happiness. Parties to a marriage that isn't working are less likely to want to stick with it today than they used to be. The view is that if you're not happy, then get out and try again, or try something else. Both parties now know that they have a good chance to make meaningful lives for themselves in some way other than by staying in a bad marriage. And with fewer children, perhaps fewer marriages are kept together "for the sake of the children." Unmarried people living together also separate. Courts have been faced with difficult cases that were essentially requests to have the legal benefits of divorce, such as community property and alimony payments, extended to people who had never married.

A recent study of husbands and wives attempted to find out what factors in a marriage were most likely to cause either party to think about divorce. Some interesting relationships were discovered. The wife's thoughts of divorce were *increased* if she had work experience, if she had a youngest child between the ages of 6 and 11, and if she felt that housework should be divided equally. Her thoughts of divorce were *decreased* according to her age at marriage (the older she was when she got married, the less likely she was to think of divorce), the length of the marriage, and her husband's increasing contribution to the housework. The husband's thoughts of divorce were *increased* if his wife worked outside the home, if there were religious differences, and if his wife believed that housework should be divided equally. His thoughts of divorce were *decreased* by the presence of a child under six, by the length of the marriage, and by being older than his wife. These findings are explained in part by the fact that women's economic roles are changing faster than are men's. This has increased her freedom and independence and has modified some of her expectations of marriage. Some husbands are having difficulty adjusting to these changes.[9]

An even more obvious example of conflict within the family is family violence. Nearly one fifth of all murders in the United States involve a person killing a member of his or her own family. Spousal abuse and child abuse are much more widespread than was earlier thought, and they may be the most underreported of all crimes. Recent research reveals that in 16 percent of the families studied, an act of violence was committed by one partner against the other within the past year.[10] If this is projected over the life of the marriage, violent acts occurred in 28 percent of the marriages. The researchers believe that this estimate is low and that the actual rate of family violence is closer to 50–60 percent of all couples. Generalizing from this study, it is estimated

The Soviet newspaper *Moskovskaya Pravda* reported in 1986 that only 37 percent of the marriages in the Soviet Union survive three years or more. Why? Drunkenness used to be the primary reason for divorce, but that seems to have changed. Housing facilities are so scarce that only one third of young married couples have an apartment of their own.

Most have to live with parents or relatives and consequently have little privacy. Vacations together are difficult to arrange because of conflicting work schedules. Also, divorces in childless marriages are easy to obtain. How to slow down the divorce rate? Some say tighten divorce laws, whereas others suggest that solving the housing shortage is the answer.

---

that 1.8 million wives are beaten by their husbands every year, and surprisingly, over 2 million husbands are beaten by their wives.

Child abuse, both physical and sexual, became an increasingly visible issue in the mid-1980s. The study by Straus and coworkers, cited in the previous paragraph, found violence toward children to be widespread in American society. Based on their interviews with 2,100 families, the researchers found that within the past year over half of the parents had slapped or spanked their children, 13 percent had hit their children with an object, and 3 percent had kicked, bit, or punched their children. The researchers estimated that some 46,000 children in the United States had a knife or gun used against them during the last year! They also found that mothers are more severe, more abusive, and more often violent with their children than are fathers. Recent headlines have reported sexual abuse of children in families, in foster homes, and in day-care centers. Children have been required to testify in court against adults, sometimes against their own parents, which has led to current controversy on two fronts: On the one hand, the child experiences the double trauma of being a victim and of then having to testify about it; on the other hand, questions have arisen about the accuracy of some of the children's testimonies and about the competence of children in general as witnesses.

Other conditions of concern in American society have roots that may be traced to the family. Eighteen percent of the arrests for violent crime and 33 percent of the arrests for property crime in 1992 were of people under the age of 18. Suicide rates among teenagers are increasing. Drug abuse and alcohol abuse by both young people and adults are serious problems. Illegitimate births increased by 50 percent during the 1970s, and by 1990, approximately 28 percent of American babies were born to unmarried mothers.

Explanations for at least some of what is happening to the family are not difficult to find. Society changes, and its institutions change as well. Fewer bonds tie family members together today, and the consequence is seen in divorce and in young people leaving home earlier. Both parents are more likely to be working, and so children may be supervised less closely than they used to be. Also, outside agencies such as day-care centers play an increasingly important role in "family" life. Less desire for having children, along with increased freedom and independence for women, has led to increasingly diverse arrangements for satisfying affectional and emotional needs. We will likely see more changes and redefinitions in the family as other aspects of society continue to change.

THE FAMILY OF THE FUTURE

And what of the family of the future? Jean Lipman-Blumen painted the following picture of the American family of the next two decades.[11] There will be fewer of the typical nuclear families that consist of two parents and one or more children. Instead, there will be more single-parent families. This will occur, she feels, because marriage and remarriage rates will continue to decline while divorce rates will continue to climb. More single-parent families will combine with others in the future for shelter, food, child care, and companionship. Lipman-Blumen feels that the smaller families of the future will be less segregated than those of today; they will get together for mutual support and cooperation. Shortened work weeks will have men at home more, which will mean that they will be playing a more real, and less ceremonial, role in the family. Women will be working more, meaning that household tasks will be shared more often in the family of the future. A growing list of companies has established paternity-leave programs, supporting the idea of increased male involvement in child rearing and increased female involvement in careers outside the family. One such program at New Jersey Bell allowed a parent of either gender to take up to six months "newborn-child leave" without pay, guaranteeing a job of the same status and salary upon return.

The institution of the family changed slowly for 200 years and now appears to be changing so rapidly that accurate predictions of the future are impossible. What are the future consequences of current changes and innovations? During 1978, the world witnessed the first human embryo transplant, in December 1981 a test-tube baby was born in the United States, and there continues to be much debate over the wisdom of manipulating the genetic code. In his book *Future Shock* (1970), Alvin Toffler warned us that more innovations and changes are taking place now than at any time in human history. What will be the effect of these changes on the family of the future? It will be interesting to see.

THEORY AND RESEARCH: A REVIEW

The study of social institutions has traditionally started from the viewpoint of *functional analysis*. If you look back at the discussion of social institutions at the beginning of this chapter, you will find institutions to be an "organized system of social relationships" that "emerges to deal with certain basic problems." Societies need to have specific basic functions performed. An institutional structure arises around attempts to supply these functions. Change is slow and the feeling of an organized and integrated system develops. Social institutions—family, organized religion, education, economics, government, and so on—show functional analysis at its clearest. This chapter focuses on structures developed and functions provided by the institution of the family.

Recently in the study of social institutions, *conflict theory* has become more prominent. In part, this is a reflection of obvious conflict—high divorce rates, crime, and violence between family members—existing in or emerging from the family. The emphasis on conflict theory is also a reflection of the increasingly popular view that scarcity, competition, exploitation, and conflict exist in all social situations. Notice the various aspects of the conflict perspective in dating, the Chinese family, and family problems discussed earlier in the chapter. The family is perhaps best seen as a small group held together by affectional ties and pulled apart by the inevitable conflicts that exist in any group of people. *Symbolic interaction* would focus on the patterns of interaction among members—how people interpret, define, and react to the behavior (symbols, gestures) of those around them. Proponents would be interested in such questions as: How is discipline handled? How do children manipulate parents and vice versa? Who exercises power and in what ways? What effect does the placement of the child in the family (firstborn, middle child, and so on) have, and why does this happen? How do aging, changing roles, social class, and other life situations affect patterns of interaction?

A number of ethnographies are referred to in this chapter in discussing family practices in other societies; most of these have been done by anthropologists, and some involve *participant observation*. The census data describing the family in the United States today were obtained through *questionnaire or interview surveys*. The study examining the relationship between presence of children and marital stability, and the study of children repeating their parents' marital experiences, were *reactive surveys*, as was Straus's study of family violence. One of the major concerns in the family-violence study was the *sample*. The sociologists selected a national probability sample of 2,100 families so that they could generalize their findings to families throughout the United States. As always happens, there were several drawbacks or problems. They chose to look at intact families, so single-parent families are not represented in the sample. Parent-child relations with children under three years of age were not included, even though very young children present an especially stressful situation. And the researchers, even after up to four visits to the family's house and letters and financial incentives, were able to reach just 65 percent of the sample. Were the families they couldn't contact different from most others in patterns of violence? The researchers hope not, but it's impossible to know.

SUMMARY

The major topics of the previous chapters in this section on social organization were *groups, categories,* and *social differentiation*. These concepts permit the study of a number of different types of collectivities of people. In this chapter we moved to another, somewhat more abstract level and focused on social institutions. Sociologists define an institution as a system of social relationships—an organized way of doing things that meets certain basic needs of society. All societies have basic needs, such as reproduction of the species and socialization of the young, for example. Consequently, many of the same institutions—family, religion, education, government, economics—appear in all societies, although the content of a given institution may vary from one society to another. As societies develop and change, other institutions, such as science and leisure, may emerge.

Sociological research and study are more apt to be concerned with specific institutions than with the concept itself. The major part of this chapter has presented an analysis of the institution of the family. We examined cross-cultural variations in courtship and the family to make the following point: Although the institution of the family exists everywhere, it is not the *same* everywhere. Constants or uniformities were discussed, and the functions that the family seems to fulfill were outlined. We took a brief look at the family in China to illustrate not only the contrast between that family system and ours, but also the idea that *change* is a constant in any institution. Finally, we examined the family in America from several perspectives: current trends and changes, problems, and predictions about the family of the future.

A major point of this chapter is that although a social institution (in this case, the family) exists in all societies, it varies tremendously from one society to the next. The group of three readings that follows describes some unique characteristics of the contemporary family in China, in Japan, and in the inner-city black community in the United States.

TERMS FOR STUDY

complementary needs theory (225)	matriarchal society (226)
compound family (227)	monogamy (226)
endogamy (226)	nuclear family (227)
exchange theory (225)	patriarchal society (226)
exogamy (226)	polyandry (226)
extended family (227)	polygamy (226)
family of orientation (227)	polygyny (226)
family of procreation (227)	process theory (225)
group marriage (226)	role and value theories (225)
incest taboo (226)	social institutions (221)
institutionalization (221)	

Reading 7.1

THEY GET BY WITH A LOT OF HELP FROM THEIR *KYOIKU MAMAS*

Carol Simons

The institution of the family fulfills many functions, but perhaps its major role is to care for and socialize children. Socialization of the young by the family trans- mits the culture and prepares the child for future participation in the adult world. Much of the child's later life is patterned on family models. This reading and the two following examine particular aspects of three contrasting family systems— the family in Japan, in China, and in inner-city America.

Carol Simons is a former associate editor for The Smithsonian *and lives in Tokyo. This reading is an excerpt from an article that appeared in* The Smith- sonian.

Two-year-old Hiromasa Itoh doesn't know it yet, but he's preparing for one of the most important milestones of his life, the examination for entry into first grade. Already he has learned to march correctly around the classroom in time with the piano and follow the green tape stuck to the floor—ignoring the red, blue and yellow tapes that lead in different directions. With the other 14 children in his class at a central Tokyo nursery school, he obeys the "cleaning-up music" and sings the good-bye song. His mother, observing through a one-way glass window, says that it's all in preparation for an entrance examination in two or three years, when Hiromasa will try for admission to one of Tokyo's prestigious private schools.

Forty-five minutes south of the capital city by train, in the small suburb of Myorenji, near Yokohama, 13-year-old Naoko Masuo returns from school, slips quietly into her family's two-story house and settles into her homework. She is wearing a plaid skirt and blue blazer, the uniform of the Sho-ei Girls School, where she is a seventh-grader. "I made it," her smile seems to say. For three years, when she was in fourth through sixth grades in public school, Naoko's schedule was high-pressure: she would rush home from school, study for a short time and then leave again to attend *juku,* or cram school, three hours a day three times a week. Her goal was to enter a good private school, and the exam would be tough.

Her brother, Toshihiro, passed a similar exam with flying colors several years ago and entered one of the elite national schools in Tokyo. The summer before the exam, he went to *juku* eight hours a day. Now, as a high school graduate, he is attending prep school—preparing for university entrance exams that he will take in March.

Little Hiromasa, Naoko and Toshihiro are all on the Japanese road to success. And alongside them, in what must surely be one of the world's greatest traffic jams, are thousands of the nation's children, each one trying to pass exams, enter good schools and attain the good jobs that mark the end of a race well run.

From *Smithsonian*, March, 1987. Reprinted with permission.

But such children are by no means running as independents. They are guided and coached, trained and fed every step of the way by their mothers, who have had sharp eyes on the finish line right from the start.

No one doubts that behind every high-scoring Japanese student—and they are among the highest scoring in the world—there stands a mother, supportive, aggressive and completely involved in her child's education. She studies, she packs lunches, she waits for hours in lines to register her child for exams and waits again in the hallways for hours while he takes them. She denies herself TV so her child can study in quiet and she stirs noodles at 11 P.M. for the scholar's snack. She shuttles youngsters from exercise class to rhythm class to calligraphy and piano, to swimming and martial arts. She helps every day with homework, hires tutors and works part-time to pay for *juku*. Sometimes she enrolls in "mother's class" so she can help with the drills at home.

So accepted is this role that it has spawned its own label, *kyoiku mama* (education mother). This title is not worn openly. Many Japanese mothers are embarrassed, or modest, and simply say, "I do my best." But that best is a lot, because to Japanese women, motherhood is a profession, demanding and prestigious, with education of the child the number-one responsibility. Cutthroat competition in postwar Japan has made her job harder than ever. And while many critics tend to play down the idea of the perpetually pushy mother, there are those who say that a good proportion of the credit for Japan's economic miracle can be laid at her feet.

"Much of a mother's sense of personal accomplishment is tied to the educational achievements of her children, and she expends great effort helping them," states *Japanese Education Today,* a major report issued in January by the U.S. Department of Education. "In addition, there is considerable peer pressure on the mother. The community's perception of a woman's success as a mother depends in large part on how well her children do in school."

Naoko's and Toshihiro's mother, Mieko Masuo, fully fills the role of the education mother, although she'd be the last to take credit for her children's accomplishments. This 46-year-old homemaker with a B.A. in psychology is a whiz at making her family tick. She's the last one to go to bed at night ("I wait until my son has finished his homework. Then I check the gas and also for fire. My mother stayed up and my husband's mother, and it's the custom for me, too") and the first one up in the morning, at 6. She prepares a traditional breakfast for the family, including *miso* soup, rice, egg, vegetable and fish. At the same time, she cooks lunch for her husband and Naoko, which she packs up in a lunch box, or *o-bento*. She displays the *o-bento* that Naoko will carry to school. In the pink plastic box, looking like a culinary jigsaw puzzle, are fried chicken, boiled eggs, rice, lotus roots, mint leaves, tomatoes, carrots, fruit salad and chopsticks. No pb&j sandwiches in brown bags for this family.

"Every morning, every week, every year, I cook rice and make *o-bento,*" Mrs. Masuo says with a laugh, winking at Naoko. "I wouldn't want to give her a *tenuki o-bento.*" Naturally. Everyone knows that a "sloppy lunch box" indicates an uncaring mother.

But Mrs. Masuo doesn't live in the kitchen. She never misses a school mother's meeting. She knows all the teachers well, has researched their backgrounds and how successful their previous students have been in passing exams. She carefully chose her children's schools and *juku,* and has spent hours accompanying them to classes. "It's a pity our children have to study so much," she apologizes. "But it's necessary." She says that someday she'd like to get a part-time job—perhaps when exams are over.

"But at the moment I must help my children. So I provide psychological help and *o-bento* help." Then she laughs.

Like most Japanese mothers, Mrs. Itoh spends most of her time with her son and her six-month-old daughter, Emi. Baby-sitters and play groups are not part of her life. She has dinner with the children well before her husband comes home from work. She takes them to the park, to swimming lessons and music, much of the time carrying her baby in a pack on her back. Indeed, a young mother with an infant in a sling and a toddler by the hand walking along a subway platform or a city street is a sight that evokes the very essence of motherhood to most Japanese.

This physical tie between mother and child is only a small part of the strong social relationship that binds members of the family together in mutual dependency and obligation. It's the mother's job to foster this relationship. From the beginning, the child is rarely left alone, sleeps with the parents, is governed with affectionate permissiveness and learns through low-key signals what is expected and what to expect in return.

Many American children are also raised with affection and physical contact, but the idea is to create independent youngsters. Discipline begins early. Children have bedrooms separate from their parents. They spend time playing alone or staying with strangers and learn early that the individual is responsible for his own actions. An American mother, in disciplining, is more likely to scold or demand; a Japanese mother is apt to show displeasure with a mild rebuke, an approach that prompted one American six-year-old to tell his own mother: "If I had to be gotten mad at by someone, I wish it would be by a Japanese."

Even a casual observer is struck by the strong yet tender mother-child connection. A Japanese senior high school teacher said that many wives, including his own, sleep in a room with their children and not their husbands. "Is it the same in America?" he asked. At a dinner party, a businessman made his wife's excuses: "I'm sorry she couldn't come tonight. My son has an exam tomorrow." Even if the excuse was not true, the use of it says a lot.

The relationship of dependency and obligation fostered in the child by the mother extends to family, school, company and country, and is the essence of Japanese society. The child is taught early that he must do well or people will laugh at him—and laugh at his mother as well. "Most Japanese mothers feel ashamed if their children do not do well at school," said one mother. "It is our responsibility to see that the child fulfills his responsibility." Bad behavior may bring shame, but good behavior has its own rewards. One woman described a friend by saying: "Her son studied very hard in order to get into a good high school and he got in. She is very clever."

This attitude is precisely what gives education mothers such as Mrs. Masuo and her *o-bento* philosophy such esteem and why they take such pride in their role, even if they don't admit it. Their goal is clear: success in entrance exams, good school, a good college and a good job. (For daughters the goal has a twist: good schools lead to good husbands.)

Today, more and more educators and parents are questioning the high-pressure system that gives rise to such popular sayings as "Sleep four hours, pass; sleep five hours, fail." Educationists speak of lost childhoods, kids never getting a chance to play, "eating facts" to pass exams, and the production of students who memorize answers but can't create ideas. They cite the cruelty of students who take pleasure when their classmates fail, increasing delinquency, and high incidence of bullying in the schools.

Not surprisingly, Japanese mothers have been among the major critics, perhaps because they bear much of the brunt and witness the effects of the pressure on their children. "My son kept getting headaches and then he didn't want to go to school," said one mother. "So I stopped the *juku*." Recently, such mothers have gained an ally in Prime Minister Yasuhiro Nakasone, whose government has been seeking ways to depressurize the education system. Nevertheless, many doubt that his efforts will have any effect in a society dedicated to hard work and competition.

Reading 7.2

THE LITTLE EMPERORS, PART 1

Daniela Deane

In an attempt to control its exploding population, China passed a one child per family policy in 1979. How has this policy affected the Chinese family? Daniela Deane explored this question while living in Hong Kong in an article she wrote for the Los Angeles Times Magazine.

Xu Ming sits on the worn sofa with his short, chubby arms and legs splayed, forced open by fat and the layers of padded clothing worn in northern China to ward off the relentless chill. To reach the floor, the tubby 8-year-old rocks back and forth on his big bottom, inching forward slowly, eventually ending upright. Xu Ming finds it hard to move.

"He got fat when he was about 3," says his father, Xu Jianguo, holding the boy's bloated, dimpled hand. "We were living with my parents and they were very good to him. He's the only grandson. It's a tradition in China that boys are very loved. They love him very much, and so they feed him a lot. They give him everything he wants."

Xu Ming weighs 135 pounds, about twice what he should at his age. He's one of hundreds of children who have sought help in the past few years at the Beijing Children's Hospital, which recently began the first American-style fat farm for obese children in what was once the land of skin and bones.

"We used to get a lot of cases of malnutrition," says Dr. Ni Guichen, director of endocrinology at the hospital and founder of the weight-reduction classes. "But in the last 10 years, the problem has become obese children. The number of fat children in China is growing very fast. The main reason is the one-child policy," she says, speaking in a drab waiting room. "Because parents can only have one child, the families take extra good care of that one child, which means feeding him too much."

Bulging waistlines are one result of China's tough campaign to curb its population.

The one-child campaign, a strict national directive that seeks to limit each Chinese couple to a single son or daughter, has other dramatic consequences: millions of abortions, fewer girls and a generation of spoiled children.

The 10-day weight-reduction sessions—a combination of exercise, nutritional guidance and psychological counseling—are very popular. Hundreds of children—some so fat they can hardly walk—are turned away for each class.

According to Ni, about 5 percent of children in China's cities are obese, with two obese boys for every overweight girl, the traditional preference toward boys being reflected in the amount of attention lavished on the child. "Part of the course is also centered on the parents. We try to teach them how to bring their children up properly, not just by spoiling them," Ni says.

Ming's father is proud that his son, after two sessions at the fat farm, has managed to halve his intake of jiaozi, the stodgy meat-filled dumplings that are Ming's particular weakness, from 30 to 15 at a sitting. "Even if he's not full, that's all he gets," he says. "In the beginning, it was very difficult. He would put his arms around our necks and beg us for more food. We couldn't bear it, so we'd give him a little more."

Ming lost a few pounds but hasn't been able to keep the weight off. He's a bit slimmer now, but only because he's taller. "I want to lose weight," says Ming, who spends his afternoons snacking at his grandparents' house and his evenings plopped in front of the television set at home. "The kids make fun of me, they call me a fat pig. I hate the nicknames. In sports class, I can't do what the teacher says. I can run a little bit, but after a while I have to sit down. The teacher puts me at the front of the class where all the other kids can see me. They all laugh and make fun of me."

The many fat children visible on China's city streets are just the most obvious example of 13 years of the country's one-child policy. In the vast countryside, the policy has meant shadowy lives as second-class citizens for thousands of girls or, worse, death. It has made abortion a way of life and a couple's sexual intimacy the government's concern. Even women's menstrual cycles are monitored. Under the directive, couples literally have to line up for permission to procreate. Second children are sometimes possible, but only on payment of a heavy fine.

The policy is an unparalleled intrusion into the private lives of a nation's citizens, an experiment on a scale never attempted elsewhere in the world. But no expert will argue that China—by far the world's most populous country with 1.16 billion people—could continue without strict curbs on its population.

Dinner time for one 5-year-old girl consists of granddad chasing her through the house, bowl and spoon in hand, barking like a dog or mewing like a cat. If he performs authentically enough, she rewards him by accepting a mouthful of food. No problem, insists granddad, "it's good exercise for her."

An 11-year-old boy never gets up to go to the toilet during the night. That's because his mother, summoned by a shout, gets up instead and positions a bottle under the covers for him. "We wouldn't want him to have to get up in the night," his mother says.

Another mother wanted her 16-year-old to eat some fruit, but the teenager was engrossed in a video game. Not wanting him to get his fingers sticky or daring to interrupt, she peeled several grapes and popped one after another into his mouth. "Not so fast," he snapped. "Can't you see I have to spit out the seeds?"

Stories like these are routinely published in China's newspapers, evidence that the government-imposed birth-control policy has produced an emerging generation of spoiled, lazy, selfish, self-centered and overweight children. There are about 40 million only children in China. Dubbed the country's "Little Emperors," their behavior toward their elders is likened to that of the young emperor Pu Yi, who heaped indignities on his eunuch servants while making them cater to his whims, as chronicled in Bernardo Bertolucci's film "The Last Emperor."

Many studies on China's only children have been done. One such study confirmed that only children generally are not well liked. The study, conducted by a team of Chinese psychologists, asked a group of 360 Chinese children, half who have siblings and half who don't, to rate each other's behavior. The only children were, without fail, the least popular, regardless of age or social background. Peers rated them more uncooperative and selfish than children with brothers and sisters. They bragged more, were less helpful in group activities and more apt to follow their own selfish interests. And they wouldn't share their toys.

The Chinese lay a lot of blame on what they call the "4-2-1" syndrome—four doting grandparents, two overindulgent parents, all pinning their hopes and ambitions on one child.

Besides stuffing them with food, Chinese parents have very high expectations of their one *bao bei,* or treasured object. Some have their still-in-strollers babies tested for IQ levels. Others try to teach toddlers Tang Dynasty poetry. Many shell out months of their hard-earned salaries for music lessons and instruments for children who have no talent or interest in playing. They fill their kids' lives with lessons in piano, English, gymnastics and typing.

The one-child parents, most of them from traditionally large Chinese families, grew up during the chaotic, 10-year Cultural Revolution, when many of the country's cultural treasures were destroyed and schools were closed for long periods of time. Because many of that generation spent years toiling in the fields rather than studying, they demand—and put all their hopes into—academic achievement for their children.

"We've already invested a lot of money in his intellectual development," Wang Zhouzhi told me in her Spartan home in a tiny village of Changping county outside Beijing, discussing her son, Chenqian, an only child. "I don't care how much money we spend on him. We've bought him an organ and we push him hard. Unfortunately, he's only a mediocre student," she says, looking toward the 10-year-old boy. Chenqian, dressed in a child-sized Chinese army uniform, ate 10 pieces of candy during the half-hour interview and repeatedly fired off his toy pistol, all without a word of reproach from his mother.

Would Chenqian have liked a sibling to play with? "No," he answers loudly, firing a rapid, jarring succession of shots. His mother breaks in: "If he had a little brother or sister, he wouldn't get everything he wants. Of course he doesn't want one. With only one child, I give my full care and concern to him."

But how will these children, now entering their teenage years and moving quickly toward adulthood, become the collectivist-minded citizens China's hardline Communist leadership demands? Some think they never will. Ironically, it may be just these overindulged children who will change Chinese society. After growing up doing as they wished, ruling their immediate families, they're not likely to obey a central government that tells them to fall in line. This new generation of egotists, who haven't been taught to take even their parents into consideration, simply may not be able to think of the society as a whole—the basic principle of communism.

Reading 7.3

Family and Peer Group in Northton

Elijah Anderson

Unemployment, racism, drug use, and lack of hope for the future are a major part of inner-city neighborhoods. How do these factors affect the family, and how are they influenced by the family? Elijah Anderson has been involved in field research in black underclass neighborhoods in Philadelphia and Chicago for more than twenty years. In this excerpt from his book, Streetwise: Race, Class, and Change in an Urban Community, *Anderson examines some of the characteristics of inner-city families in a neighborhood he calls Northton. Anderson is a sociology professor at the University of Pennsylvania.*

The young woman who becomes addicted to "rock cocaine" or crack often comes from an unstable family. The economy and drugs are often implicated in this instability. In sharp contrast, the strong ghetto family, often with both a husband and a wife, but sometimes only a strong-willed mother and her children who have the help of close relatives, seems to instill in girls a hopeful sense that they can reproduce this strong family or one even "better." Such units, when they exist, are generally regarded as advantaged. The father usually works at a regular job and has a sense that his values have paid off. Both parents, or close kin, strive to instill in the children the work ethic, common decency, and social and moral responsibility.

Young girls growing up in such situations, strongly encouraged by their mothers and fathers and other kin, are sometimes highly motivated to avoid habits and situations that would undermine their movement toward "the good life." They are encouraged to look beyond their immediate circumstances; they often have dreams of marriage, a family, and a home. The strict control their parents exert, combined with their own hopes for the future, makes them watch themselves and work hard to achieve a life better than that lived by so many in the ghetto. Because of such interests and concerns, these young women tend not to follow the social track leading to drug addiction, though even they can become victims of crack, since because of its highly addictive nature one behavioral lapse or seeking an "experience" can fundamentally change a person's life.

By contrast, young girls and boys emerging from homes that lack a strong, intact family unit, usually headed by a single mother working or on welfare and trying desperately to make ends meet, become especially vulnerable to drug addiction and unwed pregnancies. The process leading up to such events usually begins in early childhood with full participation in play groups on the streets. Children who become deeply engaged in the drug culture often come from homes where they have relatively little adult attention, little moral training, and limited family encouragement to strive for a life much different from the one they are living.

Coming from a single-parent household, often headed by a nearly destitute mother, children may have almost no effective adult supervision. As some people in the community say, "they just grow up." Many become "street kids," left largely on their own, and by the time they are preteens some are becoming "street smart," beginning to experiment with sex and drugs. The local play group may become something of a family for youths who lack support at home.

The ghetto street culture can be glamorous and seductive to the adolescent, promising its followers the chance of being "hip" and popular with certain "cool" peers who hang out on the streets or near the neighborhood school. Often such teenagers lack interest in school, and in time they may drop out in favor of spending time with their street-oriented peers.

To be sure, most adolescents dabble in the street culture, including experimenting with sex and drugs; for status and esteem, they must learn to interpret and manipulate its emblems. Those who are most successful may become invested in such pursuits to the point of becoming overwhelmed and socially defined by them. Such definitions contribute to the formation of a street identity, which then gives such youths further interest in behaving in accordance with the peer-group's norms—or even inventing them.

The morally strict and financially stable intact nuclear families, on the decline in the underclass neighborhood, with their strong emotional and social ties and their aspirations for their children, must engage in sometimes fierce competition with the peer group. With a variety of social supports, including extended kin networks and strong religious affiliations, such families can often withstand the lure of the street culture, but even they may succumb and lose control of their offspring, sometimes permanently. The much less viable family headed by an impoverished young woman who has her hands full working, socializing, and mothering stands little chance in the struggle for a child's allegiance and loyalty and often does not prevail.

Though stable families possess an outlook that considers upward mobility a real possibility and try to instill this outlook in their children, such value transfers are difficult to accomplish. One important reason is the street culture's very strong and insistent attraction. But also important is the fact that the wider culture and its institutions are perceived, quite accurately at times, as unreceptive and unyielding to the efforts of ghetto youths.

Resolving this situation depends to some degree on luck. Youths from the most socially "promising" domestic arrangements sometimes become fixated and invested in the ghetto street culture, but they may have a chance to grow out of it. . . . Those from the poorest families, with the least control over social resources, appear to be most vulnerable to the street's seductive draw.

Vulnerability to the street may be manifested in a variety of ways, from adolescent pregnancy to street crime to serious drug abuse. One may approach "hipness" through good looks and grooming, nice clothes, dancing ability, sexual activity, crime, and drug use—which may be more or less important depending upon one's particular crowd or peer group.

. . . The economic noose restricting ghetto life encourages both men and women to try to extract maximum personal benefit from sexual relationships. The dreams of a middle-class life-style nurtured by young inner-city women are thwarted by the harsh socioeconomic realities of the ghetto. Young men without job prospects cling to the support offered by their peer groups and their mothers and shy away from lasting relationships with girlfriends. Thus girls and boys alike scramble to take what they can from each other, trusting only their own ability to trick the other into giving them something that will establish their version of the good life—the best life they can put together in their environment.

We should remember that the people we are talking about are very young—mainly from fifteen years old to their early twenties. Their bodies are grown, but they are emotionally immature. These girls and boys often have no very clear notion of the long-term consequences of their behavior, and they have few trustworthy role models to instruct them.

Although middle-class youths and poor youths may have much in common sexually, their level of practical education differs. The ignorance of inner-city girls about their bodies astonishes the middle-class observer. Many have only a vague notion of where babies come from, and they generally know nothing about birth control until after they have their first child—and sometimes not then. Parents in this culture are extremely reticent about discussing sex and birth control with their children. Many mothers are ashamed to "talk about it" or feel they are in no position to do so, since they behaved the same way as their daughters when they were young. Education thus becomes a community health problem, but most girls come in contact with community health services only when they become pregnant—sometimes many months into their pregnancies.

Many women in the black underclass emerge from a fundamentalist religious orientation and hold a "pro-life" philosophy. Abortion is therefore usually not an option. . . . New life is sometimes characterized as a "heavenly gift," an infant is sacred to the young woman, and the extended family always seems to make do somehow with another baby. A birth is usually met with great praise, regardless of its circumstances, and the child is genuinely valued. Such ready social approval works against many efforts to avoid illegitimate births.

In fact, in cold economic terms, a baby can be an asset, which is without doubt an important factor behind exploitative sex and out-of-wedlock babies. Public assistance is one of the few reliable sources of money, and, for many, drugs are another. The most desperate people thus feed on one another. Babies and sex may be used for income; women receive money from welfare for having babies, and men sometimes act as prostitutes to pry the money from them.

The lack of gainful employment not only keeps the entire community in a pit of poverty, but also deprives young men of the traditional American way of proving their manhood—supporting a family. They must thus prove themselves in other ways. Casual sex with as many women as possible, impregnating one or more, and getting them to "have your baby" brings a boy the ultimate in esteem from his peers and makes him a man. "Casual" sex is therefore fraught with social significance for the boy who has little or no hope of achieving financial stability and hence cannot see himself taking care of a family.

The meshing of these forces can be clearly seen. Adolescents, trapped in poverty, ignorant of the long-term consequences of their behavior but aware of the immediate benefits, engage in a mating game. The girl has her dream of a family and a home, of a good man who will provide for her and her children. The boy, knowing he cannot be that family man because he has few job prospects, yet needing to have sex to achieve manhood in the eyes of his peer group, pretends to be the "good" man and so convinces her to give him sex and perhaps a baby. He may then abandon her, and she realizes he was not the good man after all, but a nothing out to exploit her. The boy has gotten what he wanted, but the girl learns that she has gotten something too. The baby may bring her a certain amount of praise, a steady welfare check, and a measure of independence. Her family often helps out as best they can. As she becomes older and wiser, she can use her income to turn the tables, attracting her original man or other men.

. . . In Northton there are an overwhelming number of poor, female-headed families, a significant proportion of them on welfare. As these families grow older, the children meet the street culture and must come to terms with it. Some dabble in it, experiencing enough to try to pass as hip among their friends; few want to be known as "square" or "lame." For those only loosely anchored to conventional institutions (extended families, churches, or community organizations of one kind or another), the street culture calls loudly and insistently.

Most young boys can no longer earn enough to support a family, in the regular economy or the underground economy or both. Yet they engage in sex for both biological and peer-group reasons and often father children out of wedlock. Over time such behavior becomes incorporated into sex codes that reflect the reality of the situation. The young men often adjust to their financial situation by remaining with their mothers, retaining their "freedom" and all the comforts of home; they may sometimes play the role of man of the house and aid the household through meager financial contributions. But the youth is unwilling to "play house" with the mother of his child, an adaptation his peer group turns into a virtue. The girls are often headed for adolescent pregnancy, but they are doubly unable to provide for their children, with no skills, no provision for child care, and no vision of anything else.

QUESTIONS 7.1–7.3

1. Take several concepts from earlier chapters (for example, socialization, self-concept, status, generalized other, and so on) and apply them across the three family styles described in the readings.

2. Connect family structure to behavior—how does parental conduct seem to affect children's behavior in each case?

3. How has the structure of society changed the structure of the family in each case?

4. Contrast male and female roles in the three types of families. Make hypotheses in those instances in which you cannot find specific examples.

5. How do these three family structures compare with the "typical American middle-class family"? List similarities and differences.

NOTES

1. This discussion benefits from the discussion of social institutions by S. N. Eisenstadt in the *International Encyclopedia of the Social*

Sciences (New York: Crowell, Collier & Macmillan, 1968), 14: 409–429.

2. The discussion in this paragraph is taken in part from David and Vera Mace, *Marriage:*

East and West (Garden City, N.Y.: Doubleday, 1960), pp. 141–142.

3. These two paragraphs on courtship theory benefit from the discussion by Marcia and Thomas Lasswell in *Marriage and the Family* (Lexington, Mass.: D. C. Heath, 1982), ch. 6.

4. George P. Murdock, *Social Structure* (New York: Macmillan, 1949), p. 28.

5. See "Influence of the Marital History of Parents on the Marital and Cohabitational Experiences of Children," by Arland Thornton, *American Journal of Sociology* 96 (January 1991), pp. 868–894.

6. J. Ross Eshleman, *The Family: An Introduction,* 3d ed. (Boston: Allyn and Bacon, 1981), ch. 6.

7. See "Children and Marital Disruption," by Linda Waite and Lee Lillard, *American Journal of Sociology* 96 (January 1991), pp. 930–953.

8. Numerous Census Bureau documents were used for this section on marriage and divorce. Especially see "Marital Status and Living Arrangements: March 1992," *Current Population Reports,* Series P-20, no. 468 (Washington, D.C.: United States Government Printing Office, 1991), and *Statistical Abstracts of the United States,* 1993.

9. See "Considering Divorce: An Expansion of Becker's Theory of Marital Instability," by Joan Huber and Glenna Spitze, *American Journal of Sociology* 86 (July 1980), pp. 75–89.

10. This research was done by Murray A. Straus, Richard J. Gelles, and Suzanne K. Steinmetz and is reported in their book, *Behind Closed Doors: Violence in the American Family* (Garden City, N.Y.: Doubleday, 1980), ch. 2.

11. This account is taken from Jean Lipman-Blumen, "Changing Sex Roles: Their Implications for Family Structure" (paper presented at the Pacific Sociological Association Meeting, San Diego, California, March 1976). Jean Lipman-Blumen is affiliated with the Women's Research Program at the National Institute of Education.

CHAPTER EIGHT

Institutions: Religion

TWO RESEARCH QUESTIONS

1. *Is religion related to health? Would belonging to a religious community make one more or less healthy, or are the variables unrelated?*

2. *Is there a relationship between religion and suicide? Specifically, will suicide rates vary depending on the type of church one belongs to? If so, why?*

. . . *I will give a short account of the psychology of myself when my hansom cab ran into the side of a motor omnibus, and I hope hurt it.*

 . . . *In those few moments, while my cab was tearing towards the traffic of the Strand . . . I did really have, in that short and shrieking period, a rapid succession of a number of fundamental points of view. I had, so to speak, about five religions in almost as many seconds. My first religion was pure Paganism, which among sincere men is more shortly described as extreme fear. Then there succeeded a state of mind which is quite real, but for which no proper name has ever been found. The ancients called it Stoicism, and I think it must be what some German lunatics mean (if they mean anything) when they talk about Pessimism. It was an empty and open acceptance of the thing that happens—as if one had got beyond the value of it. And then, curiously enough, came a very strong contrary feeling—that things mattered very much indeed, and yet that they*

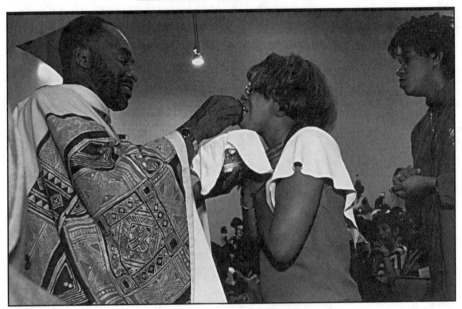

*were something more than tragic. It was a feeling, not that life was unimport-
ant, but that life was much too important ever to be anything but life. I hope
that this was Christianity. At any rate, it occurred at the moment when we went
crash into the omnibus.*[1]

IN OUR DISCUSSION of institutions in Chapter Seven, we suggested that all societies have specific constant and important central needs or problems that must
be addressed. Norms and roles emerge, and a pattern of organization or social
institution develops. In the previous chapter we looked at the family, and in
this chapter we will examine the institution of religion.

Some eternal questions face all of us: What happens after death? What is
the meaning of life? Given the misfortunes of the present, can there be hope
for the future? As people struggle with these concerns, they develop a set of
beliefs, norms, and roles that represent their attempts to answer these questions. We refer to this increasingly organized set of norms and roles that develops as the institution of religion.

The central concerns from which the institution of religion develops are
usually abstract—much more so than those we discussed in the previous
chapter on the family. One consequence is that definitions of and explanations
for the institution of religion are more varied than are those for some other
institutions. For example, Andrew Greeley suggests that religion has at its origins the human propensity to hope; that in the face of life's disappointments,

hope needs to be renewed; that a variety of experiences may renew hope; and that religion is a collectively created phenomenon that effectively renews hope. This approach focuses on what religion *does* (renews hope), whereas some other approaches focus on what religion *is* ("a set of sacred beliefs and symbols").[2]

Religion, like the family, has appeared in all societies. If one compares different societies, however, one finds tremendous variation in religious practices. To get an idea of the variation, look at Tables 8.1 and 8.2. Table 8.1 shows a

Table 8.1 Religious Population of the World, 1992

Faith	Population (millions)	Percent	Faith	Population (millions)	Percent
Christian	1,833	33	Christian		
Muslim	971	18	Catholic	1,026	56
Hindu	733	13	Protestant	374	20
Buddhist	315	6	Orthodox	170	9
Chinese folk religions	187	3	Anglican	75	4
New religionists	143	3	Other	188	10
Tribal religionists	97	2			
Jewish	18	0.3			
Sikhs	19	0.3			
Shamanists	11	0.2			
Confucian	6	0.1			
Shinto	3	0.1			
Nonreligious	876	16			
Atheist	240	4			

United States Bureau of the Census, *Statistical Abstracts of the United States, 1993* (113th ed.), Washington, D.C.

Table 8.2 Membership in Selected Religious Bodies in the United States, 1993

Religious Group	Members (thousands)	Percent
Adventist	766	0.5
Baptist	34,587	21
Churches of Christ	2,812	2
Disciples of Christ	1,022	0.6
Eastern Orthodox	2,114	1.3
Episcopal	2,471	1.5
Jehovah's Witnesses	914	0.6
Jewish	4,300	3
Latter-Day Saints	4,488	3
Lutheran	8,393	5
Methodist	14,621	9
Muslim	6,000	4
Pentecostal	10,093	6
Presbyterian	4,246	3
Roman Catholic	58,267	36
United Church of Christ	1,583	1
Other	5,225	3

The World Almanac and Book of Facts, 1994 (Mahwah, N.J.: Funk & Wagnalls, 1993).

breakdown of the religious population of the world. Table 8.2 shows recent membership figures of religious groups in the United States. Not all of America's churches are included in Table 8.2; only those with memberships of at least 500,000 are listed. This table (8.2) shows church *members,* and these represent about 60 percent of the United States population. Forty percent of the population doesn't claim any church membership. As you view the figures in these tables, keep in mind also that religious bodies count membership in different ways. For example, some churches count all baptized people, including infants, whereas others count only those people who have attained full membership, which usually includes only individuals over 13 years of age. Thus, these figures should be taken as no more than general estimates.

All countries have at least one religion, although not all people are religious. Ethnocentrism relates to our discussion of religion, as it does to our discussion of the family: The lack of acceptance of others' ways of doing things, the hostility to difference, is probably even more pronounced with regard to religion. Whereas people with peculiar practices (meaning those different from ours) in marriage, family, clothing, or eating are seen as alien, primitive, or ignorant, people with religious beliefs other than our own are frequently seen as evil, pagan, or guided by the devil. Great rivalries and conflicts have occurred between religious groups, for religious faith and conviction can be a powerful force. Knowing that "it's God's will" or that "God is on our side" has justified all kinds of behavior.

RELIGION: THE FUNCTIONAL APPROACH

Sociologists who study religion are not concerned with the possible truth or falsity of religious beliefs. They are interested in what people believe and especially in what believing does for people. Although there is great variation among religions, a discussion of some common definitions and descriptions may help our understanding of the institution of religion. **Religion** may be described as a unified system of beliefs, feelings, and behaviors related to things defined as sacred. Societies define as sacred, for example, objects (a tree, the moon, a cross, a book), animals, ideas (science, communism, democracy), and people (Buddha, Christ, saints). When things become sacred, they are endowed with a special quality or power and are treated with awe, reverence, and respect. Beliefs about the sacred entities develop, and then people often choose to lend to their beliefs a supernatural quality. A set of appropriate feelings—reverence, happiness, sadness, fear, terror, ecstasy—are established. Finally, behaviors or rituals are developed consistent with the beliefs and feelings: the confessional, the rosary, communion, dietary laws, or the content of a particular worship service.[3]

Traditionally, institutions are examined in terms of the functions they serve. What functions does religion offer for society? First, religion provides a way for people to deal with the unknown; it supplies some measure of certainty in an uncertain world. There are a number of things we cannot explain. There are many other things that can be explained and understood but are still difficult to accept, such as the certainty of death or the loss of loved ones. Religion enables people to adjust to these situations by providing a sacred, supernatural being or object to explain the unknowable; religion provides a belief system to use to deal with situations that are difficult to accept, such as the certainty of death or a life after death. Anthropologists have noted that sacred rituals and beliefs are usually found in situations that are most difficult to control, in which there is most uncertainty. Our own appeals for divine guidance and support usually come at times when uncertainty is present: when the jet accelerates for takeoff, or just before a particularly difficult test.

Religion also functions to provide people with a perspective, a viewpoint, a way of looking at the world. It provides value orientations as to how the world ought to be; it gives meaning to life. Max Weber held that people are motivated by ideologies or beliefs that stem in great part from religious origins and that these are more basic than are the economic values believed by Marx to be so important. In his classic work *The Protestant Ethic and the Spirit of Capitalism,* Weber's analysis of history led him to the conclusion that the beliefs surrounding Protestantism were central to the rise of capitalism. For example, Protestant values encourage the view that work is a calling: To work hard is virtuous, acquisition is supported, and sobriety is encouraged. Far from being a needless distraction, religion is the source, according to Weber, of many of a society's most basic values, perspectives, and orientations to the world.

Religion may serve an integrative function in that it promotes solidarity. People with similar beliefs and viewpoints are drawn together and are more unified than those without this common experience. The presence of religion does not guarantee solidarity within a society, however, as other factors are also important. Many societies—the United States, for example—have several religions, and these may conflict with each other. In this case, there may be a degree of solidarity among those of the same faith but an absence of solidarity in the society; religion may prove to be disintegrative for society even though it is integrative for the individual.

The first research question at the beginning of this chapter concerned a possible relationship between religion and health. It has been suggested that religion might affect health by regulating behavior—diet, smoking, and drinking—and by providing social support and enhancing social integration. Some time ago, sociologist Emile Durkheim emphasized the integrative and regulative functions that religion may provide for members. Whether these functions could be related to health was recently studied in an interesting way.

Two small, cohesive kibbutzim (collective settlements or communities) in the Negev region of Israel were chosen for study. They were similar in size (about 400 people) and age of members, but one kibbutz was religious and the other was nonreligious, or secular. A questionnaire given to community members included several measures of health behavior (like diet), health status (such as disabilities or sicknesses), illness behavior (for example, number of doctor visits and medication taken), and measures of religiosity (religious practice, private praying). The researchers found that the religious kibbutz practiced more health-promoting behavior and that the members of that community reported that they were more healthy and had less illness behavior

A week's schedule from a church bulletin board

THIS WEEK'S GROUP MEETINGS—7 P.M. IN THE SANCTUARY

Monday	Say No to Drugs
Tuesday	The Puzzle of Dyslexia
Wednesday	Abused Spouses
Thursday	Codependency Recovery Group
Friday	Alcoholics Anonymous
Saturday	Food Closet and Aid for the Homeless

NEXT SUNDAY 9 A.M. and 11 A.M. REVEREND CHAVOOR
"Rejoicing about America's Future"

---

than did the secular kibbutz. However, those who practiced individualistic religious activities, like private praying, were less healthy and reported more psychological distress. It seems to be the collective aspect of religion rather than the individual aspect of religiosity that is important. The authors suggest that a religious community provides additional regulation and integration (beyond that of other types of communities) that result in more healthy behavior and better psychological and physical well-being.[4]

Religion may function as an agency of social control. We noted earlier that religion provides people with a value orientation, a perspective on life, a view of the world. From shared viewpoints, common agreements arise as to what people should and should not do. Because these norms emerge from a religious base, the church becomes involved in ensuring that they are observed. They may be informally encouraged or enforced, or they may be enacted into law. For example, many religions provide for the education of the young in schools run by the church. Some churches do not recognize marriages that take place outside the church. The clergy perform marriages and engage in marriage counseling. A little more indirectly, through pressure groups and lobbies, religions are involved in ensuring the passage or rejection of laws felt to be crucial; laws dealing with liquor, pornography, abortion, and birth control are recent examples in the United States. In countries in which there is less separation of church and state, religious organizations and their representatives are more directly involved with government in determining right and wrong.

In a variety of ways, then, organized religious groups assert moral authority and attempt to provide guidance; they act as agents of social control. Problems may arise when values and viewpoints on issues change, as they have, for example, on legalized abortion and birth control in the United States. Formal church doctrines change much more slowly than do the values and attitudes of the people. The result is that the social-control function of the church is compromised when its members no longer hold its values; if this occurs on many issues, the authority of the church is likely to be seriously undermined.

A report published in 1976 provides a good example of how this can happen: In 1968, Pope Paul VI ruled against the use of artificial birth control by Catholics. Earlier in the 1960s the Second Vatican Council had made church reforms that were well accepted by Catholics in the United States. However, the Pope's 1968 ruling was, according to Andrew Greeley, who studied the consequences and authored the report, "as far as the American church goes, one of the worst catastrophes in religious history." The birth-control ruling did serious damage to the authority of the church and to the credibility of the Pope. Apparently as a consequence of the ruling, between 1963 and 1974 weekly attendance at Mass dropped from 71 to 50 percent, monthly confession dropped from 38 to 17 percent, families who favored a son becoming a priest dropped from 63 to 50 percent, parochial school enrollment dropped from 5.6 million to 3.5 million, and so on. The Pope's ruling was issued to restore faith in the church, but, in pushing a view no longer held by most members, it in fact badly weakened it.[5]

How is religion in America best explained in a functional sense? Greeley suggests a combination of three factors: belonging, meaning, and comfort. Belonging refers to the idea originated by Durkheim that religion is the source of social interaction; it provides the cement that holds society together in the face of internal and external threats; it provides unity. Meaning refers to Weber's idea that religion provides a way of looking at the world—a set of definitions and ideas that guide one's thinking. Religion gives meaning to that which is difficult to interpret. It also influences the character and direction of other institutions (economic, for example) because people's views of the world are shaped by their religions. Finally, religion provides comfort, serenity, and reassurance. It supports existing institutions and provides social stability. Greeley believes that these three factors, especially the relationship between the meaning and belonging functions, are responsible for the vigorous religious behavior in the United States.[6]

The functions we have discussed—to explain the unknown, to provide a viewpoint (meaning), to afford solidarity (belonging), and to exercise social control—represent some of the more obvious functions of religion. These functions are not mutually exclusive, and there are others as well. It is interesting to note here while we are discussing functions that *magic* in primitive societies and *science* in more advanced societies serve many of the same functions as does religion. Without careful study, it is sometimes difficult for lay people to distinguish among statements from magic, science, and religion.

RELIGIOUS ORGANIZATIONS

So far we have examined the elements of religion (beliefs, feelings, and behavior related to sacred objects) and some of the functions religion may serve. Religion may also be studied from the viewpoint of the organizational structure of religious groups. Our knowledge of bureaucracies and large organizations may be applied to churches. One way of categorizing religious bodies would be in terms of their ceremonies, their pattern of public worship. Greeley suggests the breakdown shown in Table 8.3. Religious ceremony is examined in terms of how rational or emotional it is (Apollonian or Dionysian, respectively) and by the degree of formality (simple church or high church). This breakdown provides four categories, each of which describes a fairly distinct type of religious experience. **Apollonian-simple church ceremonies** appear often in American Protestant denominations. Services are "relatively plain and simple and at the same time sober, restrained, and dignified." Even the hymns are sedate and restrained. **Dionysian-simple church ceremonies** involve "direct intervention of the spirit" and a simple and matter-of-fact approach to prayer, which at the same time encourages members to freely and openly express their emotions. **Apollonian-high church services** are "dignified, rational, and restrained, while at the same time elaborate, artistic, and stylized."

Table 8.3 Types of Religious Ceremony

	Simple Church (simple, informal ceremony)	High Church (formal, complex ceremony)
Apollonian (rational, sober, unemotional)	Traditional Protestant— Presbyterian, Congregational, Methodist	Latin mass, solemn high mass of Roman Catholic Church
Dionysian (emotional, nonrational, ecstatic)	Pentecostal, holy roller, snake handlers	Coptic mass, Javanese dance

Adapted from Andrew Greeley, *The Denominational Society* (Glenview, Ill.: Scott, Foresman & Co., 1972).

Dionysian-high church services are "ecstatic, emotional, and nonrational, and at the same time elaborate, artistic, and stylized." The latter is a difficult combination, and this category of experience is not frequently found in churches in the United States. Greeley notices several trends: Along the simple-church–high-church dimension, Catholicism in this country seems to be moving in the direction of simple church, whereas some Protestant denominations are going the other way. Along the Apollonian-Dionysian dimension, however, the direction is clear. The revolt against scientific rationalism is leading churches to become increasingly Dionysian—nonrational, emotional, and ecstatic.[7]

Terminology has also been developed to categorize churches in terms of the size and scope of the organization. One distinction that is commonly made is between church and sect. A **church,** which is large and highly organized, represents and supports the *status quo;* it is considered a respectable organization, and membership is usually automatic—one is born into the church. A **sect,** on the other hand, is smaller and less organized, and membership is voluntary. Members of sects usually show greater depth and fervor in their religious commitment than do members of churches. The sect is more closely associated with the lower classes, the church with the middle and upper classes. The sect usually arises because of protest when church members become disaffected. They may feel that the church is too compromising or religiously too conservative, or in some other way is not responding to the wishes or convictions of its membership. The disaffected, deciding they would rather switch than fight, establish a sect based on purity of belief and on the individual religious needs of its membership.[8]

Milton Yinger sees church and sect as two types on a continuum of religious organizations. He adds several other types, including denomination and cult, as shown in Table 8.4. The **denomination** appeals to a somewhat smaller category of people—a racial, ethnic, or social-class grouping—than does the church. Yinger defines **cult** as a small, short-lived, often local group frequently built around a dominant leader. Cults and sects arise, it is argued, because people feel deprivation in some form—poverty, ill health, value conflict, anomie. This deprivation leads to the formation of new religious movements. Once formed, the movement may later die out, or it may become larger, more established, and conservative, and move to another category of organization—from cult to sect to denomination, for example.

RELIGION IN AMERICA

To a sociologist looking at religion in the United States, particular aspects stand out. In America, as in other countries, there is *differential involvement* in religious activities. If we take church attendance as a measure of religious involvement, we find that women are more active than men, Catholics more than Protestants, college-educated more than those with a high-school education or less, people over 30 more than those under 30, and people in the East most and those in the West least. Middle- and upper-class people attend and participate in church activities more than those of the lower class. Lower-class

Table 8.4　Religious Organizations

Type of Organization	Characteristics	Examples
Church	Large; universal appeal; very respectable; members are born into it; supports the status quo and existing nonreligious (secular) institutions; formal structure	Roman Catholic Church of the Middle Ages
Denomination	Less universal appeal but organized around class, racial, or regional boundaries; conventional and respectable; less emotionalism and fervor than is shown in the sect; mainly middle class; stable, organized, and at peace with secular institutions	Presbyterian Church, Lutheran Church
Sect	Smaller; arises out of protest; membership is voluntary; uncompromising on religious doctrine; more emotional; less formal organization; more withdrawn from society and at war with secular institutions; it alone "has truth"; mainly lower class	Pentecostal and evangelical movements, Jehovah's Witnesses
Cult	Smallest; less organized; somewhat formless and often short-term; frequently centered around a charismatic leader; concerned mainly with problems of the individual (loneliness, alienation); more exotic ritual; may focus on a religious theme (salvation) or on a more secular theme (self-improvement, self-awareness)	Snake worshipers, People's Temple, urban storefront churches

Adapted from J. Milton Yinger, *Religion, Society and the Individual* (New York: Macmillan, 1957), ch. 6.

people tend to show greater intensity and emotion in religious feeling. Religious groups also seem to be stratified according to social class. Episcopalians, Congregationalists, and Jews tend to be of higher social class than Presbyterians and Methodists, who in turn tend to rank higher than Baptists and Catholics. This ranking in great part reflects ethnic or national origin and recency of immigration into the United States. The religious and social-class picture is complicated by the apparent tendency of Americans to change their religious affiliation as they move up the social-class ladder.[9]

Great *religious diversity* characterizes this country. The United States has no official state-supported religion and, unlike some countries, America has attempted to maintain a separation between church and state. Broadly speaking, three religious bodies—Protestant, Catholic, and Jewish—dominate. However, there are numerous Protestant denominations, of which the largest are Baptist, Methodist, Lutheran, and Presbyterian, as well as a multitude of sects and cults. Some authorities have noted this diversity in American religious groups and have argued that, on the contrary, the differences are in name only. They believe that there is a growing consensus on religious beliefs and that we are approaching a common religion in America that is eroding the traditional differences among Protestantism, Catholicism, and Judaism. Charles Glock and Rodney Stark tested this idea by asking several members of Protestant and Catholic churches about their beliefs. They concluded that great diversity *does* exist, not only between Protestants and Catholics but among various Protestant denominations as well. In fact, Glock and Stark believe that the differences among various Protestant denominations are so great that it is inappropriate to speak of "Protestants"—there is no such thing as a unified Protestant religion. Some of the results of Glock and Stark's research, shown in Table 8.5, indicate that members of different Protestant denominations have marked differences in beliefs. These differences are reflected in numerous topics and seem to indicate that a common American religion, at least in terms of beliefs, does not now exist.

On the other hand, Protestantism, Catholicism, and Judaism in the United States are probably peculiarly American; often they differ markedly from their antecedents in other countries. A *Time* magazine poll of American Catholics conducted in August 1987 on the eve of a visit to the United States by Pope John Paul II seemed to support this. Often, American Catholics differed with the official church: American Catholics favored permitting women priests (52 percent), married priests (53 percent), and remarriage in the church for the divorced (76 percent). Only 24 percent viewed artificial birth control as wrong. Seventy-eight percent of those polled felt it was permissible for Catholics to make up their own minds on such moral issues as birth control and abortion. Often, the viewpoints of Catholics and Protestants were similar: 27 percent of Catholics and 34 percent of Protestants felt that women should have the right to abortion on demand; 43 percent of Catholics and 41 percent of Protestants

Table 8.5 Percentage of Members of Various Christian Religious Groups Who Held Particular Beliefs to Be "Completely True"

	Congrega-tionalists	Meth-odists	Presby-terians	Missouri Lutherans	Southern Baptists	Catholics
Belief in the existence of God	41%	60%	75%	81%	99%	81%
Belief that Jesus was born of a virgin	21	34	57	92	99	81
Belief in life after death	36	49	69	84	97	75
Belief in the existence of the devil	6	13	31	77	92	66
Belief in original sin	2	7	21	86	43	68

Adapted from Charles Glock and Rodney Stark, *Religion and Society in Tension* (Chicago: Rand McNally, 1965), ch. 5.

agree that the Pope is out of touch with Catholics in the United States. The melting pot has modified traditional religions so that they all probably reflect the cultural values of the United States.

The second question at the beginning of this chapter dealt with the possible relationship between religion and suicide. Durkheim started the research on this issue in 1915, and much has been done since his time. A study on suicide is relevant to our discussion here on religious changes in the United States. This study, which examined suicide rates within different religious groups, found that religious affiliation was associated with suicide rates in complicated ways. Catholics, Jews, and evangelical Protestants had lower suicide rates, whereas Presbyterians, Methodists, and Lutherans had higher suicide rates. Religions that might be called "mainstream" and liberal or moderate had higher rates. The authors suggest that people belong to different "networks" in society. People's membership in groups or organizations and their affiliations are examples of networks. Some of these networks are tightly knit, more controlling, and provide more energy and support for members; others are more loosely structured and provide less support for day-to-day behavior. Some religions (Catholicism, evangelical Protestant denominations) are more tightly involved in their members' lives and provide additional social support and guidance in times of personal crises. Other religions (mainstream Protestant denominations) have responded to changes in society and have become looser and consequently less central to members' lives. They may provide less energy and support and be less able to help in times of crisis, perhaps leading to higher suicide rates.[10]

In the late 1970s, the White House was occupied by a president who was a born-again Christian, and numerous people on the national scene, from Watergate conspirators to athletic heroes, publicly became religious. This trend continued through the 1980s and into the 1990s, with religious groups like the Moral Majority exerting political pressure at high levels, and with the spread of electronic religion, in which radio and television sermons reach a wide audience. The United States appears to be going through a religious revival of sorts, with renewed interest in fundamentalist, pentecostal, and evangelical types of experiences. Evangelical movements have a high profile and are continually in the news. Television ministries ("televangelism"), a phenomenon of the cable and satellite technologies of the 1980s, are reaching a large audience, attracting much money, and exerting political influence. Some traditional churches have discovered to their dismay that many of their members have found themselves attracted to a different message, perhaps to more emotional and exotic religious conduct like faith healing, speaking in tongues, mass revival meetings, and so on. Membership trends in the mid-1990s have shown declines in old-line Protestant churches (Methodist, Episcopal, Presbyterian) and increases in the conservative Protestant and Evangelical churches. Switching from the religious tradition in which one was raised is more common, especially among liberal and moderate Protestants. It also seems to have become more acceptable and common to state no religious affiliation and to have no attachment to any organized church.

A passing parade of recent religion-related events includes the following: scandals involving financial and sexual improprieties of religious leaders;

strong interest in Eastern religions and mysticism; members of Hare Krishna chanting in the streets; government investigations of Scientology; Reverend Moon's Unification Church holding mass rallies and mass weddings; religious groups being charged with brainwashing converts while parents attempt to kidnap their children from the group; a gun battle that left four ATF agents and six members of the Branch Davidians dead, followed in April 1993 by a fire in which sect leader David Koresh and 78 of his disciples die.

Religious behavior in general seems to be fragmenting and taking many different forms. Overall church attendance has not changed much between 1970 and the present—it is the *form* of religious expression that is undergoing change. Great diversity continues to characterize religion in America.

Religion in America is *losing functions:* Other institutions are taking over the traditional functions that religion has served. Education and socialization of the young is handled for the most part by educational institutions rather than by the church. More dramatic is the rise in the importance of science. Today we tend to put our faith in science rather than in religion when dealing with the unknown or uncertain. Scientific knowledge increasingly forces a redefining of religious beliefs. How do you describe heaven when rockets and spaceships are racing through the universe to the moon and beyond? What is the soul when hearts and possibly even brains can be transplanted from one body to another? Today it seems that we look to science for the answers rather than to religion; the scientist is more of a folk hero than is the minister. We might predict that either religion will change to adapt to a scientific world or that religion will have ever decreasing influence. Some religious organizations, in a spirit of "if you can't lick 'em, join 'em," are attempting to develop new functions to replace those they have lost. Usually this means becoming more "this-worldly" and less "other-worldly." The new church looks like a social-work agency, and religious leaders frequently lead the way in speaking out on social issues. Clergy put their bodies on the line in the 1960s in civil rights demonstrations in the South. Individual representatives of religions spoke out on the draft and on the war in Southeast Asia among other things, and have been active in the farm-labor movement in California and in other parts of the Southwest. Religious organizations have provided food and shelter for youth "doing their own thing." It is frequently noted that the clergy in the pulpit sound more as though their college degrees were in psychology or sociology than in theology.

These trends may not be widespread or even widely accepted in the very churches in which they are taking place. Some argue that these changes are certain guarantees of the decline of religion. After all, churches were formerly offering something different, a special way of looking at the world unattainable elsewhere. Now the church is no different from other secular organizations and therefore has lost any unique appeal it might have had. Others believe, however, that in a changing world, religion too must change or run the risk of becoming irrelevant to modern people.

Finally, the **secularization** of American religion bears discussion. By secular we mean nonspiritual, nonsacred, and this-worldly. Americans seem to be religious and nonreligious at the same time. Millions of Bibles are sold yearly,

THE TNEVNOC CULT

The 1970s have witnessed a profusion of new religious movements, and as these groups have grown in size and wealth, there has been a parallel spread of alarm at the tactics by which they recruit and hold members. Allegations of deception, seduction, hypnosis, brainwashing, and other forms of mind control are made.

David Bromley and Anson Shupe argue that characteristics of this "cult menace" are not new but have been a persistent feature in our history. They describe the Tnevnoc cult, a once-powerful and widespread religious movement of the 19th century. Young women at school were openly recruited and induced to commit themselves totally to the cult. Once they had joined the Tnevnocs, they had to surrender all aspects of their former lives: All possessions were taken away; all luxuries and even basic amenities were eliminated; members had their hair cut off; they could not own or look into mirrors; they were permitted no outside visitors during the first year; they could write no more than four letters a year; they could not associate with the opposite sex; and they had to accept cult leaders in parental roles. Cult life was relentless and consuming: Members were awakened at 4:30 A.M., and they spent a long day at manual labor mixed with meditation, mind-numbing chanting, and compulsory religious ceremonies. Like the Hare Krishna, they carried prayer beads, which they used in their repetitive, monotonic chanting. In a bizarre ceremony involving ritualistic cannibalism, Tnevnocs consumed food that symbolically represented parts of the dead founder's body. All loyalty had to be channeled to the cult and its leaders.

In one macabre nuptial ceremony, members were required to become the living brides of the dead cult leader.

The group that Bromley and Shupe are talking about is not all that obscure. It is a Roman Catholic convent (*Tnevnoc* spelled backwards). The description doesn't mention the other side of the picture—the sense of majesty, purpose, and personal fulfillment that nuns feel, or the order, stability, and integration of the convent as a social organization. The methods may seem strange, but socialization and resocialization take place in elite military units, communes, convents, religious groups, and social movements advocating radical social change. These groups seek the total commitment of their members, which means detachment from former lifestyles and attachment to new ones. Society's response to what it sees groups doing depends on the group's *degree of legitimacy*. The less the degree of a group's legitimacy, the more objections a society raises to its behavior. Much as the Unification Church, Hare Krishna, and Children of God are currently labeled "cults," the Catholic Church once was pejoratively lumped with groups such as the Mormons and Masons despite enormous doctrinal and organizational diversity. Bromley and Shupe suggest that groups that seek to initiate sweeping social change become the focus of hysterical reactions well out of proportion to the real threat they present, and that most religious movements through American history have gradually accommodated to the larger society.

Adapted from "The Tnevnoc Cult" by David G. Bromley and Anson D. Shupe, Jr., with the permission of the authors.

and polls indicate that over 90 percent of Americans say they believe in God, and yet 53 percent of the people asked could not name one of the four authors of the first four books of the New Testament, less than 50 percent could name at least four of the Ten Commandments, and 46 percent of those polled by Gallup in 1980 felt that religion was losing its influence. Americans believe in prayer, in heaven and hell, and in life after death, and yet church attendance has dropped from 49 percent in a given week in 1958 to 40 percent in 1971, and it remains at that level today. In 1957, 97 percent of Americans identified with a religious denomination. Yet at about the same time a panel of outstanding Americans, when asked to rate the most significant events in history, gave first place to Columbus's discovery of America. Christ's birth or crucifixion came in 14th, tied with the discovery of the X ray and the Wright brothers' first flight.[11] Religion used to be of vital significance; it guided the pioneers' way of life. People read the Bible regularly at family gatherings. It was the law, and its teachings were carefully followed. But as society changes in other ways, so does religious observation. Today people join churches and say they are religious, but perhaps the meaning of religion and the reasons for believing have changed. Although many remain religious in the traditional sense, many others are religious for social and secular reasons—it looks right and it makes one respectable. Religion is used to obtain other desired results. Belonging to the right church will help the aspiring businessman make the right contacts, meet the right people, and have the right friends. Athletic teams pray before games; even the United States Marine Corps holds special religious services before shooting the rifle on qualification day. We seem to feel that if we have religion—belong, or go to church—other good things will happen to us. The result is an increasing secularization of American religion. As Will Herberg puts it, contemporary American religion is *man-centered*. People, not God, are the beginning and end of the spiritual system of much of present-day American religiosity. The result is a religiousness without religion, a way of sociability or belonging rather than a way of orienting one's life to God.[12]

Making categorical statements about religion in America is risky business, however. In 1992, 75 percent of Americans polled were members of churches or synagogues, over half said religion was very important in their personal lives, and the public consistently says that it has more confidence in the church and organized religion than in other key institutions. Although less than half of the American population attends church on a given Sunday (41 percent in 1992), this percentage is higher than in most other predominantly Protestant nations in the world. Interest in television ministries and evangelical churches is strong. Next to the secular strain of religion is another strain emphasizing its spiritual and emotional aspects. This seems to be the keynote of religion in America: great variation in belief and practice.

THEORY AND RESEARCH: A REVIEW

As mentioned in the previous chapter, the study of social institutions usually starts with *functional analysis*. Institutions imply order, stability, and a unified system with integrated parts, the whole thing working together to make society run—an apt description of the functional approach. *Conflict theory* provides another perspective by focusing on the role religion plays in social conflict. Contrasting religious beliefs have often been a source of conflict; take, for example, the battles between Protestants and Catholics in Northern Ireland over the past decades. Religious differences may also become the basis for political differences, as currently is the case in the Middle East. Karl Marx was critical of religion because of people's tendency to use it in times of uncertainty or stress. Marx described religion as the opiate of the masses because he felt that it took their minds off their real problems in favor of the otherworldly answers of Christianity. The working class's real problem in Marx's view was its manipulation and exploitation by the wealthy industrialists—the bourgeoisie—in capitalist economic systems. Religion was a needless distraction that pacified the oppressed; it could only slow the workers' revolt and subsequent freedom from oppression.

Religion plays an important role in social conflict, according to conflict theorists, because the dominant religion in society usually represents the dominant social class. The dominant religion supports the status quo, offers justifications for existing class lines, and operates to make working-class and lower-class people satisfied with their condition. The conflicts among social classes may lead to the emergence of specific religious groups that represent specific class interests. One example is the People's Temple of the late Reverend Jim Jones, which reportedly drew much of its membership from the ranks of the disaffected and alienated in society. According to the conflict perspective, a consequence of the development of class-based religions is that often what appear to be religious differences between opposing religious groups (for example, Protestants and Catholics in Northern Ireland) actually turn out to be class differences. One religious group represents money and propertied people and the other represents the working class. Conflict, then, is a central element in the discussion of religion, according to this viewpoint.

General information about church membership and religious opinions and beliefs comes from *reactive surveys* (public opinion polls) and *nonreactive surveys* (analysis of membership records and historical data). Glock and Stark's study was a *reactive questionnaire survey* of a *random sample* of Northern California church members. Because only 50 percent of the Catholics and 70 percent of the Protestants returned their questionnaires, the researchers compared what they knew about those who didn't return the questionnaires with what they knew about those who did, and they concluded that the bias represented by the nonreturns was negligible. Are there other ways in which the sample might be biased?

SUMMARY

Religion has been the subject of this second chapter on social institutions. We have suggested that institutions emerge because of societies' important needs. In the case of religion, these needs involve renewing hope, giving meaning to life, explaining the unknown, and dealing with death. Religion may be involved in other functions such as socialization, social control, and the promotion of solidarity and unity. Religion also plays an important role in social conflict: Marx felt that religion distracted people from their real problems, making it less likely that these problems would ever be corrected. Religious groups may battle each other or support one social class or ethnic or racial group to the disadvantage of another.

After examining the functions of religion, we dealt with the diversity of religious groups by suggesting several ways to distinguish among various types of religious organizations. Finally, we suggested that religion in America is characterized by differential involvement in religious activities, much religious diversity, an apparent loss of functions, and some interesting trends on the topic of secularization.

Religious groups may be similar in their beliefs and ceremonies, or they may be vastly different. In the following reading, Michael Watterlond describes the Holy Ghost People of Appalachia, a religious group that is somewhat off the beaten track.

TERMS FOR STUDY

Apollonian ceremony (259)	Dionysian ceremony (259)
church (260)	religion (255)
cult (260)	sect (260)
denomination (260)	secularization (264)

Reading 8.1

THE HOLY GHOST PEOPLE

Michael Watterlond

Writer Michael Watterlond's description of the Holy Ghost People is interesting on several counts. The practices of this religious group appear quite bizarre, something most of us haven't experienced. Comparing their activities with the formal ritual of the Catholic Church makes us wonder if both could possibly be part of the same social institution. And yet, if we study the Holy Ghost People carefully, we should be able to discern certain common elements that this religious group has in common with the mainstream religions with which we are more familiar.

For the Holy Ghost people of Appalachia, it is a fragile fabric that separates this world from the next.

They are serpent handlers.

They take up poisonous snakes in church—timber rattlers, copperheads, even cobras—and they drink strychnine. They handle fire, using coal oil torches or blow torches. And they speak "with new tongues." They lay hands on the sick, trusting that—God willing—the sick shall recover. And when necessary, they cast out devils in Jesus' name.

"We live in the world, but we are not of the world," they say repeatedly during their twice-weekly church services, as though this were the statement that most distinctly defines them from us.

That these people are of another plane is unquestionable. They exist wholly in the thin, almost dimensionless region of "hard doctrine," of "getting right with Jesus" and most importantly of what they offhandedly call "this thing" or "this."

"This" is the featureless article of language they use to encompass the fiery, dramatic, even deadly practices that set their religion, their lives, and their select social grouping apart from most of Christian culture.

They belong to a variety of independent, fundamentalist sects called loosely Jesus Only. And they belong specifically to churches that subscribe to a doctrine based on what they call "the signs" or "the signs following."

The basis for these doctrines is a stone-hard reading of the last few verses of the Gospel according to Mark and other passages in the New and Old Testaments as presented in the King James version of the Bible, an English translation published in 1611.

In the pertinent section of Mark, Jesus has already risen from the tomb and appeared to several characters, including his disciples. As he is about to ascend, he issues these final pronouncements:

Mark 16:17 "And these signs shall follow them that believe; In my name shall they cast out devils; they shall speak with new tongues;

Mark 16:18 "they shall take up serpents; and if they drink any deadly thing, it shall not hurt them; they shall lay hands on the sick, and they shall recover."

Reprinted from the May issue of *Science 83*.

The Book of Daniel, in which Shadrach, Meshach, and Abednego are cast into the fiery furnace, is the basis for fire handling.

The fundamentalist serpent handlers take these passages as absolute. That these verses in Mark do not appear in the earliest extant texts is of no importance to them at all. Only the King James Bible counts. It acts as a major initiator of personal action, behavior, and decision making in their lives. It defines them.

Mary Lee Daugherty, formerly a professor of religion at the University of Charleston, estimates that there are now about 1,000 members of serpent-handling sects in West Virginia. She says numbers have dwindled recently because high unemployment has forced many members to migrate to urban areas. Since no organization links these congregations, a total member count would be guesswork.

Though there are local laws against handling dangerous animals in public in many states where serpent-handling churches exist, the laws aren't always strictly enforced. There are congregations in most areas of Appalachia, stretching from West Virginia south through Kentucky, Tennessee, the Carolinas, and into Georgia. Migrations from hill country into some of the urban areas of the Midwest have led to serpent-handling churches in nonrural areas such as Columbus, Cleveland, Flint, Indianapolis, and Detroit. The Full Gospel Jesus Church of Columbus, under the leadership of Willie Sizemore, has purchased and renovated a building in one of the city's industrial areas.

George Hensley, who began the practice of snake handling in rural Tennessee around 1909, could not have predicted its spread to urban environments. Hensley, a Holiness circuit preacher, died of a snake bite near Atha, Florida, in 1955. He is not revered by these people and is not considered to have been a prophet of any sort. His death, however, does illustrate a point one frequently hears in Jesus Only congregations: The Bible says to take up serpents; it doesn't say they won't bite.

"Make sure it's God," a deacon warns. He holds the microphone close to his lips like an entertainer. His hefty coal miner's shoulders straighten as he whips the mike cord away from his feet. "Only God can do it. Not me. Not you."

The loud, mechanical sounding buzz from the white pine serpent box makes his point unmistakable. The box this afternoon contains four timber rattlers and a rosy, velvet-textured copperhead.

"Death is in there," he says. "Don't go in the box unless it's with you."

By "it," the deacon means "the anointing"—the protection and spiritual direction of God that is manifest in what believers report as physical and emotional sensations.

"It's different for everyone," according to Sizemore. "Some people get a cold feeling in their hands or in their stomachs. Some don't."

Investigators in the past have reported that members say the anointing has a different and, for them, recognizably distinct sensation for each sign. The anointing to handle serpents may appear as a tingling, chilled feeling in the hands. An anointing to drink strychnine may appear as a trembling in the gut. The reports vary just as the sensations vary.

The deacon's warning about acting on the signs only if the anointing is present is a typical part of services, and the saints, as the members call each other, enthusiastically applaud. As he turns from the pulpit, the underlying rhythmic drum beat that has throbbed subliminally in the background becomes distinct and powerful. It is joined by electric organ, electric guitar, tambourine, and a room full of the clapping hands, stomping feet, and hearty, frantic voices of 40 saints.

As the service gets underway, the saints stand, clapping with the music, and walk slowly toward the open area between the pulpit and the first row of pews. The movement forward and together is called a "press." There is some feeling that the spiritual power of the group is concentrated by this gathering together. Should a member be bitten by a snake or "get down on strychnine," the gathering of saints around the victim is also called a press. Members refer to this communal praying and support of the stricken person as a "good press."

The services themselves have an informal structure that begins with loud, insistent singing. Most songs are belted out in four-four time, and each lasts as long as 20 or 30 minutes. After a few songs there will be requests for prayers or healings and testament. The prayer requests may be for "sinner children" or for members of the church who are sick or injured. There will be more songs before the sermon. Services last from two to five hours depending on the time of year and work schedules in the mines.

The signs may show themselves at any time during the service but are most apt to be manifest in association with the driving rhythms of the music. The enthusiasm may verge on violence. One woman slumps to the floor, "slain in the spirit," they say. As she trembles in front of the pulpit, other saints close in on her screaming, "JESUS! JESUS! JESUS!" in her ears.

"I saw one man get up and run around the inside of the church more than 50 times," says sociologist Michael Carter of Warner Southern College in Lake Wales, Florida, an experienced observer of the sect.

Members wail and shake and lapse into the unintelligible, ecstatic "new tongues" of glossolalia. Each member has his or her own style of speaking. As one saint sails into a new oration of tongues, another clamps his hands to her head and speaks in tongues himself; the ecstasy spreads like contagion. "He's translating," Carter observes.

There are two camps of psychological explanation and description of such services. Some researchers theorize that the wild activities are attempts to transcend reality, while others believe the point is self-actualization. Possibly both are right. It takes only a brief conversation with participants to learn that their daily life is one ongoing seance. They see God's movement and directives in every action, object, or thought.

Jerking as though his head were being battered by some unseen opponent, one brother moves across the floor toward the canister of coal oil next to the pulpit. He lights it quickly and thrusts his hands into the fire as though he were washing himself in flame. Next to him a young woman has been dervishlike for several minutes. She continues spinning under the tent of her long brown hair for nearly half an hour. In the back, the children play tic-tac-toe, practice spelling exercises, or sleep, unconnected to the proceedings.

Investigation into psychological and sociological aspects of fire and serpent handling has been conducted sporadically since the 1940s. A paper presented last year by Carter and sociologist Kenneth Ambrose of Marshall University in Huntington, West Virginia, at the Fifth Annual Appalachian Studies Conference, attempted to determine the satisfaction church members derive from such activities. They interviewed members of an urban congregation to compare the rewards of church versus non-church activity.

Results of the study indicate that taking part in the signs gives these people personal reward equalled in no other aspect of their lives. Participation in "signs of the spirit" were statistically evaluated by Ambrose and Carter to reveal the highest level of satisfaction for members when they spoke in tongues. Handling serpents and handling fire were also rated highly, while drinking strychnine—clearly the most deadly sign—provided more shallow levels of satisfaction.

"When you drink strychnine," members say, "you're already bit"—your fate has been decided. Strychnine, commonly a white or colorless powder, causes a warm feeling in the gut, tingling sensations, and muscle spasms. In poisonous doses—15 to 30 milligrams for a human—the spasms can be severe enough, as the serpent handlers say, to "snap muscles right off the bone" and stop the heart. However, it is possible to ingest a considerable amount of strychnine and not exhibit the symptoms of poisoning. The drug does not accumulate in the body because it is rapidly oxidized in the liver and excreted in urine. The likelihood of developing a tolerance to it is remote.

While members of serpent handling churches—as well as most Jesus Only churches—forgo worldly diversions such as films, television, and politics, they do hold down jobs. Sizemore, for example, is a factory worker. But they consider mingling with the world on the job part of their earthly burden.

Since most members have had to face stern criticism and even ostracism by friends and relatives—including husbands and wives at times—they are even more strongly pushed into this cluster of supportive friends at church, where they speak of themselves as the chosen people. The social and psychological bonds are reinforced more or less constantly by hugging, touching, kissing. This creates a rich, meaningful world for them—a sense of being special.

Also, the pastor rides herd spiritually on the congregation. During service he will approach a member he suspects (or "discerns," as they would say) is backsliding, or going to other churches. He will preach to that member eye-to-eye, only the microphone between them, and talk in generalities about worldliness, sin, or traipsing around.

"I don't hold with people going from one church to another," says Brother Sizemore.

It is sometimes a strenuous task for the outsider to pry himself loose from all this music and high spirits and remind himself that these people take poisonous reptiles out of the countryside, bring them into their churches and drape them over their bodies, wear them in bundles on their heads. While the churchgoers are enthusiastic and emotional and committed, they do not appear to be disturbed and certainly not suicidal.

"I'm just as afraid of serpents as anybody," one 22-year-old West Virginia man says. He has attended these churches since childhood and has left the church his mother attends in recent years because that congregation has stopped handling serpents. "I'm afraid when I am in the flesh," he explains, "but when it is the spirit, there's nothing. I'm just not afraid."

He pulls up his sleeve to illustrate the critical point he wants to make: that these serpents are real and that their venom is real.

"It bit me here," he says, pointing to his wrist. "And it swelled up so much that the skin just pulled apart up here."

"When Richard died," one member says, "his whole arm split open from the shoulder down to the elbow."

Richard was Richard Williams. He died in 1974 after being viciously bitten by a huge eastern diamondback. Like most members who suffer bites, Williams refused medical aid and waited for fate to reveal itself.

The people of this church still talk about Richard Williams' anointing and about the serpent-handling feats he performed. Pictures on the walls of the church show him with his face in a mound of snakes; lying with his head resting on them like pillows, stuffing them into his shirt next to his skin. In a photo of Williams and the snake that killed him, it appears as if he is holding the felled limb of a tree rather than a snake.

"What really killed him was that the serpent got his vein when it hit the second time," one member says.

"You just can't tell," Willie Sizemore says, "you just have to make sure you have the anointing."

The folk myth that individuals who suffer repeated snakebites develop an immunity to the venom is viewed skeptically by Sherman A. Minton, Indiana University School of Medicine microbiologist and toxicologist. Minton says that such immunity is developed rarely and only when regular, gradually increased doses of the venom are administered. He points out that many people have allergies to venom, making successive bites more painful and causing greater swelling, asthma, and other symptoms.

As one snake expert puts it, however, the chances of dying by snakebite in the United States are comparable to the probability of being struck by lightning. One 10-year study showed that 8,000 venomous bites occur each year resulting in an average of only 14 deaths.

There are many theories about why snakes strike the saints infrequently. It is possible that given many warm-bodied targets pressing closely around it, a snake lapses into a sort of negative panic, a hysteria that makes it unable to single out one target.

Some observers have reported that church members' hands feel cold to the touch after handling fire or snakes. "I have felt their hands after serpent handling or fire handling," says Ambrose, who has observed these services for 15 years. "Their hands are definitely cold, even after handling fire." This would correspond with research in trance states involved in other religious cultures. It would also account for the vagueness of memory, almost sensory amnesia, that researchers have reported in serpent handlers as well as fire handlers. In his doctoral dissertation at Princeton University in 1974, Steven Kane reported that in North Carolina serpent handling congregations, young women were known to embrace hot stovepipes without injury or memory of the event. It has also been suggested by some observers that cold hands on the body of the snake would camouflage the touch and prevent it from feeling the handler.

Retired sociologist Nathan Gerrard, formerly of the University of Charleston, observed serpent handlers for seven years in the 1960s. By administering portions of a psychological test to measure deviate personalities, Gerrard concluded that serpent handlers had healthier attitudes about death, suffered less from pessimistic hypochondria, and generally seemed better adjusted in certain ways than "conventional" churchgoers he used as a control group.

However, one of Gerrard's conclusions at the time was also that serpent handling represented a sop to desperation. He called them the "stationary working class" and attributed their stern doctrine to highland ignorance and poverty—a fatalistic creed that offered death and salvation as the only way out of West Virginia. Since that time, however, Jesus Only churches have spread out of Appalachia as economic hardships pushed followers into urban centers perched on the edge of the Midwest.

Also, the members of the Full Gospel Jesus Church of Micco, a creek-bank town about 10 miles from Logan, West Virginia, are distinctly not the archetypal mountain folks. They come to church in well-polished, late-model automobiles and dress like middle-class people in most parts of the country.

While it may be difficult for outsiders to say that it is simply common faith that keeps these groups together, that would be the first thing they would say themselves. And this is a self-definition that has been immensely important to Western civilization.

"You can clearly see early Christianity as having very strong sectarian overtones," says Robert Bellah, a sociologist at the University of California, Berkeley. "Without that at the beginning, there would not have been any Christian church."

"Sect religions," Bellah says, "are most apt to occur in relatively low-status groups, relatively low-educated groups where a combination of intense religious experience and rather high group discipline create a kind of separate world in which the people in the sect live.

"They largely reject the surrounding culture," Bellah explains, "rejecting many of the prevailing cultural forms. In other words, the whole round of life tends to be bound up' in the sect itself. Social contact is limited by the sect, and the meaning of most things that one does derives from the sect."

Bellah makes the distinction that churches involve a structure which attempts to en-compass the whole of society. They include a range of social classes and do not op-pose the dominant social power but view themselves as having influence on how that power is exercised.

"The church accepts the culture while working within it," Bellah says. "A sect is ex-clusive."

It is the current public concern about total-commitment sects that has many people wondering about what have come to be called cults these days.

"'Cult' is really not a sociological category," Bellah explains. "It is a pejorative, popu-lar term that we use for groups that are unfamiliar to our culture." Generally, the word is used to describe groups that are non-American in origin or are aberrantly individualis-tic—"in that they have been created by some 'kook,'" Bellah says. "Like Jim Jones."

"We are not a cult," Sizemore says forcefully. "If Jim Jones had been right with Jesus, those people wouldn't have died."

And still, outsiders want to know why anyone handles serpents.

"Because it is written," says Bishop Kelly Williams of the Jesus Only Church of Micco, West Virginia. "The main purpose in our doing this is to obey the Word of God." He points out that it really does not matter that the snakes symbolize Satan.

"Now, we don't think that everyone has to take up serpents," Williams says. "It doesn't say that all signs will follow all believers.

"You," he says to a visitor. "You might only speak in tongues. It depends on how God moves."

"Everyone knows that it is the nature of those serpents to bite you," he nods toward the buzzing pine box of copperheads and rattlers. "But you saw last night that God gave us a victory over those serpents. They were new serpents. They'd never been handled."

In fact, in a quiet evening before, Williams had gone to the box, lifted out a rattler and held it close to his body. He stared down at the serpent in his palms, smiled calmly for about three minutes, and set the snake back down in the box.

"You've seen that there is no one way to handle serpents," he says. "You been around enough to know that. There's no trick."

The styles of serpent handling are as various as the vocabulary of new tongues. On one Sunday afternoon, several men lift up the box, shake it hard and dump a mass of rattlers and copperheads onto the wooden floor without much caution. One man reaches into the tangle of scales and rattles and pulls up a timber rattler about four feet long. The other snakes, as well as a microphone cord, are tangled up with it, so he just shakes the whole mess until the excess snakes drop off. They coil stunned on the floor, as if paralyzed by the bad manners of it all.

The serpents are passed between hands. Loose copperheads wander around the box, uncertain of any particular route of escape. The heavy beat of gospel music picks up and the floor vibrates, the congregation claps and wails and sings.

In the back row of pews a girl about 11 turns back over her shoulder to talk to the visitor. She is playing hairdresser with another girl and holds an unfinished braid in her hands. She smiles halfheartedly and rolls her eyes heavenward as though her patience were limitless and unconditional.

"Boring," she says, "isn't it?"

QUESTIONS 8.1

1. Which of the functions of religion discussed in this chapter are supplied by the religious practices of the Holy Ghost People?

2. What would Karl Marx say about the Holy Ghost People? How would he explain their behavior?

3. Analyze the Holy Ghost People in terms of religious organizations—which category are they and why?

4. It is obvious that the Holy Ghost People are different from most, but what *similarities* do you find between them and other religious groups?

NOTES

1. From "An Accident" by G. K. Chesterton in *Tremendous Trifles* (New York: Sheed and Ward, 1955), pp. 29–33.

2. Andrew Greeley, *Religion: A Secular Theory* (New York: The Free Press, 1982), ch. 1.

3. Much of our discussion in this section is drawn from Charles Glock and Rodney Stark, *Religion and Society in Tension* (Chicago: Rand McNally, 1965). This definition of religion is close to that of Glock and Stark. In their book, Glock and Stark trace the background of their definition, which includes the works of Durkheim, Yinger, Parsons, Nottingham, and Williams.

4. See "Religious Community, Individual Religiosity, and Health: A Tale of Two Kibbutzim," by Ofra Anson, Arieh Levenson, Benyamin Maoz, and Dan Bonneh, *Sociology* 25 (February 1991), pp. 119–132.

5. This project was researched by the National Opinion Research Center in Chicago, directed by Andrew Greeley. See Andrew Greeley, William McCready, and Kathleen McCourt, *Catholic Schools in a Declining Church* (Mission, Kan.: Sheed and Ward, 1976).

6. This comes from Andrew Greeley's book *The Denominational Society* (Glenview, Ill.: Scott, Foresman & Company, 1972), ch. 2.

7. The categories and quotations discussed in this paragraph and in Table 8.3 come from Andrew Greeley's book *The Denominational Society*, pp. 23–24 (see Note 6).

8. These paragraphs on types of religious organizations make use of material from Glock and Stark, *Religion and Society in Tension*, ch. 13 (see Note 3), and from J. Milton Yinger's book, *Religion, Society and the Individual* (New

York: Macmillan, 1957), ch. 6, especially pp. 142–155.

9. George Gallup conducts a yearly poll on church attendance.

10. See "Durkheim, Suicide, and Religion: Toward a Network Theory of Suicide," by Bernice Pescosolido and Sharon Georgianna, *American Sociological Review* 54 (February 1989), pp. 33–48.

11. Some of the statistics in this and the following paragraph come from Gallup's study of religion in America in 1983. Also see Will Herberg's interesting book, *Protestant-Catholic-Jew* (Garden City, N.Y.: Doubleday, Anchor Books, 1960), ch. 1. Also see Glock and Stark, *Religion and Society in Tension*, ch. 4 (see Note 3).

12. Herberg, *Protestant-Catholic-Jew,* ch. 11, especially p. 268 (see Note 11).

CHAPTER NINE

Institutions: Economic and Political

TWO RESEARCH QUESTIONS

1. Do large U.S. companies (like IBM, Exxon, Ford, and General Motors) make more of their money in the United States or abroad?

2. From what social class come the people who run the country, who have the highest positions in government? Are they selected from all walks of life, from the middle class predominantly, or from some other group?

IN CHAPTER SEVEN we defined social institution as the norms, roles, and patterns of organization that develop around societies' central needs or problems. In Chapters Seven and Eight, family and religious institutions were examined. In this chapter we turn to political and economic institutions.

Institutions are interrelated, with change in one inevitably affecting another. The relationship between political and economic institutions is perhaps closer than most. Looking at their central functions may help us understand this better. Economic institutions arise out of societies' efforts to produce and distribute goods and services. Political institutions refer to the social relationships surrounding the assignment and use of power. And the connection between the two? Goods and services are limited and unequally distributed in society. The distribution of goods and services is determined by those with

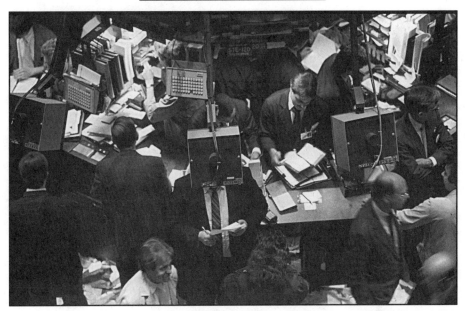

Elliott Erwitt/Magnum Photos, Inc.

the power to make such decisions. Who has the power to make such decisions? Probably those who benefit most from the existing distribution of goods and services. So we see that the political distribution of power and the economic distribution of goods and services are closely related. Remember Murphy's Golden Rule: Whoever Has the Gold Makes the Rules.

ECONOMIC INSTITUTIONS

Economic institutions arise out of societies' efforts to produce and distribute goods and services. Important issues surface: What is the meaning and value of work? How does a society deal with the situation in which only limited supplies of goods are available and demands for those goods by citizens are unlimited? Is unequal distribution of goods and wealth in society beneficial or detrimental? Several quite contrasting views exist regarding the most appropriate economic system to deal with such issues.

Capitalism

Capitalist systems cherish the ideas of private property, the profit motive, and free enterprise. *Private property* means that property, especially that used to produce goods and services, is held by private individuals, not by the state. *Profit motive* means that individuals are motivated by the idea of making a profit; if they produce a commodity that people want and that is a better product at a lower price than their competition's, they will capture the market. *Free competition* means that each private entrepreneur is free to compete with others in a marketplace unencumbered by any rules and regulations other than fair play. The market and the economy are self-correcting: Buyers make free choices among goods and services; bad products aren't purchased; workers paid too little go elsewhere or do poor work, whereas those paid too much cause goods to be priced too high to be sold. Good businesses succeed, bad ones fail. Build a better mousetrap at a good price, and the world will beat a path to your door.

However, not many—if any—capitalist systems really work this way. Take the United States, for example. Private property? Some property is publicly held, and many privately held businesses are owned by many scattered stockholders who have little say in ongoing activities. Further, the government has a great, although indirect, influence; many businesses are dependent on the government aid they receive through subsidies and tax breaks. Profit motive? The profit motive is certainly strong. However, there are many unprofitable but necessary businesses that are supported by the government through tax dollars and are not allowed to fail. Free enterprise? Laws regulate prices, minimum wages, and fairness in hiring. Government regulations limit or control entry to occupations, and buyers' "free" choices are limited by what they can afford. Domination of whole segments of the economy by one or two large corporations allows them to "create" markets.

Socialism

In the 1850s, Karl Marx saw much about capitalism that he didn't like. Workers were exploited, and capitalism seemed to benefit only the capitalists, not the rest of society. The apparent evils of capitalism led Marx to suggest a socialist system in which property is collectively, not privately, owned. Goods and services are produced according to people's needs, regardless of profitability. Centralized planning determines what is produced and how much. Wages and benefits should be equally distributed, and there should be no accumulation of private wealth. As with capitalism, a pure socialist system is difficult to find. In the former Soviet Union, for example, there was some private property and some private free enterprise, and income differences between the managerial level and workers created unequal distribution of wealth. In fact, most economic systems represent a mix of the extremes of capitalism and socialism. In many societies you find private property and the profit motive, government regulation of the economy, and government ownership of key industries such as mass media, transportation, and health care.

Economic Change and the American Economy

The evolution of economic systems has been remarkable. In early societies, people spent all of their time finding food. There was little specialization or division of labor—everybody did most everything. Specialization gradually increased as societies became larger and more complex. By the time of the Industrial Revolution in the 19th century, the division of labor was extensive; workers would perform one small task continually for the whole of their workday. One of Marx's concerns, however, was the alienation, the meaninglessness, and the powerlessness a worker must feel when he or she is a powerless cog with no meaningful input in the final product. Work is a major and important part of life, but if one has no sense of accomplishment—if there is no meaning or creativity in work—then it may end up being unsatisfying and demoralizing. Some businesses have approached this problem by allowing workers to change jobs or follow the product through from start to finish, by humanizing the workplace, and by giving workers greater voice in management.

The increasing specialization and division of labor caused businesses to become bigger and bigger. Does this also mean better? Specialization and size did mean that tasks earlier unimagined could now be performed. Can you imagine, for example, primitive hunters and gatherers reaching the moon? Or even wanting to? But it has also led to some criticisms and problems (as we discussed in Chapter Four in the section on large organizations). One of the dominant features of the American economy in the mid- and late-20th century is bigger and bigger businesses and corporations that buy up other corporations and become conglomerates. Recently, mergers have become more extensive than ever before. Nine out of the ten largest mergers in United States history have occurred since 1980. And one or two of these large corporations can influence or dominate large segments of the economy. The dominance of big organizations such as IBM, AT&T, and General Motors can be illustrated in the following ways: First, they employ many individuals and have great wealth. Table 9.1 shows the sales figures and number of employees of the 15 largest U.S. companies in 1993, 5 of which, incidentally, are oil companies. The 500 largest industrial corporations employed 11.5 million people in 1993 and had sales of $2.37 trillion. There is clearly a tremendous concentration of wealth at the top.

Second, many of the large corporations are called **multinationals** because they go beyond national boundaries and have connections throughout the world. Many are truly huge in scope. The total revenues from just the top 100 United States–based multinationals in 1992 were about $1.8 trillion. Table 9.2 shows the ten largest United States–based multinationals in 1992. Note that the multinationals are huge; Citicorp's assets, for example, are $106 billion. Notice also that many of the multinationals make a substantial part of their profit (especially when compared with assets) *outside* the United States. This relates to the first research question asked at the beginning of the chapter. Oil companies like Exxon and Mobil make more money abroad than at home, as does computer giant IBM. Car companies like Ford and General Motors face much greater competition from companies in other countries, and their foreign sales are a smaller percentage of their total sales.

The behavior of the multinational corporations raises interesting issues. Their wealth is often greater than that of the countries in which they do business. Multinationals may encourage development in underdeveloped coun-

Table 9.1   The Fifteen Largest U.S. Companies, 1993

Company	Sales (billions)	Profits (billions)	Assets (billions)	Employees
General Motors	$134	$2.4	$188	710,800
Ford	109	2.5	199	322,200
Exxon	98	5.3	84	91,000
IBM	63	−8.1	81	267,196
General Electric	61	4.3	252	222,000
Mobil	57	2.1	41	61,900
Philip Morris	51	3.1	51	173,000
Chrysler	44	−2.5	44	128,000
Texaco	34	1.1	27	32,514
Du Pont	33	0.5	37	114,000
Chevron	32	1.3	35	47,576
Procter & Gamble	30	−0.6	25	103,500
Amoco	25	1.8	28	46,317
Boeing	25	1.2	20	134,400
Pepsico	25	1.6	24	423,000

Adapted from *Fortune*, April 18, 1994.

Table 9.2   The Ten Largest U.S.–Based Multinationals, 1992

Company	Foreign Sales (billions)	Foreign Sales as % of Total	Total Assets (billions)	Foreign Assets as % of Total
Exxon	$79	77%	$48	57%
General Motors	42	32	44	23
IBM	40	62	47	54
Mobil	39	68	23	56
Ford	36	36	49	27
Texaco	26	53	11	37
Chevron	21	44	12	34
Citicorp	19	60	106	47
Du Pont	17	47	16	41
Procter & Gamble	15	50	10	43

Adapted from *Forbes*, July 19, 1993.

tries, but whether their efforts are welcome or not is much debated. They have been charged with meddling in local politics to encourage a friendly climate for their investments. They often bring their own employees rather than using native workers, which is of little help to the local employment situation. They often produce products for their own economy that may be of little use to the local society.

POLITICAL INSTITUTIONS

The social relationships surrounding the assignment and use of power is the central issue in the study of political institutions. **Power** refers to the ability of one party, either an individual or a group, to affect the behavior of another party. (We discussed power in Chapter Five as one of three variables, along with wealth and prestige, that determine one's socioeconomic status or rank in society.) Here we find that power is the main force in political institutions. Power is probably a factor in all social situations: The doctor has power over the patient, the teacher over the student, the parent over the child; and even in dating situations, one party often has superior bargaining power. In this chapter, however, we focus on the power that has a more general impact—power that influences actions and decisions at the societal level.

There are three major types of power: *authority, influence,* and *resource control.*[1] These types of power generally overlap; having one type means you'll probably have another type as well. **Authority** refers to socially approved power. It has legitimacy, meaning it is socially acceptable. People believe it is right and appropriate to drive their cars on the right-hand side of the road and to stop at stop lights, and they would probably do so even if no one were around. Authority is often impersonal in the sense that it is linked to position rather than to the character of the occupant. Students pay attention in class not because of the instructor's fantastic personality, but because the "teacher" position holds a particular authority. We said that authority was viewed as legitimate—people are socialized to obey the rules. At base, however, authority rests on the threat of force and coercion. The relationship is complicated: Does one go along because one wants to, or because of fear of the consequences? Would you stop at the stop light if no one were looking?

Authority comes in different shapes, and Max Weber has identified three specific types: *traditional, legal-rational,* and *charismatic.* **Traditional authority** means that power is granted according to custom. Leaders are selected by inheritance, and people go along because "it has always been done this way." Hereditary monarchies and tribal chieftainships are examples. The leaders appear to have much power, laws tend to be understood but unwritten, and there is a sacred quality to the whole operation. Historically this has been a common type of authority but is increasingly difficult to find today. **Legal-rational authority** is more typical of modern societies. Leaders are selected to fill positions, the boundaries of which are carefully designated by a formal set of rules and procedures. The *position* has the power rather than the person occupying it. If one goes beyond the boundaries, something bad may result. The example of Watergate and Richard Nixon is typically used to show how the legal-rational system is supposed to work; he went beyond the legitimate boundaries and was forced to leave office. Legal-rational authority is typical not only of governments but of businesses and other types of organizations as well.

Charismatic authority is based on the unique personality of a particular person. This person—Mao, Gandhi, Martin Luther King, Jr., or Castro, for example—has authority over others based on his or her personal appeal. Charismatic authority is very unstable; it tends to emerge out of crisis and is tied to a particular person. **Pseudocharisma** is a recent addition to Weber's idea.[2] The term refers to public figures who appear to have charisma but in fact have a packaged and carefully created image manufactured by public-relations techniques. Making politicians into TV celebrities is a clear example of the development of pseudocharisma.

Influence refers to subtle, informal, indirect power based on persuasion rather than coercion and force. Lines of influence are very widespread and less easily recognized than those of authority. Parents influence behavior through a look or tone of voice, or because their children should "know how they would feel." Admirers of rock stars and other entertainment personalities may be influenced to ingest particular restricted substances or wear wild and peculiar garments. Friends exercise influence, and so do the mass media. Holding an important position like "member of the board of trustees" or "chair" naturally enhances one's influence. It follows that authority and influence often go together, but occasionally their boundaries may differ. A person may have narrow authority but much influence, or vice versa.

Power as **resource control** refers not to things people do, but to their possession or control of valued items. In this case, power often represents a capacity, potential, or threat to act rather than actual decision-making behavior. If one possesses great wealth, control of a major energy source, or ownership of a string of radio and TV stations, one would influence others' behavior whether one intended to or not. Think of the amazing power shift in the direction of the oil-rich Middle East in recent years.

Distribution of Power

Power is stratified in society; it is distributed unequally. Those few at the top who have it are called elites; those without it are called the masses. But beyond this, how is power distributed? What type of people have it? What is the relationship between elites and the masses? Three theories provide differing answers to these questions, and as with most competing theories, there is evidence to support and refute each of them. These three approaches are called the **pluralistic model**, the **elite model**, and the **class model**.

The Pluralistic Model   Pluralists believe that power is diffuse, spread across many diverse interest groups. An individual's interests are protected because everyone is potentially represented by one or more interest groups. Continuous competition and conflict for power among these interest groups leads to a balance—an equilibrium—and thus no radical or rapid social change takes place. The groups resolve differences by bargaining and compromise, and because they are specialized, no single interest group has absolute authority. Some have political power, some have economic power, and some have military or religious power. Political parties, voluntary associations such as the PTA and the League of Women Voters, all levels of government, and finally even the voters—all exercise power.

This traditional pluralist approach, which sees the masses producing interest groups that deal with the state, has been modified in a modern version called elite-pluralism. **Elite-pluralism** recognizes the fact that mass political participation is impossible in modern complex societies and that small elite groups do develop and make major decisions. It is suggested, however, that these elites have competing interests and tend to balance or neutralize each other so that no one elite group gains dominance.

Does the model fit the facts of modern society? Critics have some doubts. They suggest that most people are not represented by interest groups, and those who are don't have much voice unless they are part of the leadership. Interest groups that do exist tend to be dominated by business and monied interests. Finally, the balance of power suggested by the pluralists just doesn't exist, argue the critics; the power differences among various interest groups are vast.

The Elite Model The elite model, which is sharply different from the pluralist model, sees relatively unrestrained power in the hands of a few who rule the masses, who, in turn, are apathetic and incapable of self-rule—incapable because they are poorly trained, uninformed, and unorganized. Of the early elite theorists, Pareto viewed elites as the highest achievers in society; they govern through physical coercion (the "lions") and through cunning and intellectual persuasion (the "foxes"). Mosca felt that elites gain and keep power because they are organized (whereas the masses are not), they have particular personal attributes (strength of personality or intelligence), and they control valued resources or "social forces." Michels viewed rule by a few as an inevitable consequence of large complex organizations. Because a large number of people can't make decisions efficiently, a few are given the power to make decisions on behalf of the whole, and a leadership group develops. Once in power, the elite spends more and more of its time and energy maintaining its power rather than concentrating on other organizational activities. For example, think of how much of a politician's time (from the president on down) is spent on activities designed to keep himself or herself in office.

Modern elite theorists see the masses as manipulated and exploited by an elite motivated by self-interest. Whereas the earlier theorists felt the development of an elite to be inevitable, modern theorists think that such a condition can and should be avoided. Theorists today see the elite as a cohesive group of people bolstered by their control of key resources, including wealth, government authority, and communication facilities. In sociologist C. Wright Mills's view, the top layer of power in society is in the hands of "the warlords, the corporation chieftains, and the political directorate," who work together to form a power elite in America.[3] According to Mills, decisions are made in this country by a few people, and they govern a fragmented mass of people who have no power. Between the two is a group whose power is semiorganized and stagnant. The middle is there, but it neither represents the masses nor has any real effect on the elite. The elite is highly organized and made up of military leaders (admirals and generals), politicians, and business leaders. Mills suggests that a system in which so much power is held by a few who are not responsible to anyone but themselves is both immoral and irresponsible.

Does the model fit the facts? Elites unquestionably exist, and the forces in large groups and organizations work much as Michels predicted. But critics raise specific questions: Must elites inevitably emerge? Are elites cohesive, and do they cooperate to the extent suggested by Mills and others? Are elites only self-serving or do they work in the public interest? And are the masses as totally powerless as the elite model predicts?

The Class Model The class and elite models are similar in that they see power concentrated in the hands of a few. They differ in their description of the elite group and its origins. The class model brings us back again to the ideas of Karl Marx. Marx starts with the idea that economic institutions are the key to society; all else stems from such economic activities as finding food, shelter, and the other necessities of life. Classes emerge based on the means of production, and in industrialized societies, two classes will dominate: the bourgeoisie, or capitalists, who own the means of production, and the proletariat, who own only their own labor. Classes become political groups when class members realize that they need to gain and use power to protect their common interests. Continual domination and exploitation of the proletariat by the bourgeoisie means continuing class conflict. Because economic matters are so important, economic domination is the key to power and control in other noneconomic elements in society.

Contemporary class theorists grant that the workers' revolution hasn't happened and that the classes are not as polarized as Marx predicted. In fact, the classes have become increasingly complex and diverse; in many societies the growing middle class has become a dominant feature. In spite of these deviations from Marx's predictions, contemporary class theorists say that the current corporate elite has the power of a ruling class.

Answering the second research question posed at the beginning of this chapter—Who runs the country?—is no problem for William Domhoff. In his book *Who Rules America?*, Domhoff describes a governing class that "owns a disproportionate amount of a country's wealth, receives a disproportionate amount of a country's yearly income, and contributes a disproportionate number of its members to the controlling institutions and key decision-making groups of the country." Excessive wealth and income are the key; having these elements, which other classes do not, allows the upper class to *control* other individuals and institutions. This upper-class elite has well-established ways of training and preparing new members for future service. Bright young people are processed into the upper class by means of "education at private schools, elite universities, and elite law schools; through success as a corporate executive; through membership in exclusive gentlemen's clubs; and through participation in exclusive charities."[4]

Domhoff holds that the leaders of the executive branch of the federal government are either members of the upper class or former employees of institutions controlled by members of the upper class. He cites an impressive array of evidence that describes who has the wealth and how it is used to control power. Domhoff also traces the occupants of various government positions and finds, for example, that members of the upper-class power elite dominate the president's cabinet, especially in the departments of State, Treasury, Defense, Commerce, and even Labor. Of the 13 men who were Secretary of Defense or Secretary of War between 1932 and 1964, eight have been listed in the *Social Register*, and six of the eight people who were Secretary of State in the same period were members of the upper-class power elite. Through the CIA and other agencies, the upper class has manipulated intellectuals by giving money to individuals, schools, and foundations to exert influence, to start organizations, to hold conferences, and to publish magazines and books. As a consequence, the elite manages ideas, co-opts potential critics, and spreads its influence internationally. Do they act for themselves, or do they act for the good of the country? Domhoff's view on this point is clear, for he states that no matter how much they may plead otherwise, "they are primarily self-interested partisans, their horizons severely limited by the ideologies and institutions that sustain and justify their privilege, celebrity, and power."[5]

Does this model fit the facts? As we mentioned previously, Marx missed on several points. The workers' revolution hasn't happened, and classes have developed in ways other than those Marx predicted. Critics also suggest that the class model overemphasizes class and economic factors and ignores the fact that elites continually emerge in modern complex societies regardless of their relationship to the means of production. At the same time, impressive data support the class viewpoint.

The major aspects of the three theories of power are summarized in Table 9.3. Each theory makes important points about the nature of power and the relationship between power and other segments of society. Each model speaks to particular characteristics of society, and each is criticized, usually from the vantage point of another model.

Political Systems and American Politics

The **state** is the political body organized to govern; it is the center of legitimate power within a society. States may exercise their power in various ways. Look at several contrasting approaches: In a **totalitarian system,** the state has abso-

Table 9.3 Societal Power as Seen by Three Models

Model	Chief Source of Power	Key Power Group(s)	Role of Masses	Function of State
Class	Control of society's productive resources (wealth)	Ruling class (owners and controllers of the corporate system)	Manipulated and exploited by the ruling class	Protect capitalist class interests; reproduce class system
Elitist	Control of key institutions, primarily of the corporation and of the executive branch of the federal government	Relatively cohesive power elite, made up of top corporate and government leaders	Manipulated and exploited by the power elite	Protect interests of dominant elites and their institutions
Pluralist	Various political resources, including wealth, authority, and votes	Elected political officials; interest groups and their leaders	Indirectly controls elites through competitive elections and interest-group pressures	Referee the arena of interest groups; create political consensus

Martin Marger, *Elites and Masses* (New York: Van Nostrand, 1981), p. 112.

lute or total control, and power is highly centralized in one party or group; dissent or conflicting viewpoints are not allowed; and all parts of society—education, mass media, and so on—are controlled to maintain purity of viewpoint. In a **democratic system,** the power of the state is limited or modified by the wishes of the people. The people are seen as having direct input into the decision-making process. They may themselves be involved, or they may participate indirectly through their elected officials. Ideally, there are restraints on power—"checks and balances"—that keep the state in line. If the government goes too far, the people have an obligation to oppose it and make their feelings heard. Democratic systems seem to be more inclined to succeed in economically developed societies that have a well-educated and well-informed citizenry and in societies in which there is diffused or decentralized power and an absence of major political conflicts or differences of opinion over issues. As is usually the case with ideal or pure types, the complete totalitarian or democratic state is seldom, if ever, seen; systems generally fall somewhere between the extremes.

The political system in the United States has several curious characteristics. It is surely a democracy, yet where is the participation? In 1992, 55 percent of the voting-age population voted in the presidential race. In the off-year elections of 1990, the voter turnout was 33 percent. This low turnout is not unusual in the United States, where perhaps individuals feel that a single vote in a mass society is of little consequence. However, voter participation is much higher in western European countries. The **political party,** a number of people organized to gain legitimate governmental power, represents another path into participation. It was assumed for a time that American political parties were divided along class, racial, and ethnic lines. The Republican Party was the party of the middle class and of big business, and the Democratic Party represented the working class and the common man. In recent elections, however, this seems to have broken down, with much crossing of party lines. The importance or influence of parties is muddled by the fact that both parties have liberal and conservative wings and by the tendency in many situations for people to vote for candidates rather than for issues. The **interest group,** a number of people organized to influence decisions on issues important to them, is another type of participation in American politics. A regular alphabet soup of interest groups flourishes on the political scene—AMA, NRA, NAACP, Right to Life, and so on. Once a significant bill is proposed, the interest groups possibly affected by that bill immediately become geared up. For example, a major tax revision was suggested in May 1985: Real estate investors had depreciation concerns, corporations didn't like the minimum tax, industries worried about loss of investment credits, and others focused on loss of state-tax deductions. They hired lobbyists to aid their cause. **Lobbyists** are people in the business of knowing, influencing, and persuading the right people in government. The final product of any governmental decision-making process is inevitably greatly influenced by the pressures of interest groups and lobbyists. The final product is a compromise, manipulated by the concerns of politicians who wanted to be faithful to their party and get reelected by the citizens and by the pressures exerted by interest groups and lobbyists on politicians—an interesting example of the democratic process in action.

THE POLITICAL ECONOMY: SEPARATE INSTITUTIONS?

The connection between economics and politics is clear: One of the three major sources of power is resource control, and the economic theme runs through both the class and elite theories of power. The huge corporations, with the wealth and work force they control, have tremendous power to influence governmental decisions. And the concentrated power of the corporation is further strengthened by other factors. Corporations are often linked to each other through interlocking directorates. An individual sitting on the board of one organization is likely a director of several others as well. This encourages cooperation, reduces competition among groups, and smooths the path toward greater profits. In addition, although large companies appear to be publicly owned by stockholders, ownership is in fact highly concentrated. Most stock is owned by a select few, which are often banks and other corporations. Because corporate leaders sit on the governing boards of universities, higher education naturally responds to the needs of business and industry. Corporations influence the mass media either by direct ownership or because they purchase advertising. In summary, the power of corporations is great and concentrated; their influence extends to all segments of society.[6]

One might expect that government—the political sector—would act as a restraint, but in fact the contrary seems to be the case: The two institutions seem to work together. Corporations influence political decisions through lobbying and campaign financing and by having their leaders move freely back and forth between the two job worlds. For example, in 1971 a professor was appointed head of an insurance regulatory agency in Pennsylvania. His task, with the support of the powers of the state government, was to reform insurance-company practices. He was ambitious, interested in reform, and had power, but he ran into difficulty. Nearly all the government staff and managerial people he worked with had a background in the insurance industry. Furthermore, the insurance industry had a large lobbying operation; lobbyists attended legislative committee meetings and helped draft legislation favorable to the industry. Finally, good jobs in insurance companies were promised to state employees who were loyal to insurance-industry needs. During the professor's three years in office (he resigned seven months early to be replaced by a candidate acceptable to the insurance industry), no legislation put forth by his department passed if it was opposed by the insurance industry.[7]

Corporate leaders serve as presidential advisors, cabinet officials, and heads of regulatory agencies who later move back to the corporate world. Government, in return, supports a tax structure and gives tax breaks that are favorable to individuals and corporations. Louis B. Mayer, founder of MGM, saved $2 million in taxes when he retired by hiring a Washington lawyer who obtained a special tax provision that applied only to him. Government is also the corporation's best customer as a major consumer of goods and services;

the Department of Defense, for example, is the largest single consumer in American society. And government pursues a foreign policy favorable to corporate interests: Home markets are protected, foreign markets encouraged, and continued supply of necessary foreign raw materials assured. Government and economic elites clearly have mutual interests; as one prospers, so does the other.

Members of the government and economic elite are generally wealthy, male, college-educated with advanced degrees, and "well connected." Some from lower social classes occasionally sneak in, but not often. "The outstanding fact of elite recruitment in the United States and other Western industrial societies is that leaders are chosen overwhelmingly from socially dominant groups, and have been for many generations."[8] These elites differ on some issues but agree on the important ones, which should be expected given their common interests, backgrounds, and needs. The masses have little authority or influence in selecting or controlling the elites. There is potential control through the political process by voting or working for candidates, but few participate significantly, and those who do tend to be from the higher social classes. Social movements (see Chapter Thirteen) might be effective in controlling elites, but they are seldom a direct challenge.

THEORY AND RESEARCH: A REVIEW

The study of social institutions starts with *functional analysis:* Economic institutions function to organize production and distribution of goods and services in society. Governmental institutions function to maintain some semblance of order in a complex society, to enforce norms that people feel are important, to settle disputes, and to deal with other states. All are important and necessary functions. Again we see the emphasis on the basic elements of functional analysis—order, stability, structure, and function. *Conflict theorists* see other aspects of these institutions: Economic and governmental institutions operate to benefit powerful groups. Economic institutions clearly illustrate the basic battle between capitalists and workers that Marx saw long ago. Governmental institutions aid tremendously by helping the privileged to insulate and protect their position. The two institutions work together for the benefit of the established order. Both theories seem to "fit the facts"; both provide important insights toward understanding these institutions.

As for the data cited in this chapter, Domhoff used records in a *nonreactive approach*. He used information available in *Who's Who* and the Social Register, and he obtained lists of the directors, trustees, and partners from records of the largest companies, banks, foundations, and law firms. The economic data on corporations and multinationals were collected by business magazines from information collected and reported by the businesses themselves. Should we be suspicious? What would a conflict theorist say?

SUMMARY

In this third chapter on social institutions we have focused on economic and political institutions. We have discussed the economic systems of capitalism and socialism, and we have looked at some of the distinctive features of the American economy—large corporations and multinationals. After turning to political institutions, we have defined power, discussed types of power, and explored different models explaining the distribution of power. In addition, we have described totalitarian and democratic political systems and have discussed some of the characteristics of politics in America. We then followed Martin Marger's argument from his book *Elites and Masses*, in which he suggests that the key factor in understanding modern industrialized societies is their domination by elites from the political and economic realms. These elites have common interests, come from common backgrounds, and work together to their mutual benefit. The masses are relatively powerless and only marginally involved in the political process. There is considerable debate about the power and influence of elites, and one has to decide which of the various models and explanations best fits the facts of modern society.

To understand the American political economy, one needs to look at the power and influence of large corporations. The first of the readings that follow is a description of Exxon, the world's largest industrial company. Class and elite theorists believe that a powerful group of people, well known to each other, run this country. In the second reading, William Domhoff describes Bohemian Grove, making the point that members of the elite do not only work together; in the process they play together and make important decisions as well.

TERMS FOR STUDY

authority (283)

capitalism (279)

charismatic authority (283)

class model (286)

democratic system (288)

elite model (285)

elite-pluralism (284)

influence (284)

interest group (288)

legal-rational authority (283)

lobbyist (288)

multinationals (281)

pluralistic model (284)

political party (288)

power (282)

pseudocharisma (283)

resource control (284)

socialism (280)

state (287)

totalitarian system (287)

traditional authority (283)

Reading 9.1

THE NO. 1 INDUSTRIAL COMPANY

Douglas Martin

Large corporations are different from the rest of us—Exxon spent $100 million on changing its name! In this article by New York Times *writer Douglas Martin, we get a look at one of the large corporations that dominate the American economic scene.*

Mr. Rockefeller was right.

Although we pay homage to Jeffersonian democracy, Americans have built an economy that relies on goliaths to get the job done. And the biggest of all is Exxon, the most prominent descendant of the vast empire that John D. Rockefeller, who made no bones about the end of individualism, put together piece by piece.

The Exxon Corp., the world's largest industrial company, shuffles billions of dollars like pocket change. Its decisions shape nations and shake governments.

Consider the Colony project, a $5 billion oil shale development in the Colorado high plains. Until recently, it was the centerpiece of U.S. involvement in synthetic fuel; it was also an undertaking where Exxon, paying its own way, was a kind of equal partner with Washington, to whom lesser companies looked for financial help. Exxon withdrew from the venture, 2,000 workers got their last paychecks; 1,000 more were on their way out. A new town died and, practically unnoticed, a $100 million pipeline project to ship the shale oil was shelved.

The scale of this thing called Exxon—which opened the celebration of its 100th birthday at its annual meeting 10 days ago—is boggling. It consists of more than 400 corporate entities operating in 100 countries. Each day, six million motorists stop at Exxon's 65,000 service stations worldwide. The company's oil and gas pipelines would more than circle the earth. The cost of a single offshore drilling platform, towering higher than a 50-story building, might be four times the size of Harvard University's annual operating budget. Each week, Exxon spends $200 million on capital projects. Last year it earned $176.53 a second. So rich is Exxon that it still gives away road maps free.

"There is no wad of cash like this anywhere on earth," said Jack Blum, a Washington energy lawyer and former Senate investigator. "This is a wad of cash to break banks, even governments."

Take the high-speed elevator to the executive dining room on the 53rd floor of Exxon's regal white skyscraper on the Avenue of the Americas, and one finds the founder's rolltop desk still prominently displayed. Two floors below, Clifton C. Garvin Jr., chairman and chief executive officer, plus the seven men who make up his executive team, walk corridors lined with chamois and brightened by original paintings and freshly cut flowers.

Twenty-seven floors lower, technicians in what outsiders like to term "the war room" direct the far-flung movements of Exxon's tankers, a fleet on a par with the Royal Navy. In office after office, globes and world maps suggest the breadth of Exxon, the company that was a multinational 80 years before the term slipped into popular usage.

Size, indeed, sometimes seems to blind the company to all but the gigantic. What other corporation would claim that $55 million illegally doled out to Italian politicians was too small a sum to notice?

Exxon has perhaps been most deft at making its way in a changing world. From the 19th century, when companies began moving out into the world to lift other nations' oil, through the 1970s, as governments wrenched back title to their resources, Exxon has proved its skill at what amounts to statecraft, an ability to endure as a power on the world scene. Perhaps most notable, its grasp of global politics and of enlightened self-interest enabled it during the 1973–74 Arab oil embargo to shuffle oil supplies around like checkers. No nation was denied oil, but the companies nevertheless honored the letter of the embargo. "Exxon had the clearest vision of balancing all the interests and the need not to push anyone to the wall," says Jerome Levinson, former staff counsel to the Senate multinational subcommittee.

One might argue—as Exxon does—that colossal, largely undemocratic enterprises such as itself are precisely what enable America to afford democracy. It quietly points out that the $3 trillion economy of the United States, which generates one of the world's highest standards of living, rests on giant foundations—General Motors, General Electric, International Business Machines.

"Things in this country are done on a large scale," Garvin said, "and yet each one of us likes to think about things in a small way. It's always been a strange thing to me."

But what is Exxon? To begin, it is a corporation, a legal entity allowing the organization to transcend the lives of men. "A corporation doesn't have a soul to damn or a butt to kick," Henry Banta, a left-leaning energy lawyer, observes.

Exxon sprouted as the Standard Oil Co. of New Jersey with the granting of a charter to Rockefeller on Aug. 5, 1882. Originally the operating arm of John D. Rockefeller's Trust, it became the vehicle for controlling the entire skein of Standard Oil companies in 1899, after New Jersey abolished its antitrust laws and otherwise encouraged companies to incorporate there.

Then, with the Supreme Court's busting of the trust in 1911, Jersey Standard became one of 34 new oil companies, a collection including some of today's biggest—Mobil, Socal, Indiana Standard, Sohio, and Arco.

One obvious definition of Exxon, therefore, is that it is a child of Rockefeller. But only up to a point. Rockefeller ceased being an active manager in 1897, and his descendants are believed to own less than 1 percent of Exxon stock. Today, Exxon is owned by 804,000 shareholders, up more than 100,000 over the last year, largely because of the company's hefty 10 percent dividend.

Another definition of Exxon is people, those dauntless employees on the television commercials who seem to be searching for energy everywhere but the hall closet. Like others in the capital-intensive oil industry, Exxon is a relatively small employer, considering its clout. Some 180,000 people work for Exxon. By contrast, the Bell system has more than one million employees, and General Motors, more than 700,000.

Exxon's comparatively small army may make up in expertise what it lacks in numbers. Fully one-third are in managerial, professional, or technical slots, and they seem consumed by arcane details of highly technical tasks, speaking a language of undersea production systems, satellites, and laser technology.

Visits with Exxon hands over the years—the drilling supervisor in the Louisiana bayous, top executives in posh restaurants, world-weary Middle East experts—indicate that Exxon's people, almost uniformly, are nice folks.

But they are also very much Exxon: After lunch, groups will surround television sets to watch the in-house production of "This Week at Exxon"; even in New York City, participation in the Exxon Club's sweater sales and baseball expeditions is keen. A silver-haired functionary says that after decades of cocktail parties, his wife recently remarked Exxon men tend to be smart, friendly—and boring.

Exxon is also money. In addition to leading industrial companies in sales, assets, and stockholders' equity, the company last year made more than the combined profits of—to pick a smattering of giants in other fields—International Telephone and Telegraph, Dow Chemical, Procter & Gamble, Eastman Kodak, Union Carbide, Boeing, R.J. Reynolds, Goodyear, and General Motors combined.

Still, it was not the world's most profitable company; American Telephone and Telegraph, whose return is set by regulators, earned over $1 billion more than Exxon and claimed twice the assets.

And what's in a name? Jersey Standard became Exxon in 1972, after other Standard companies successfully sued to limit use of Jersey's "Esso" brandname (derived from Standard Oil's initials). Exxon came from 10,000 names dredged up by a computer—to then be studied by lawyers, linguistic experts, and other specialists to make sure the new name would have no adverse connotations in any of the multitude of languages the company deals in. Total cost: more than $100 million.

If Exxon is history, owners, people, cash, and a name, it is also unmistakably a bureaucracy. Decision-making is so marked by committees, hierarchies, and set processes that one employee believes the only comparable organization may be the Roman Catholic Church.

Garvin is said to devote more time to judging personnel than any other topic. Last year, he sat down with his seven top advisers 39 different times to discuss the matter.

Indeed, except for the newest and lowliest, everybody at Exxon is continuously rating somebody else, planning his future. The result: The company, fairly early in careers, has identified those who may someday lead it.

"Exxon ends up with 300 guys who have board potential by their late 30s or early 40s," an observer says. (And guys it is, so far. Exxon's highest-ranking woman is Carol C. Tatkon, treasurer of Exxon U.S.A.)

The "Exxon system," says John Buckley, who left Exxon to become vice president of the Northeast Petroleum Corp., results in "dog-eat-dog competition under the patina of working together."

From another viewpoint, it can also stunt entrepreneurship and personality. "The individual bureaucrat cannot squirm out of the apparatus in which he is harnessed," Max Weber wrote.

The rigid discipline of Exxon's personnel system—with its unrelenting pull toward Exxon's center—also characterizes other aspects of the company's business. In part, this reflects an international oil company's imperative, that need for control of a disorderly commodity first voiced by Rockefeller. But it also is a direct function of Exxon's size, prestige, and predisposition to do things its way.

The nature of control at Exxon has shifted in recent years, oldtimers say. The reason is the opening of the board of directors to outsiders in 1966, something Exxon long resisted. Outside directors—necessarily knowing less about the energy business— now outnumber insiders, 11 to 8. This has had the perverse effect of increasing the power of Garvin, the chief executive, while decreasing the influence of other senior management.

"The CEO used to be the first among equals, who could and did outvote him," said one of the few executives who has left the company over the years. "Now the chairman dominates the board." Garvin partly accepts the point, but notes that board meetings can be feisty affairs. In particular, the deliberations regarding the $1.2 billion purchase of Reliance Electric in 1979 were characterized by long debates and some dissenting votes by outside directors.

The acid test of the Exxon system, as with any organization, is results. And the results in recent years have principally been a matter of good luck. Thanks to the Organization of Petroleum Exporting Countries, and the 15-fold price increase it spurred over the last decade, seven of the top 10 industrial companies on *Fortune* magazine's 1982 list are now oil companies, up from three in 1955.

But what of Exxon's success on projects undertaken at its own volition, such as finding oil and natural gas or buying Reliance? Sometimes only sheer size seems to protect it: Exxon can afford to lose.

QUESTIONS 9.1

1. Show how politics and economics might be connected in the Exxon case. How and why might government help Exxon? Would it be difficult for government to control Exxon? Why or why not?

2. Show how Exxon would fit, in turn, into pluralistic, elite, and class explanations of power. Which makes the most sense?

3. What characteristics of bureaucracy (from Chapter Four) do you see in Exxon?

4. What types of power does Exxon have (from the text), and what type of authority (Weber, from the text) best describes Exxon?

Reading 9.2

POLITICS AMONG THE REDWOODS

G. William Domhoff

Is there an elite running society? William Domhoff thinks so. He describes a ruling class "which owns a disproportionate amount of a country's wealth, receives a disproportionate amount of a country's yearly income, and contributes a disproportionate number of its members to the controlling institutions and key decision-making groups of the country." They work together to exert tremendous power and influence, nationally and internationally.

They also play together, as this article on Bohemian Grove illustrates. William Domhoff is a professor of psychology and sociology at the University of California, Santa Cruz.

Ronald Reagan is many things to many people—radio announcer, actor, union leader, rancher, governor, and now president-elect of the United States. But to high-level members of the corporate business community he is just another fellow Bohemian—a member of a unique California social club whose yearly retreat into its redwood grove is a major social event on the calendar of the nation's power elite.

Perhaps the dozens of detailed accounts of Reagan's ascent to the White House that are sure to appear in the next few years should explore this little-known affiliation, unmentioned in Reagan's biography in *Who's Who in America,* for the Bohemian Grove has provided the setting for major events in the political careers of three former Republican Presidents and has played a role in making Reagan known on a first-name basis within the small circles of the social and corporate elites. Several of his Bohemian campmates are likely to be major advisers or officers in the new Administration.

The Bohemian Grove is a 2,700-acre campground in a virgin redwood forest on the meandering Russian River, seventy-five miles north of San Francisco. Owned and operated by the Bohemian Club of San Francisco, it has been host since the 1890s to a three-weekend respite beginning in the middle of July. Members and their guests number anywhere from 1,500 to 1,900 men on the weekends, but as few as 300 or 400 are in attendance during the week—even most of the corporate rich work during the summer, a considerable change since the more leisurely pace that prevailed before World War II.

Bohemian campers are treated to plays, skits, symphonies, band concerts, lectures, and political commentaries by top entertainers, scholars, musicians, and government officials from the local to the national level. They also trap shoot, canoe, swim, drop by the Grove art gallery, and take guided nature tours into the outer fringe of the mountain forest. But mostly a stay at the Bohemian Grove is a time for relaxation and drinking in the modest tents, lodges, huts, bunk houses, and even tepees fitting unobtrusively into the landscape along two or three unpaved roadways that join the few "developed" acres within the Grove.

Bohemian Grove resembles nothing so much as a summer camp for overgrown Boy Scouts, or maybe a collection of university fraternities displaced to an outdoor setting. Herbert Hoover, who became a Bohemian in 1913 and held forth as the encampments' final speaker every year from 1935 until his death in 1964, once called it "the greatest men's party on Earth." It provides a respite from establishment wars and from conflicts with the various activist groups within the underclasses.

One member in four, approximately, is a major business executive, director, or lawyer within the big-business community that is outlined for list lovers in the annual *Fortune* line-up of the largest firms. Such reasonably well-known multimillionaires as David Packard of Hewlett-Packard, Ray Kroc of McDonald's, and America's current richest man, Daniel K. Ludwig, are among the business members. There are celebrity members, too—such well-known but fading entertainers as Art Linkletter, Phil Harris, and Ray Bolger, and such traditional writers as Irving Stone and Herman Wouk.

Then there are the associate members, several hundred strong—lesser mortals on the status ladder, but talented men nonetheless who write the skits, act in the shows, sing in the chorus, design stage sets, play in the band or orchestra, and do the paintings and sculptures that are on sale in the Grove gallery. It costs them much less to be members, and most of them come from the Bay Area around San Francisco.

There are no women in the Bohemian Club, and there are no female employees at the Bohemian Grove. Last October, the Club had to defend its employment discrimination at the Grove before California's Fair Employment and Housing Commission. The proceedings revealed something of the atmosphere at the encampments, giving outsiders an insight into the fabled spirit of Bohemia. Corporate lawyer Del Fuller, secretary of the Club, argued that the presence of women would destroy the "intimacy" of the occasion because women would "distract" the more "flirtatious" of the men from the concern with just plain fellowship. He explained that the men like to let their hair down and become "boisterous," and that they would be "inhibited" by the presence of women.

Besides, continued Fuller, the men who have to dress up as women for parts in Bohemian plays would be embarrassed to do so if women were present. Under questioning, Fuller added that he had played the part of a wood nymph in the Club's centennial celebration in 1972, wearing wings and a body stocking.

When president-elect Reagan visits this enchanted Grove—he has been a member since 1975 and was a frequent guest before that—he stays at Owl's Nest, one of 130 or so little camps of ten to thirty members in which most of the men gather during their stay. Other campsites have equally strange names—Zack, Stowaway, Woof, Sons of

Toil, Cave Man, Mandalay, Toyland, and even Parsonage. Most camps are organized around a small building that serves as a kind of lodge or tavern, often housing a grand piano and an unusual contraption for mixing drinks, and cluttered with photos, drawings, and memorabilia from past encampments. The sleeping quarters are close to the main lodges.

Some camps are noted for special drinks, brunches, or luncheons. Jungle Camp talks up mint juleps, free for the asking. Halcyon has a three-foot-high martini-maker constructed out of chemical glassware. Poison Oak is remembered for a Bull's Balls Lunch, featuring the testicles from the castrated herds of a central California cattle baron. The specialty at Reagan's camp is a gin-fizz breakfast—about 100 Bohemians from other camps are invited one morning during each encampment for eggs Benedict, gin fizzes, and assorted trimmings.

Reagan's fellow Owls at the Nest are a group to be reckoned with in the corporate world. Only a few other camps can claim the concentration of corporate power that resides in this twenty-two-person campsite. They include the chairmen or presidents of United Airlines, United California Bank, Dart Industries, Carter-Hawley-Hale Stores, Dean Witter Reynolds & Co., and Pauley Petroleum, along with retired chieftains from Pacific Telephone, General Dynamics, and United Airlines. And, of course, several of these campmates sit on each other's boards of directors and on many other boards as well.

From a Reagan-watcher's point of view, Justin Dart of Dart Industries is the most important member of the camp, for he is probably Reagan's closest friend and political sponsor within the business establishment. No stranger to the Grove, Dart has been a Bohemian since 1951, and he has played a central role through the years in introducing Reagan to other business leaders, whether Bohemians or not.

Mandalay Camp, perched on the hillside about 100 feet above the Grove floor and housed in beautiful redwood buildings whose sleeping quarters are called Condemned Row, is one of the few camps that tops Owl's Nest in overall corporate connections. However, the chairmen and presidents it boasts from General Electric, Bankers Trust, Bank of America, and Utah Mining and Construction (now merged with General Electric) have retired from their offices in the past five or six years, so Mandalay is no longer the unequivocal Number One it used to be. Nonetheless, there are still Stephen D. Bechtel, Jr., chairman of Bechtel Construction; Richard P. Cooley, chairman of Wells Fargo Bank; and Jack K. Horton, chairman of Southern California Edison, among others.

Two Reagan intimates also reside at Mandalay, which is known among Bohemians for its gin-and-lemon-juice drink, its Welsh Rarebit dinners, and plush furnishings well beyond what other camps have to offer. One of these friends is lawyer William French Smith of the Los Angeles firm of Gibson, Dunn & Crutcher. The other is George P. Shultz, vice chairman of Bechtel Construction Company and a secretary of both labor and treasury during the Nixon Administration.

Smith, who along with Dart is Reagan's closest social friend with wide connections throughout the entire range of the power elite, is a quintessential interlocking overlapper. He serves as a director of Pacific Lighting, Pacific Telephone, Pacific Mutual Life, and Jorgensen Steel, but also finds time to be a regent of the University of California, a member of the executive committee of the California Roundtable, and a member of the advisory board for the Center for Strategic and International Studies at Georgetown University, from which Reagan will draw several foreign policy advisers. Shultz, who became a top Reagan adviser only after the Republican convention, is somewhat more modest in his affiliations, serving as a director of Morgan Guaranty Trust, Sears Roebuck & Co., the Alfred P. Sloan Foundation, and that favorite *bete noire* of Reagan's ultra-right supporters, the Council on Foreign Relations, which has about thirty-five members in common with the Bohemian Club.

Another prominent Republican at Mandalay who came to be a Reagan adviser is former President Gerald R. Ford. A close friend of Leonard K. Firestone of the tire manufacturing fortune, Ford was Firestone's guest at Mandalay several times before becoming a member of the Bohemian Club in 1977.

Hill Billies is another camp with a strong core of business executives, and it is the camp of the vice president-elect of the United States, George Bush. The corporate parallels between the Bush and Reagan camps are instructive. Reagan rubs shoulders with the chairman of United Airlines, but Bush can lay claim to being a chum of the president of Eastern Airlines. Reagan's camp has the chair of United California Bank, but the Hill Billies include among their twenty-three members the president of the Bank of America. The president of Dean Witter Reynolds may be at Owl's Nest, but the first vice president of Blyth Eastman Dillon is under the Hill Billies' tent. When it comes to overall corporate connections, however, Hill Billies seems to have the edge, for it houses executives and directors from General Motors, Southern Pacific, Westinghouse Electric, B.F. Goodrich, Morgan Guaranty Trust, Mutual Life Insurance of New York, Superior Oil, and Metromedia.

One other member of Hill Billies, aside from Bush, recently appeared in the news: Alden W. Clausen, president of the Bank of America, was nominated by Jimmy Carter to be the next head of the World Bank. According to press reports, Carter's aides approached George P. Shultz to clear Clausen's acceptability should Reagan win the forthcoming election. Reagan gave his approval, but it is not recorded whether he was chagrined to learn from a member of Mandalay that someone from Hill Billies had gained Carter's nod instead of one of his fellow Owls.

The guest list of about 400 for each year's retreat only adds to the impression that the Grove is a playground for the powerful. Among the guests at Owl's Nest in 1980, for example, was Charles F. Luce, chairman of Consolidated Edison of New York, hosted by Edward Carlson, chairman of United Airlines. Also at Owl's Nest was William E. Simon, former Secretary of the Treasury and now a Reagan adviser, invited by Justin Dart. Guests at Mandalay included Donald M. Kendall, chairman of Pepsico and a Nixon intimate; Reginald Jones, chairman of General Electric and an adviser to Carter; and Walter B. Wriston, chairman of Citicorp and an adviser to Reagan. David Rockefeller, chairman of Chase Manhattan Bank, stayed at Stowaway as a guest of William A. Hewitt, chairman of Deere & Co. Willard C. Butcher, president of the same Chase Manhattan, was at Cave Man with John E. Swearingen, chairman of Standard Oil of Indiana. Robert D. Stuart Jr., chairman of Quaker Oats, was at the Isles of Aves with Reagan adviser Caspar Weinberger, a vice president at Bechtel Corporation and a director of Quaker Oats.

Simon at Owl's Nest and Wriston at Mandalay were not the only Reagan advisers on the 1980 guest list. Campaign chairman William Casey, a Wall Street lawyer, was staying at the Parsonage, a guest of Darrell M. Trent, himself an academic adviser to Reagan. Senator Paul D. Laxalt, at one point in charge of keeping Reagan from saying anything off the cuff, dropped in at Mandalay as the guest of a retired Anheuser-Busch executive, John Flanigan. Jack Kemp, the tax-cutting member of Congress who urged Reagan to appeal to blue-collar workers in the North, was at Toyland, the guest of George Lenczowski, a political scientist of the University of California, Berkeley, who also serves as an adviser to a Reagan-oriented think tank, the American Enterprise Institute.

How much does all this matter? What difference does it make that a socially chummy corporate community likes to do some drinking and joking with lots of Republicans and a few Democrats in rustic little campsites in the California redwoods? After all, the official motto of the club is "weaving spiders come not here," and most members agree that 80 percent of those in attendance are so overwhelmed by the fine food, strong drink, and sterling entertainment that they couldn't remember the details of a new policy or a diabolical conspiracy from one day to the next even if they had tried to plan one. The Bohemian Grove is not a place of power in the sense of attempting to formulate new policies. Corporate board rooms, charitable foundations, and such discussion groups as the Conference Board, Committee for Economic Development, Business Council, Business Roundtable, and, yes, even the Council on Foreign Relations, are the institutional settings for those kinds of activities.

But if the Bohemian Grove is not a place of power, it is nonetheless a place where powerful people congregate, and as such it has provided the setting for new developments in Republican politics at the presidential level ever since 1927, when Herbert Hoover was sitting by his tent in Cave Man camp just as President Calvin Coolidge made his cryptic announcement, "I do not choose to run for president in 1928." And it was then that the Hoover campaign began. "Within an hour," Hoover wrote in his memoirs, "a hundred men—publishers, editors, public officials, and others from all over the country who were at the Grove—came to my camp demanding that I announce my candidacy."

Dwight D. Eisenhower made what is thought to be his first general pitch to the Republican establishment with a speech at the Grove in 1950, and then became an honorary member in the same year. It was a short speech, delivered without notes, and it drew applause when Ike declared he didn't see why someone who wouldn't sign a loyalty oath should have the right to teach in a state university.

Before the talk, Eisenhower had lunch at Cave Man with Herbert Hoover and his campmates, most of whom supported even more conservative candidates than Eisenhower. Among those present as a guest was Richard Nixon, who had to wait until 1953, after he was vice president, to become a Bohemian and a Cave Man. Nixon's memoirs suggest that a serious political discussion took place at Cave Man after the Eisenhower speech:

"After Eisenhower's speech we went back to Cave Man Camp and sat around the campfire appraising it. Everyone liked Eisenhower, but the feeling was that he had a long way to go before he would have the experience, the depth, and the understanding to be president. But it struck me forcibly that Eisenhower's personality and personal mystique had deeply impressed the skeptical and critical Cave Man audience."

Nixon himself launched his first successful presidential campaign at the 1967 encampment by means of a Lakeside talk not far from the Owl Shrine.

He called this talk "the speech that gave me the most pleasure and satisfaction of my political career"—and one that "in many important ways marked the first milestone on my road to the presidency . . . an unparalleled opportunity to reach some of the most important and influential men, not just from California, but from across the country." The speech was important because it unveiled a "new" Nixon to the corporate establishment, a Nixon who had decided to take the high road and rise above his usual anticommunist harangues. His new tack was what he called a "sophisticated" hard line, and apparently it met with approval from the assembled Bohemians.

The Grove also was helpful to Nixon in another way in 1967, for it was there that he and Reagan made a deal calling for Nixon to go first in Republican primaries against the more moderate opposition, with Reagan jumping in only if Nixon faltered. The ubiquitous William F. Buckley, Jr., is the source of this story, but Nixon acknowledges in his memoirs that he and Reagan had a "candid discussion of the political situation as we sat outdoors on a bench under one of the giant redwoods."

It seems worthwhile to keep an eye out for the Bohemian Connection in understanding both the rise of Ronald Reagan and his forthcoming presidency, for the Bohemian Grove provides an ideal setting for informal politics, a place to take an off-the-record look at potential candidates and advisers and to gain a first-hand impression of a person's style and personality.

And if George P. Shultz, William French Smith, Caspar Weinberger, and a few other of the many Bohemians who supported Reagan end up in his new Administration, the only question will be whether to hold the July cabinet meetings at Mandalay, Hill Billies, or Owl's Nest.

QUESTIONS 9.2

1. How would a pluralist argue with Domhoff over the significance of Bohemian Grove?

2. Domhoff notes a "quintessential interlocking overlapper." What is it? What is its importance for the political economy?

3. "So certain people get together recreationally once in a while—what's wrong with that?" Discuss this statement from both sides of the issue.

NOTES

1. The major source for this section is *Elites and Masses* by Martin Marger (New York: Van Nostrand, 1981). I have followed Marger's organization and viewpoint on most topics. His discussion of power and the three theoretical models comes from his chapters 2–5.

2. The idea of pseudocharisma comes from Joseph Bensman and Bernard Rosenberg in *Mass, Class, and Bureaucracy: An Introduction to Sociology* (New York: Praeger, 1976), pp. 431–433 and is discussed by Marger (see Note 1) on pp. 20–21.

3. See C. Wright Mills, *The Power Elite* (New York: Oxford University Press, 1959).

4. Both quotations in this paragraph come from *Who Rules America?* by G. William Dom-

hoff (Englewood Cliffs, N.J.: Prentice-Hall, 1967), p. 5.

5. This is from Domhoff's later book, *The Higher Circles* (New York: Random House, 1970), p. 275.

6. The source for these paragraphs is Marger, *Elites and Masses* (see Note 1), especially chapters 6–8 and 11.

7. This case is described in *The Powers That Be* by G. William Domhoff (New York: Vin-

tage, 1978), pp. 32–36. The original source is David L. Serber, "Regulating Reform: The Social Organization of Insurance Regulation," *The Insurgent Sociologist*, Spring 1975. The Mayer example in the next paragraph is also from Domhoff, p. 29.

8. Marger, *Elites and Masses* (see Note 1), p. 207.

CHAPTER TEN

Institutions: Education

TWO RESEARCH QUESTIONS

1. Is the high-school dropout rate in a state related to the amount of money spent on education in that state? In other words, if a state has high expenditures on education, will it have a lower high-school dropout rate?

2. Do American elementary and secondary students spend more, less, or about the same number of days in school each year as students from other countries?

IN CHAPTER SEVEN we defined social institutions as the norms, roles, and patterns of organization that develop around central needs or problems that societies experience. In previous chapters we discussed the family, religion, and economic and political institutions. In this chapter we look at education.

One basic problem that all societies face is socialization and training of the young. The young must *learn* how to become adults, to count to 100, to drive cars, and to become rocket scientists. Who is going to teach them these things? Early parts of the socialization process are carried out primarily in the family. In primitive societies, parents and those close to the family probably carried most of the load. But as societies become more complex and the path to becoming a rocket scientist becomes more convoluted, other elements are necessary to continue the process. All of these elements, then—from small groups like a family, a kindergarten class, or a graduate seminar, to larger organizations like the University of California or the Boston school system—are part of the institution of education.

As we have learned, there are overlaps among institutions. Socialization and training of the young take place in the institutions of the family and religion, and perhaps in others as well. However, by focusing on education in this chapter, we may be able to identify some of the problems and issues that are unique to this institution.

EDUCATION AND SCHOOLING DEFINED

A distinction should be made between education and schooling. **Education** is the more general term and refers to the process of socialization—learning the knowledge, skills, and appropriate ways of behaving in order to exist in a larger social world. Education (socialization) is thus most intense for the very young, but it continues to occur throughout the life span. **Schooling** is the narrower term and refers to formal instruction in a classroom setting. Our major focus in this chapter will be on schooling.

The United States pioneered the creation of the "common school," a publicly supported elementary school for the education of lower- and middle-class children. The comprehensive high school grew out of the elementary-school program. We started what has been called a "schooling revolution" in the modern world, a revolution that represented a fundamental change in how young people were prepared for life in the adult world.[1] No longer did the child receive informal teaching in the community; the school became the setting for most of the child's preparation for the adult world. Following the lead of the rest of society, the small, folk-society, gemeinschaft, isolated-village approach to education was left behind; thus the "schooling revolution" was but one aspect of our move to a more complex, industrial and scientific, secondary, gesellschaft type of society.

THE FUNCTIONAL APPROACH TO EDUCATION

The functional approach makes several assumptions about schooling that require some attention.[2] Functionalists see modern society as a system in which success is *achieved*; ability and hard work are more important than being given something or inheriting it. Children of the poor should have the same opportunities to get high-status jobs as do the children of the rich. Functionalists see society as an *expert society,* one that needs people who are highly trained and who have developed specialized skills. In fact, modern society is becoming more and more complex, and this implies an ever-increasing need for expertise and technical know-how. Functionalists also see modern society as a *democratic society* in which an educated citizenry is an informed citizenry composed of people who are more likely to make responsible and informed decisions, are likely to be less prejudiced and more tolerant, and will be more concerned with the quality of life and social justice.

In this context, functionalists view schooling as providing a number of obvious and important functions:

passing on the culture of a society

facilitating the mixing and merging of different classes and ethnic groups

creating the informed citizenry necessary in a complex society

producing cognitive skills

screening, sorting, and selecting of talent and ability

developing new knowledge and providing scientific discoveries

Obvious functions like these could be called *manifest functions.* Perhaps equally important but less obvious consequences (*latent functions*) of schooling might include the following:

delaying young people from entering the job market

"baby sitting" young children, thereby freeing parents to work outside the home

teaching habits of discipline and obedience to authority that are necessary for survival in a bureaucratic society

How good is the fit between what the functionalist approach proposes and the current facts? Christopher Hurn suggests that it is not as good as it might be. According to the functionalist approach, modern society needs an ever-increasing number of highly skilled and trained individuals. Those who perform best will reap the highest rewards in modern society. And yet in studies of college graduates, grade point average does not consistently predict either occupational status or future earnings! Actually, it may be that the presence of a *degree* or *credential* is the crucial factor for being hired, not the knowledge or skill that is assumed to go with it. This suggests that rather than teaching skills, the education system more likely rewards *persistence*—the student who sticks it out to the B.A., M.A., or Ph.D. gathers the benefits. As job candidates become more educated, entrance standards for the job seem to get raised. Does this mean that the job has become more complex, or that entrance standards are used as ways of controlling admission to the occupation?

Many schools use a procedure called **tracking** (or curriculum tracking) by which students are placed in certain programs based on curricular needs or students' abilities. They are assigned to these tracks based on test scores and on input from teachers, parents, and school administrators. Some students are placed in a college-preparatory track, others in a vocational-career track. Some schools have three or four curricular tracks; other schools may have more depending on such factors as school size, budget, type of community being served, and diversity of faculty. According to the functionalist view, people should *achieve* success, not be given it. People should get ahead according to their ability, not their birthright. Tracking makes sense to functionalists because it provides equal opportunity for all and rewards ability and hard work. Functionalists hope for a "sorting mechanism" that is truly class- and color-blind and in which the sharper and harder-working students are rewarded by being placed in the higher (college-prep) classes. The question, of course, is whether tracking is truly color-blind and in fact treats everybody equally, separating only on the basis of ability and hard work. Research shows a relationship between I.Q. (ability) and later earnings, but it also shows that one's social class is the overpowering factor in the equation. In other words, the evidence seems to indicate that one's social class is a better predictor of future economic success than is one's measured I.Q. With these criticisms of the functionalist approach in mind, let's turn to a discussion of the conflict approach.

THE CONFLICT APPROACH TO EDUCATION

Conflict theorists see education and schooling quite differently than the functionalists do. First, conflict theorists see a struggle among competing groups for the control of schooling. The struggle for control represents each group's attempts to advance its own set of interests and instill its own values in the education system. Second, the struggle among groups is an unequal one. The elite—the "haves," the privileged class with its superior power and resources—always has the upper hand. Third, conflict theorists are convinced that even though employers recognize the need for a certain minimum level of skill, they are far more interested in having workers who are loyal, obedient, and submissive. What the employers really want from schooling, then, is workers who have "appropriate" attitudes and values.

Conflict theorists emphasize the role schools play in maintaining the existing social-class system. Tracking is one way of doing this because, from a conflict perspective, there is clearly no equality of opportunity. Higher-class children are tracked into colleges and universities and thus into positions of economic and political dominance like those their parents hold. Children with lower-class or working-class origins tend to be placed in tracks that will move them toward blue-collar work in adulthood.

Another way in which schools reinforce the existing class system is by teaching the values and personality characteristics necessary in certain positions in society. A **hidden curriculum** exists in schools whereby these important cultural values and attitudes are taught. These values could be either compliance, punctuality, and conformity, or independence, self-reliance, and nonconformity. A hidden curriculum is not part of the regular or intended curriculum, and teachers may not even be aware it exists, but in subtle and effective ways students come to understand their social place in school and in society as a whole.

Conflict theorists Samuel Bowles and Herbert Gintis suggest specifically that schooling serves the interests of capitalist society.[3] They propose that for capitalist societies to work efficiently, production-line workers must have a different approach to their jobs than scientists, heads of corporations, or presidential advisors need to have to theirs. Schools recognize this, and schools whose graduates enter lower-level occupations stress following rules, being punctual, having respect for authority, having fewer choices, and showing less creativity. Schools whose graduates enter higher-level occupations are likely to encourage independent thinking, creativity, less obedience to authority, and more freedom and flexibility. If you were to compare lower-track high-school and junior-college courses with college-prep high-school courses and courses at elite universities, you would probably find that the former had more assignments, less flexibility, more supervision, and less encouragement of initiative or creativity. According to Bowles and Gintis, then, schools are conferring on students the values and attitudes they will need for their place in society, a place not much different from the place their parents hold.

In much the same vein, Randall Collins suggests in his book *The Credential Society* that educational credentials that are largely unnecessary are used to control access to desirable jobs. Collins believes that most jobs are not particularly complex, that they are not becoming more complex, and that the skills most jobs require can easily be learned on the job. He contends that schools are not particularly good at teaching intellectual skills and that in fact much of what is taught is quickly forgotten. Collins concludes that various groups use the credentialing that education provides as a way of elevating their own status, *not* because the complexity of the job requires it. For example, many police departments now want their candidates to have a college education, when not so long ago a high-school education (or even less) was seen as adequate. Has the job gotten that much more complex? Is a college-educated police officer a significantly better officer than a high-school–educated officer? Or are the new requirements a way of elevating the profession—a way of giving it higher status and controlling admission to the job? Collins agrees with other conflict theorists who say that much of what schools teach are good manners and how to behave, not cognitive skills.

You might think of the discussion in the previous several paragraphs in terms of a debate between the relative strengths of two models: the social reproduction model and the democratic model. The **social reproduction model** refers to the tendency of the institution of education to reproduce the social class of the student. The **democratic model** refers to the struggle to provide equal opportunity for all and to reward ability and hard work regardless of one's origin. The constant tension between these two models is nicely illustrated in the two competing theories, functionalist and conflict.

The conflict perspective sees tension and conflict as normal in social situations, so it is no surprise that they exist in schools. Teachers organize into unions and educational associations for negotiating salaries and work conditions. Teachers use punishment (grades, detention) to ensure conformity. Students may resort to informal subversive activities to counter discipline and classwork demands made by teachers. Accordingly, a recent high-school student recalls seeing graffiti on walls, bringing a walkman or books to class to help pass the time, figuring out clever ways to get out of class or to cut class, vandalizing equipment (clogging the toilets, stuffing paper or pencils into gas outlets, or breaking lockers), and carving or spray-painting slogans or initials on various surfaces. It is estimated that 130,000 children bring guns to school, and many schools now have security guards and metal detectors.

Another type of behavior that may be a form of resistance or rebellion is to drop out of school entirely. It is estimated that about 600,000 youngsters drop out of school each year—in many urban schools, up to half of the enrollment drops out.[4] The next stop for many is the correctional system. Table 10.1 shows percentages of high-school dropouts for two different years and for different groups. Why do students drop out? When one study asked students about it, the most frequently given reasons for dropping out were: "School was not for me"; had poor grades; was offered a job and chose to work; got married or planned to get married; couldn't get along with teachers; had to support family; and was pregnant.

High-school dropout rates vary across the country, as can be seen in Table 10.2. Minnesota has the highest graduation rate, and Louisiana has the highest dropout rate. The first research question at the beginning of this chapter asked whether the amount of money spent on education might be related to the dropout rate: Do states that spend less per student have higher dropout rates? Expenditures on education may well be part of the explanation, but as the data in Table 10.2 indicate, expenditures don't appear to be the whole answer. Some states that spend the least have low dropout rates (North Dakota, South Dakota, Iowa, and Nebraska), and some states that spend the most have high rates (New York and Florida). It's more likely that dropout rates are related to the demography of a state. States with large cities and a predominantly urban population are likely to encounter the problems that inner-city schools often have, which means that they will likely have much higher dropout rates than will more rural states that need not contend with these issues.

Conflict theorists suggest that one reason for students' resistance or rebel-

Table 10.1 Percentages of High-School Dropouts Aged 14 to 24 Years

	1980	1991
U.S. (all groups)	12.0	10.5
White	11.3	10.5
Black	16.0	11.3
Hispanic	29.5	29.5

United States Bureau of the Census, *Statistical Abstracts of the United States, 1993* (113th ed.), Washington, D.C.

Table 10.2 Top Ten States in High-School Graduation and Dropout Rates and Educational Expenditures, 1988 (U.S. Average = $4,209)

Most Graduates	Expenditures (per student)	Most Dropouts	Expenditures (per student)
Minnesota	$4,513	Louisiana	$3,211
North Dakota	3,353	Florida	4,389
Iowa	3,846	Georgia	2,939
South Dakota	3,159	Texas	3,462
Nebraska	3,641	South Carolina	3,075
Montana	4,061	New York	6,864
Wisconsin	4,991	Arizona	3,265
Kansas	4,262	Mississippi	2,760
Connecticut	6,141	North Carolina	3,911
Hawaii	3,894	California	3,994

The World Almanac and Book of Facts, 1991 (New York: Pharos Books, 1990) and United States Bureau of the Census, *Statistical Abstracts of the United States, 1989* (109th ed.), Washington, D.C.

lion, especially dropping out of school, is that some students, particularly working-class students, detect the "hidden curriculum" and decide that they won't go along with it. They refuse to accept the school's view that conformity to the school's rules and regulations will lead to economic opportunity. They see that the system is stacked against them, that there "Ain't no making it," and they rebel.

Read the box entitled "The Typical Dropout" and decide whether this is an example of the functionalist or conflict interpretation of why students drop out.

EDUCATION IN AMERICA

Are we schooling our young people as well as we should be? We hear a lot about Scholastic Aptitude Test (SAT) scores. Both math and verbal SAT scores were lower in 1992 than scores of 25 years earlier. At the same time, curiously, the number of students doing very well on the test increased. This led one expert to comment that we have an "educational elite" and an academic "underclass" that is "ill-prepared for the demands of college or the workplace." The International Association for the Evaluation of Educational Achievement (IEA) has been trying to gauge the quality of education in various countries. In their comparison of eighth-grade math students in 20 countries in 1981–

THE TYPICAL DROPOUT

Compared with those who will graduate, the future dropout shows a clear indication of academic problems by the third grade. Achievement test scores are below the scores of his or her classmates and also below the level one would expect given the student's ability. The poor attendance and underachievement increases as the student goes into middle school, and by the seventh grade failing grades are present. By the ninth grade a pattern of high absences, failing grades, and a low overall GPA is well established, and it continues until the student drops out of high school.

Regarding family background, the dropout's father tends to have a job of lower social status, although not significantly lower, than that of the graduate's father (about 1.5 points on the 9-point Hollingshead scale). Although a lower percentage of the dropout group lived in families with two parents present than did the graduates, more than two thirds (69 percent) were living at home with both parents at the beginning of the ninth grade. This finding evidently questions the assumption that the typical dropout is from a "broken home." . . .

The typical dropout was no more likely to be either male or female than the student who graduated, and similarly being in a more flexible modular scheduled high school or following a more traditional schedule had no significant effect on the likelihood of being a dropout.

The high accuracy with which the elementary school data identified potential dropouts probably reflects family attitudes toward education. The elementary school student who is not in school is absent with the parent's knowledge and at least tacit consent. The parents who are uninterested in their children's attendance in elementary school, and probably their achievement, likely not only convey their values to the child, but also are willing to agree when the child later decides to leave high school.

From "Differentiating Characteristics of High School Graduates, Dropouts, and Nongraduates," by B. L. Barrington and B. Hendricks, Journal of Educational Research 82 (July/August, 1989), pp. 309–319.

1982, American students ranked 10th in arithmetic, 12th in algebra, and 16th in geometry.[5] Japan ranked first in all three of these categories. The IEA examined the science knowledge of ten-year-olds in 15 countries from 1983 to 1986; American students ranked eighth. Other studies that suggested that American students fall farther behind each year were confirmed when the IEA examined 14-year-olds and found them tied for 14th (with students from Singapore and Thailand) out of 17 countries. Most countries don't try to educate as many of their young people as we do, so perhaps these comparisons aren't completely valid. However, in another study IEA compared 12th-grade students who were engaged in the serious study of math (were in their fourth year of high-school math), clearly a more elite group of students. A portion of the results of this international comparison is shown in Table 10.3.

Perhaps we don't take schooling seriously enough. A current debate involves a simple variable like the amount of time spent in school. The second

Table 10.3 Rankings of Student Achievement in
Two Math Subjects in Various Parts of the World

Advanced Algebra	Geometry
1. Hong Kong	1. Hong Kong
2. Japan	2. Japan
3. Finland	3. England/Wales
4. England/Wales	4. Sweden
5. Flemish Belgium	5. Finland
6. Israel	6. New Zealand
7. Sweden	7. Flemish Belgium
8. Ontario	8. Scotland
9. New Zealand	9. Ontario
10. French Belgium	10. French Belgium
11. Scotland	11. Israel
12. British Columbia	12. United States
13. Hungary	13. Hungary
14. United States	14. British Columbia
15. Thailand	15. Thailand

"The Case for More School Days," by Michael J. Barrett, *The Atlantic Monthly* 266 (November 1990), pp. 78–106.

research question asked at the beginning of this chapter concerned the number of days children spend in school in different countries. It may surprise you to learn that our children spend *fewer* days in school than do the children in many other countries—especially in those countries with which we increasingly find ourselves competing (see Table 10.4). This fact would seem to explain the tendency of American students to get farther behind students from other countries with each passing year. For over 40 years, Gallup has polled Americans on their attitudes about lengthening the school year, and the response has always been negative. The gap has been closing recently as more people realize that some nations' students spend up to 25 percent more time in school than do students in this country, but people are still not enthusiastic about the idea.

Does the problem lie with teachers? We hear that they burn out, that they are bored, that they aren't happy in their work. It is probably true that the best college-bound high-school seniors aren't headed for careers in education. But why should they be? Education is not widely seen as a challenging, interesting, or valuable career. It's difficult to get the best people to go into a profession in which the pay isn't great, there isn't much prestige, and the working conditions are often far from ideal. Table 10.5 shows starting salaries for a number of occupations. You would need to be quite dedicated to take the lowest starting salary on the list, and you won't get the best and the brightest to go into teaching under these circumstances. Education is always squeezed for money, and the failure of yet another bond measure to raise funds for schools seems to be an annual occurrence.

Maybe our values need reviewing. A T-shirt making the rounds locally

Table 10.4 Number of Days in a Standard School Year in Various Countries

Japan	243	Finland	190
West Germany	226+	New Zealand	190
South Korea	220	Nigeria	190
Israel	216	France	185
Soviet Union	211	Ireland	184
Netherlands	200	Spain	180
Scotland	200	Sweden	180
Thailand	200	United States	180
England/Wales	192	French Belgium	175

"The Case for More School Days," by Michael J. Barrett, *The Atlantic Monthly* 266 (November 1990), pp. 78–106.

Table 10.5 Average Starting Annual Salaries of Public-School Teachers and Selected Private Industries, 1992

Teaching	$22,171
Sales/marketing	26,532
Liberal Arts	27,324
Business administration	27,156
Economics/finance	27,708
Accounting	28,404
Mathematics/statistics	29,472
Computer science	30,888
Chemistry	30,360
Engineering	34,620

United States Bureau of the Census, *Statistical Abstracts of the United States, 1993* (113th ed.), Washington, D.C.

says, "It will be a great day when our schools get all the money they need and the air force has to hold a bake sale to buy a bomber." Maybe we are complacent and take education for granted. A few years ago a study compared the performance and attitudes of students and parents from Japan, Taiwan, and the United States. The American fifth graders performed poorly (the worst of the three groups) on mathematics tests. When the mothers were asked how their children were doing in school, the American mothers were more likely to report that they were "very satisfied," whereas the Chinese and Japanese mothers were more likely to say that they were "not satisfied." Ninety-one percent of the American mothers said the schools were doing an "excellent" or a "good" job, whereas only 42 percent of the Chinese mothers and 39 percent of the Japanese mothers were that positive.[6] The author of the study despaired at the practicality of pushing educational reform if the parents and citizens who provide the money and the force for change aren't interested enough to urge change.

Much public discussion has focused on the declining quality of American education, and many ideas have been offered about how to change it. Leon Botstein, president of Bard College, suggests some adjustments in the box entitled "How to Change the System."

HOW TO CHANGE THE SYSTEM

How do we go about changing the system? First of all, I would simplify the curriculum and focus on a few very basic things: a command of language, reading and writing, mathematical reasoning. Science and mathematics, as opposed to the humanities, should be at the center of the curriculum from the beginning. I take this radical position because what is taught in the humanities—in reading and writing and history—has become so politicized that science is the last common ground we have. It's also the easiest curiosity to sustain; every child wants to know how the world works. College is too late to develop a curiosity in science and mathematics; it can only be done early in life.

I would eliminate the American high school, because it is not fixable in its current design. High school is a nineteenth-century structure that has run its course: it is no longer congruent with the rapid development and autonomy of adolescents, their physical maturation, and the independence that society now permits them. I would divide schooling into five or six years of elementary school and then middle school. At the end of what is now the sophomore year of high school, students would feed into the community college system, which works very well by comparison, doing what high schools should be doing.

I think we also have to face the fact that we must make teaching a desirable profession. In order to do this, we have to raise teachers' salaries. But more importantly, we have to change the working conditions in our public schools. Why do good people go to private schools to teach, even though salaries are considerably lower than at public schools? Because in public schools the teaching profession is bureaucratized, teachers are not treated as professionals, there is enormous regulation, the conditions of work are horrendous, there is no professional community. The administration and organization of schools need a radical overhaul.

From "Educating in a Pessimistic Age," Harper's, *August 1993.*

IS CHANGE COMING IN EDUCATION?

So far this chapter has sounded quite negative. Is there another side to the story? There continues to be tremendous interest in educational reform. We continually hear about individual teachers, particular schools, and cities in which innovative programs are working or being developed. Traditional policies and procedures are being reviewed. For example, there is increasing interest in getting away from the memorization of facts and focusing instead on developing critical thinking skills, even in the lower grades. California is getting away from A, B, C grading in elementary schools; instead, teachers are using written reports. This approach eliminates some of the competition present in the early years and probably gives a more complete evaluation of the child. This reminds me of the time several years ago when the parents of a beginning student proudly told me that their child had made the prekindergarten honor roll. . . .

The types of tests being given have come under question. What do the SAT, ACT, or I.Q. tests actually measure? Their ability to predict future academic or job success is being increasingly questioned. It has been recognized for some time that they are culture bound and class linked. The tests are being revamped and revised to accommodate cultural diversity.

Tracking is currently in disrepute and is widely criticized, especially by those with the conflict perspective. There is a growing interest in looking at educational stages rather than tracks and in emphasizing developmental age rather than chronological age. We have tended to place children in school based on **chronological age**— kindergarten at 5, first grade at 6, and graduation from high school at 18. The Gesell Institute has long held that children should enter school and be promoted based on their actual behavior, or **developmental age,** rather than on their chronological age. What parents should ask themselves is, "Is my child (regardless of his or her actual age) *ready* for kindergarten?" Starting a child too early may be a disservice; many contend that overplacement is a major factor in a child's subsequent decision to drop out of school. Or the overplaced child may develop perfectly well in one area (intellectually, for example) but be left hopelessly behind in other areas (socially, for example). Emphasis on developmental age recognizes that people don't all mature at the same rate.

Finally, we might look back at some of the ideas that functionalist and conflict theorists debate. Is a mass of college-educated people needed for our society to work well? The current job market contains mostly service (nontechnical) types of jobs that don't need a high level of proficiency in math and science. Perhaps we need to focus our resources on work-force needs rather than on educating everyone to the greatest extent possible. Or, if we want to continue with our goal of educating everyone, a task few societies attempt, we must consider the possibility that, given that goal, perhaps we are doing the best we can do.

THEORY AND RESEARCH: A REVIEW

This chapter has been structured around a comparison of the two dominant theoretical approaches, *functional analysis* and *conflict theory*. Functional theorists see society as a meritocracy in which all have an equal chance to achieve success. Ability and hard work will be rewarded, and the educational system will produce the number of highly skilled and trained people that an expert society needs. Functionalists see schooling as providing a number of important functions: selecting people with talent and ability, creating an informed citizenry, and passing on the culture of a society to the next generation. They also recognize the presence of other (latent) functions: baby sitting young children so parents can work, and teaching habits of discipline and obedience.

Conflict theorists ask, as they usually do, who benefits and who is deprived. They see a social-class bias in our educational system, which means that privileged children benefit and working- and lower-class children are the losers. They believe that the social reproduction model operates in that schooling works to reproduce the social class of the student. Children of working-class parents go to schools that teach them to be obedient and follow rules, and students of the elite go to schools that encourage creativity and flexibility. Conflict theorists believe that working-class students drop out at least in part because they see that the system does not benefit them, and they give up. Conflict theorists would be skeptical about educational reform; they fear that given the type of changes suggested and the continuing lack of funding, the beneficiaries of reform will be those who are already privileged.

Much of the data discussed in this chapter comes from records: *reactive* and *nonreactive surveys* of numbers of days in school, salary by occupation, and high-school dropout rates by state and ethnicity. Some data came from *reactive questionnaires* and *interviews*—for example, asking students why they drop out of school. The cross-cultural studies on student performance were clearly *reactive* and were probably *experiments.*

SUMMARY

The chapter focused on education, the fifth of five institutional areas explored in this book. Previous chapters examined the family, religion, and economic and political institutions. All societies must socialize and train their young—the central issue for the institution of education. We distinguished between education and schooling. The major focus of the chapter was on the unique and contrasting ways in which functional analysis and conflict theory view education and schooling. Finally, we discussed some current issues in education in the United States.

The two readings that follow look at opposite ends of the schooling continuum: Studs Terkel interviews a college faculty member teaching a group of honor students, and Harry Gracey describes a day in kindergarten.

TERMS FOR STUDY

chronological age (315)

democratic model (309)

developmental age (315)

education (305)

hidden curriculum (308)

schooling (305)

social reproduction model (309)

tracking (307)

Reading 10.1

SCHOOL DAYS

Studs Terkel

In his book The Great Divide: Second Thoughts on the American Dream, *author Studs Terkel tells his story through the words of other people. In this piece, Professor Robert Franke makes some observations about students in his class at the University of Arkansas–Little Rock.*

For the past several years, [Robert Franke] has been teaching an honors course at the University of Arkansas–Little Rock: Science and Society.

Most of his students are from Arkansas, urban as well as rural. Almost all come from fundamentalist families; politically conservative and middle-class.

They have learned how to take college tests. They score high, especially in math. They are quite verbal. They give the impression of being bright. Encouraged by their families, they come with the conviction that education is something they want, something they need. But their definition of education is something else. To them, education is not treated as an opportunity to explore new ideas, to sort out what other people think. For most of these young people, to be educated is to better articulate what they stand for, why they're right.

Early in the course, we talked about the nature of science. We look at it as a process, as a way of discovering ways of looking at our world and universe. They look at facts as something you memorize and tell your kids about, and that's what science is. We're telling them it's a voyage during which we sometimes take a wrong turn, and new findings help us get back on course. Trial and error. They work from a preconceived set of ideas.

They've been taught that the university is a dangerous though necessary place. It's something you need if you want to live comfortably. It's good to be educated, but it's a necessary evil.

You don't challenge any doubts you may have. There are some truths in the world and you know what those truths are, because you've already been taught in church and by parents.

First and foremost, there are religious truths. The assumption is that everyone does or should hold these truths. Almost a hundred percent of my students do not believe in evolution. I ask, "What is your source of information about evolution?" Two-thirds say, Home. The other third says, Church. They are assigned reading and they're very conscientious about it. They come to class prepared, their chores done. As though girded for battle. The enemy is any part of society out there that isn't quite what they are. The enemy is anybody who is not a fundamentalist. The enemy is anybody who is not a Reagan Republican. They would never say, I don't like you. They are attractive young people who even like their enemies. They come to school to find out more about those others, to learn how to cope with them. A student told me: "I'm glad to hear you talk about evolution. I really know what it's all about, but I haven't changed my views. I can answer you better."

Surprisingly, with these young people, communism is a bit dated as an enemy. They know it was something to worry about in the fifties. Today, it's sort of vague and historical. They have no conception of anything historical. It is only now that counts. We exist for ourselves. The past has nothing to do with us. There's a puzzlement about Russia. They sort of respect Russia because it's developing in technology. They, the Russians, are trying to make it, like everybody else. There's almost a kinship.

We frequently go to the original works. They've read *The Origin of Species*. They become torn because Charles Darwin is so logical. He is also very gentle and persuasive. As you get into his work, how do you respond to a kindly old man writing something interesting? It's very hard to get mad at him. That's what makes this book so diabolic. They're thrown into a turmoil and become very uptight, very angry.

I don't lay this on them very hard at all. I try to be gentle. I tell them I don't expect them to change their point of view, but I do expect them to be receptive to another's so they may be able to contrast the two.

For them, it's a new way of looking at things. I represent a peculiar sort of openness. They're not used to it. They're disturbed. I think they appreciate my respectfulness toward them, but they suspect me as they do anybody who deals with science. I tell them that we need to look at ideas for their own sake and see how they fit into the way they look at the world. It doesn't mean we dislike each other because we have different ideas.

A student, early in the course, will tell me, "Charles Darwin didn't go over too well at home." What probably happened is he took this idea home, took a point of view similar to mine, and threw his parents into an absolute dither. He tried it out just to see how it would go. So when he said, "You won't believe how upset my father got," I realized he was taking the course seriously.

Amazingly enough, the parents are very respectful toward us teachers. We represent authority and knowledge. Their children have been taught that we have some of it and they should get it. Knowledge is something you buy, you own, and then pass on to your children. A piece of property.

I have a hunch, because of their native intelligence, some of these students will do pretty well. They'll probably be managers, directors of laboratories, or work their way up in computer firms. Believe it or not, I have a feeling we may get a couple of writers and possibly a poet. Though these few haven't wholly abandoned the teaching of their parents and church, they're taking into account other ideas, and one day these will fall into place with their earlier views. Perhaps when they're in their thirties, they'll make a radical breakthrough.

Remember, in most cases, they've never been out of Arkansas. Church and family is all they have. It's out of this milieu came their values of what is right and wrong. They believe in Authority. Authority is knowledge. Authority is government. Authority is parent. Authority is church. You do not question Authority.

Robert Franke taught at Iowa State during most of the sixties. "They had their Bible Belt, too. The students were not that much different in upbringing."

Before the Reagan era, they were more willing to grapple with what was right and wrong. They were willing to assume responsibilities of moral choice. I know it's been referred to as a decadent era. I found it just the opposite. There was a search. The great unrest among my students came out of a dissatisfaction with some of the presidents, with what was happening in our society, with wars. I saw it as a time of the young searching, examining alternative ways of doing things, better ways. Always questioning.

The students in my class today are very reluctant to assume responsibility for their own moral choices. If I were to tell them persuasively something patently false, they'd probably go along. It doesn't really matter to them.

In the sixties, there was beneath all the turmoil a hopefulness, a genuine hopefulness that the world would be better. That gave spirit to the time. It is false to describe the young then as lost souls.

I took a poll in my class recently: How many of you feel there will be a nuclear war? Seventy-five percent of the class raised their hands. I was appalled, thinking of their absolute hopelessness. Assuming the Vietnam War were still going on, I have a terrible hunch these students would have said there's nothing to be done about altering the government's stand. What's the point if nuclear war is inevitable? When they all saw *On the Beach* on TV—the world destroyed by the Bomb—I tried to make it personal. They'd have none of it.

With absence of hope, I find an absence of generosity. We had a course on disease. We contrasted today with the time of the bubonic plague: how medicine developed, how attitudes changed, how superstition was challenged. Inevitably we came to a discussion of AIDS, of its devastating effects, of our fears. The prevailing attitude in class threw me for a loop: people with AIDS deserve to die. Others would say, I want nothing to do with them. There was hardly any show of compassion. They were so cold. Of course, if you're going to be blown up, you have to be dispassionate and not worry about people with disease. I don't think that was true in the sixties.

Yet when I showed them a film on AIDS and the dying, they were disturbed. Some in the class cried openly.

QUESTIONS 10.1

1. We discussed the hidden curriculum in this chapter. Can you find examples of it in this essay? If so, describe it.

2. How do the family institution and the education institution interact in this essay? Which would you say is more powerful? Why?

3. Analyze this essay from two viewpoints: first, from the perspective of functional analysis; second, from the perspective of conflict theory.

Reading 10.2

LEARNING THE STUDENT ROLE: KINDERGARTEN AS ACADEMIC BOOT CAMP

Harry L. Gracey

Education is part of the process of socialization, the process whereby one learns the expectations, skills, beliefs, and other attributes necessary for effective participation in social groups. It involves reading, writing, and arithmetic. It also involves a "hidden curriculum." In this article, Harry Gracey compares kindergarten to military basic training—a boot camp. He first describes a day in kindergarten; then he analyzes what the students are really learning—how to play important roles that they will play throughout their lives.

. . . Kindergarten is generally conceived by educators as a year of preparation for school. It is thought of as a year in which small children, five or six years old, are prepared socially and emotionally for the academic learning which will take place over the next twelve years. It is expected that a foundation of behavior and attitudes will be laid in kindergarten on which the children can acquire the skills and knowledge they will be taught in the grades. A booklet prepared for parents by the staff of a suburban New York school system says that the kindergarten experience will stimulate the child's desire to learn and cultivate the skills he will need for learning in the rest of his school career. It claims that the child will find opportunities for physical growth, for satisfying his "need for self-expression," acquire some knowledge, and provide opportunities for creative activity. It concludes, "The most important benefit that your five-year-old will receive from kindergarten is the opportunity to live and grow happily and purposefully with others in a small society." The kindergarten teachers in one of the elementary schools in this community, one we shall call the Wilbur Wright School, said their goals were to see that the children "grew" in all ways: physically, of course, emotionally, socially, and academically. They said they wanted children to like school as a result of their kindergarten experiences and that they wanted them to learn to get along with others. . . .

The unique job of the kindergarten in the educational division of labor seems rather to be teaching children the student role. The student role is the repertoire of behavior and attitudes regarded by educators as appropriate to children in school. Observation in the kindergartens of the Wilbur Wright School revealed a great variety of activities through which children are shown and then drilled in the behavior and attitudes defined as appropriate for school and thereby induced to learn the role of student. Observations of the kindergartens and interviews with the teachers both pointed to the teaching and learning of classroom routines as the main element of the student role. The teachers expended most of their efforts, for the first half of the year at least, in training the children to follow the routines which teachers created. The children were, in a very real sense, *drilled* in tasks and activities created by the teachers for their own purposes and beginning and ending quite arbitrarily (from the child's point of view) at the command of the teacher. One teacher remarked that she hated September, because during the first month "everything has to be done rigidly, and repeatedly, until they know exactly what they're supposed to do." However, "by January," she said, "they know exactly what to do [during the day] and I don't have to be after them all the time." Classroom routines were introduced gradually from the beginning of the year in all the kindergartens, and children were drilled in them as long as was necessary to achieve regular compliance. By the end of the school year, the successful kindergarten teacher has a well-organized group of children. They follow classroom routines automatically, having learned all the command signals and the expected responses to them. They have, in our terms, learned the student role. The following observation shows one such classroom operating at optimum organization on an afternoon late in May. It is the class of an experienced and respected kindergarten teacher.

An Afternoon in Kindergarten

At about 12:20 in the afternoon on a day in the last week of May, Edith Kerr leaves the teachers' room where she has been having lunch and walks to her classroom at the far end of the primary wing of Wright School. A group of five- and six-year-olds peer at her through the glass doors leading from the hall cloakroom to the play area outside. Entering her room, she straightens some material in the "book corner" of the room, arranges music on the piano, takes colored paper from her closet, and places it on one of the shelves under the window. Her room is divided into a number of activity areas through the arrangement of furniture and play equipment. Two easels and a paint table near the door create a kind of passageway inside the room. A wedge-shaped area just inside the front door is made into a teacher's area by the placing of "her" things there: her desk, file, and piano. To the left is the book corner, marked off from the rest of the room by a puppet stage and a movable chalkboard. In it are a display rack of picture books, a record player, and a stack of children's records. To the right of the entrance are the sink and cleanup area. . . .

At 12:25 Edith opens the outside door and admits the waiting children. They hang their sweaters on hooks outside the door and then go to the center of the room and arrange themselves in a semicircle on the floor, facing the teacher's chair which she has placed in the center of the floor. Edith follows them in and sits in her chair checking attendance while waiting for the bell to ring. When she has finished attendance, which she takes by sight, she asks the children what the date is, what day and month it is, how many children are enrolled in the class, how many are present, and how many are absent.

The bell rings at 12:30 and the teacher puts away her attendance book. She introduces a visitor, who is sitting against the right wall taking notes, as someone who wants to learn about schools and children. She then goes to the back of the room and takes down a large chart labeled "Helping Hands." Bringing it to the center of the room, she tells the children it is time to change jobs. Each child is assigned some task on the chart by placing his name, lettered on a paper "hand," next to a picture signifying the task—e.g., a broom, a blackboard, a milk bottle, a flag, and a Bible. She asks the children who wants each of the jobs and rearranges their "hands" accordingly. Returning to her chair, Edith announces, "One person should tell us what happened to Mark." A girl raises her hand and when called on says, "Mark fell and hit his head and had to go to the hospital." The teacher adds that Mark's mother had written saying he was in the hospital. . . .

At 12:35 two children arrive. Edith asks them why they are late and then sends them to join the circle on the floor. The other children vie with each other to tell the newcomers what happened to Mark. When this leads to a general disorder Edith asks, "Who has serious time?" The children become quiet, and a girl raises her hand. Edith nods and the child gets a Bible and hands it to Edith. She reads the Twenty-third Psalm while the children sit quietly. Edith helps the child in charge begin reciting the Lord's Prayer, while the other children follow along for the first unit of sounds and then trail off as Edith finishes for them. Everyone stands and faces the American flag hung to the right of the door. Edith leads the pledge to the flag, with the children again following the familiar sounds as far as they remember them. Edith then asks the girl in charge what song she wants and the child replies, "My Country." Edith goes to the piano and plays "America," singing as the children follow her words.

Edith returns to her chair in the center of the room, and the children sit again in the semicircle on the floor. It is 12:40 when she tells the children, "Let's have boys' sharing time first." She calls the name of the first boy sitting on the end of the circle, and he comes up to her with a toy helicopter. He turns and holds it up for the other children to see. He says, "It's a helicopter." Edith asks, "What is it used for?" and he replies, "For the army. Carry men. For the war." Other children join in, "For shooting submarines." "To bring back men from space when they are in the ocean." Edith sends the boy back to the circle and asks the next boy if he has something. He replies "No" and she passes on to the next. He says "Yes" and brings a bird's nest to her. He holds it for the class to see, and the teacher asks, "What kind of bird made the nest?" The boy replies, "My friend says a rain bird made it." Edith asks what the nest is made of and different children reply, "mud," "leaves," and "sticks." There is also a bit of moss woven into the nest and Edith tries to describe it to the children. They, however, are more interested in seeing if anything is inside it, and Edith lets the boy carry it around the semicircle showing the children its insides. Edith tells the children of some baby robins in a nest in her yard, and some of the children tell about baby birds they have seen. Some children are asking about a small object in the nest which they say looks like an egg, but all have seen the nest now, and Edith calls on the next boy. . . .

At 1:30 Edith has the children line up in the center of the room; she says, "Table one, line up in front of me," and children ask, "What are we going to do?" Then she moves a few steps to the side and says, "Table two over here, line up next to table one," and more children ask, "What for?" She does this for table three and table four and each time the children ask, "Why, what are we going to do?" When the children are lined up in four lines of five each, spaced so that they are not touching one another, Edith puts on a new record and leads the class in calisthenics, to the accompaniment of the record. The children just jump around every which way in their places instead of doing the exercises, and by the time the record is finished, Edith, the only one following it, seems exhausted. She is apparently adopting the president's new "Physical Fitness" program in her classroom.

At 1:35 Edith pulls her chair to the easels and calls the children to sit on the floor in front of her, table by table. When they are all seated she asks, "What are you going to do for worktime today?" Different children raise their hands and tell Edith what they are going to draw. Most are going to make pictures of animals they saw in the zoo. Edith asks if they want to make pictures to send to Mark in the hospital, and the children agree to this. Edith gives drawing paper to the children, calling them to her one by one. After getting a piece of paper, the children go to the crayon box on the right-hand shelves, select a number of colors, and go to the tables, where they begin drawing. Edith is again trying to quiet the perpetually talking girls. She keeps two of them standing by her so they won't disrupt the others. She asks them, "Why do you feel you have to talk all the time?" and then scolds them for not listening to her. Then she sends them to their tables to draw. . . .

At 2:15 Edith walks to the entrance of the room, switches off the lights, and sits at the piano and plays. The children begin spontaneously singing the song, which is "Clean up, clean up. Everybody clean up." Edith walks around the room supervising the cleanup. Some children put their toys, the blocks, puzzles, games, and so on back on their shelves under the windows. The children making a collage keep right on working. A child from another class comes in to borrow the 45-rpm adaptor for the record player. At more urging from Edith the rest of the children shelve their toys and work. The children are sitting around their tables now, and Edith asks, "What record would you like to hear while you have your milk?" There is some confusion and no general consensus, so Edith drops the subject and begins to call the children, table by table, to come get their milk. "Table one," she says, and the five children come to the sink, wash their hands and dry them, pick up a carton of milk and a straw, and take it back to their table. Two talking girls wander about the room interfering with the children getting their milk and Edith calls out to them to "settle down." As the children sit, many of them call out to Edith the name of the record they want to hear. When all the children are seated at tables with milk, Edith plays one of these records called "Bozo and the Birds" and shows the children pictures in a book which go with the record. The record recites, and the book shows the adventures of the clown, Bozo, as he walks through a woods meeting many different kinds of birds who, of course, display the characteristics of many kinds of people or, more accurately, different stereotypes. As children finish their milk they take blankets or pads from the shelves under the windows and lie on them in the center of the room where Edith sits on her chair showing the pictures. By 2:30 half the class is lying on the floor on their blankets, the record is still playing, and the teacher is turning the pages of the book. The child who came in previously returns the 45-rpm adaptor, and one of the kindergartners tells Edith what the boy's name is and where he lives.

The record ends at 2:40. Edith says, "Children, down on your blankets." All the class is lying on blankets now. Edith refuses to answer the various questions individual children put to her because, she tells them, "it's rest time now." Instead she talks very softly about what they will do tomorrow. They are going to work with clay, she says. The children lie quietly and listen. One of the boys raises his hand and when called on tells Edith, "The animals in the zoo looked so hungry yesterday." Edith asks the children what they think about this and a number try to volunteer opinions, but Edith accepts only those offered in a "rest-time tone," that is, softly and quietly. . . .

At 2:50 Edith sits at the piano and plays. The children sit on the floor in the center of the room and sing. They have a repertoire of songs about animals, including one in which each child sings a refrain alone. They know these by heart and sing along through the ringing of the 2:55 bell. When the song is finished Edith gets up and coming to the group says, "Okay, rhyming words to get your coats today." The children raise their hands and as Edith calls on them, they tell her two rhyming words, after which they are allowed to go into the hall to get their coats and sweaters. They return to the room with these and sit at their tables. At 2:59 Edith says, "When you have your coats on, you may line up at the door." Half of the children go to the door and stand in a long line. When the three o'clock bell rings, Edith returns to the piano and plays. The children sing a song called "Goodbye," after which Edith sends them out.

Training for Learning and Life

The day in kindergarten at Wright School illustrates both the content of the student role as it has been learned by these children and the processes by which the teacher has brought about this learning or "taught" them the student role. The children have learned to go through routines and to follow orders with unquestioning obedience, even when these make no sense to them. They have been disciplined to do as they are told by an authoritative person without significant protest. Edith has developed this discipline in the children by creating and enforcing a rigid social structure in the classroom through which she effectively controls the behavior of most of the children for most of the school day. The "living with others in a small society" which the school pamphlet tells parents is the most important thing the children will learn in kindergarten can be seen now in its operational meaning, which is learning to live by the routines imposed by the school. This learning appears to be the principal content of the student role.

Children who submit to school-imposed discipline and come to identify with it, so that being a "good student" comes to be an important part of their developing identities, *become* the good students by the school's definitions. Those who submit to the routines of the school but do not come to identify with them will be adequate students who find the more important part of their identities elsewhere, such as in the play group outside school. Children who refuse to submit to the school routines are rebels, who become known as "bad students" and often "problem children" in the school, for they do not learn the academic curriculum and their behavior is often disruptive in the classroom. Today, schools engage clinical psychologists in part to help teachers deal with such children.

In looking at Edith's kindergarten at Wright School, it is interesting to ask how the children learn this role of student—come to accept school-imposed routines—and what, exactly, it involves in terms of behavior and attitudes. The most prominent features of the classroom are its physical and social structures. The room is carefully furnished and arranged in ways adults feel will interest children. The play store and play kitchen in the back of the room, for example, imply that children are interested in mimicking these activities of the adult world. The only space left for the children to create something of their own is the empty center of the room, and the materials at their disposal are the blocks, whose use causes anxiety on the part of the teacher. The room, being carefully organized physically by the adults, leaves little room for the creation of physical organization on the part of the children.

The social structure created by Edith is a far more powerful and subtle force for fitting the children to the student role. This structure is established by the very rigid and tightly controlled set of rituals and routines through which the children are put during the day. There is first the rigid "locating procedure" in which the children are asked to find themselves in terms of the month, date, day of the week, and the number of the class who are present and absent. This puts them solidly in the real world as defined by adults. The day is then divided into six periods whose activities are for the most part determined by the teacher. In Edith's kindergarten the children went through serious time, which opens the school day, sharing time, play time (which in clear weather would be spent outside), work time, cleanup time, after which they have their milk, and rest time, after which they go home. The teacher has programmed activities for each of these times.

Occasionally the class is allowed limited discretion to choose between proffered activities such as stories or records, but original ideas for activities are never solicited from them. Opportunity for free individual action is open only once in the day, during the part of the work time left after the general class assignment has been completed (on the day reported, the class assignment was drawing animal pictures for the absent Mark). Spontaneous interests or observations from the children are never developed by the teacher. It seems that her schedule just does not allow room for developing such unplanned events. During sharing time, for example, the child who brought a bird's nest told Edith, in reply to her question of what kind of bird made it, "My friend says it's a rain bird." Edith does not think to ask about this bird, probably because the answer is "childish," that is, not given in accepted adult categories of birds. . . .

While children's perceptions of the world and opportunities for genuine spontaneity and creativity are being systematically eliminated from the kindergarten, unquestioned obedience to authority and rote learning of meaningless material are being encouraged. When the children are called to line up in the center of the room they ask "Why?" and "What for?" as they are in the very process of complying. They have learned to go smoothly through a programmed day, regardless of whether parts of the program make any sense to them or not. Here the student role involves what might be called "doing what you're told and never mind why." Activities which might "make sense" to the children are effectively ruled out, and they are forced or induced to participate in activities which may be "senseless," such as the calisthenics.

At the same time the children are being taught by rote meaningless sounds in the ritual oaths and songs, such as the Lord's Prayer, the Pledge to the Flag, and "America." As they go through the grades children learn more and more of the sounds of these ritual oaths, but the fact that they have often learned meaningless sounds rather than meaningful statements is shown when they are asked to write these out in the sixth grade; they write them as groups of sounds rather than as a series of words, according to the sixth-grade teachers at Wright School. Probably much learning in the elementary grades is of this character, that is, having no intrinsic meaning to the children but rather being tasks inexplicably required of them by authoritative adults. . . .

The kindergarten has been conceived of here as the year in which children are prepared for their schooling by learning the role of student. In the classrooms of the rest of the school grades the children will be asked to submit to systems and routines imposed by the teachers and the curriculum. The days will be much like those of kindergarten, except that academic subjects will be substituted for the activities of the kindergarten. Once out of the school system, young adults will more than likely find themselves working in large-scale bureaucratic organizations, perhaps on the assembly line in the factory, perhaps in the paper routines of the white collar occupations, where they will be required to submit to rigid routines imposed by "the company" which may make little sense to them. Those who can operate well in this situation will be successful bureaucratic functionaries. Kindergarten, therefore, can be seen as preparing children not only for participation in the bureaucratic organization of large modern school systems, but also for the large-scale occupational bureaucracies of modern society.

QUESTIONS 10.2

1. Give examples of the hidden curriculum from this article.

2. Is the process described in kindergarten still going on in your college classes? If so, give examples.

3. Analyze this article from the perspective of the two theories discussed in this chapter. What would a functionalist see? What would a conflict theorist see?

NOTES

1. This discussion comes from chapter 3 of Christopher Hurn's book, *The Limits and Possibilities of Schooling: An Introduction to the Sociology of Education* (Boston: Allyn and Bacon, 1985).

2. This section on functionalism and conflict theory follows from Hurn's discussion in his chapter 2 (see Note 1).

3. See *Schooling in Capitalist America* by Samuel Bowles and Herbert Gintis (New York: Basic Books, 1976).

4. Estimates on guns and dropouts come from "America Skips School," by Benjamin Barber, *Harper's Magazine* 288 (November

1993), p. 39. Other information on dropouts in this paragraph comes from "High School Dropouts; Descriptive Information from High School and Beyond," U.S. Department of Education, National Center for Educational Statistics, Bulletin NCES 83 221b, November 1983.

5. "The Case for More School Days," by Michael J. Barrett, *The Atlantic Monthly* 266 (November 1990), pp. 78–106.

6. This work by Harold Stevenson is discussed in Michael Barrett's article (Note 5), p. 94.

CHAPTER ELEVEN

Population and Ecology

TWO RESEARCH QUESTIONS

1. *Some states are gaining population rapidly; others are losing population. How is population being redistributed in the United States? Are people moving to the West and South or to the Midwest and East, or is there any pattern?*

2. *Is there really a world population explosion? If so, what is the cause? Is it fertility (increases in birth rate), mortality (declines in death rate), or migration (population relocation)?*

PREVIOUS CHAPTERS IN this section on social organization have examined concepts dealing with groups, categories, aggregations, and institutions. In this chapter, we will show how social organization may be analyzed through the study of population and human ecology. Population refers to the number of people in a given unit, as in a state, society, world, or universe. **Human ecology** refers to the adaptation of people to their physical environment, their location in space. **Demography** is the study of human population—its distribution, composition, and change.

As an introduction to some of the important techniques and variables, let's try a demographic analysis of a college sociology class. What does the professor see as he or she looks out at the mass of eager young faces? The total population of the class selected is 130, of which 80 are females and 50 are males. There are 4 African Americans, 11 Asian Americans, and 115 Caucasian Americans. The age distribution of this "society" is overwhelmingly in the 19 to 23 year category, with trace amounts of ages up to 45. Both birth and death rates for this "society" are exceedingly low. Apparently this represents a very healthy but nonfertile tribe. The life expectancy for the majority will be one academic term, although for some it may be at least double that. Migration variables are peculiar for this tribe. Temporary immigration ("in-migration") occurs on those occasions when the class is planning to discuss sex or deviant behavior. Permanent emigration ("out-migration") occurs just before tests are scheduled or papers are due.

Ecologically, this class is also interesting. After several meetings they arrange themselves in space consistently and predictably. There are occasional shifts as dating alliances change, but generally they sit in the same seats throughout the term. Part of the patterning may be explained by classical ecological processes. Cooperation: X takes great notes, so she is constantly surrounded by A, B, and C (at least I think that's the reason). Competition: When asked why they sit in front, several have said they pay more attention to what is being said and feel the instructor is more likely to remember them; consequently, they will get better grades. Films of migrating apes have shown ecological processes at work: Older males proceed in front, younger males at the rear, and females are protected in the middle. Do these principles apply in your classroom?

POPULATION

Demographers study populations from several viewpoints. They may want to look at population growth or its general characteristics—a statistical portrait, so to speak. A number of different variables may be used. For example, if the issue is population growth, the important variables are births, deaths, and immigration. (Population growth = births − deaths + immigration.) If the concern is a general description of the population, then age, gender, race, ethnicity profiles, and perhaps patterns of internal migration may be examined. Growth potential and general characteristics are related; the number of young and old people in a population as well as the number of women in childbearing ages will have an effect on birth and death rates. In the following paragraphs we will look at some of the major population variables and show how they are used in the United States and in some other societies.

Fertility

Fertility, which refers to the number of children born, may be measured in several ways. The **birth rate** (or crude birth rate) is the number of live births per 1,000 people in the population. These data are often used because they are easily collected, but they ignore the age structure of the population. The **general fertility rate,** which is a better measure, is defined as the number of births per 1,000 women aged 15 through 44. The **total fertility rate** is an estimate of the average number of children born to each woman based on current rates.

Fertility has both a biological and a social aspect. The physical or biological ability to reproduce, also known as **fecundity,** is present in most of us, and yet actual fertility rates differ greatly. A 19th-century Russian woman gave birth to 69 children; the Hutterite women (of North and South Dakota and Canada) were averaging more than 12 children each in the 1930s; currently, American women are averaging about two children each. The social influences on fertility are factors that vary from one society to another. For example, fertility would be affected by age at marriage, the proportion of people who never marry, frequency of sexual intercourse, and whether birth-control techniques are used and how effective they are. These factors are influenced by tradition, lifestyle, standard of living, and countless other social values.[1]

Trends in birth, general fertility, and death rates in the United States are shown in Figure 11.1. Our birth rate has been around 15 per 1,000 since the mid-1970s (16.2 in 1991). Remember that although birth and general fertility rates are similar, the general fertility rate is the better measure because it focuses on the number of women in the childbearing ages (15 to 44). The rates follow the same pattern, but the fluctuations are seen more clearly in the general fertility rate. The rates were high in the early 1900s, dropped in the 1920s

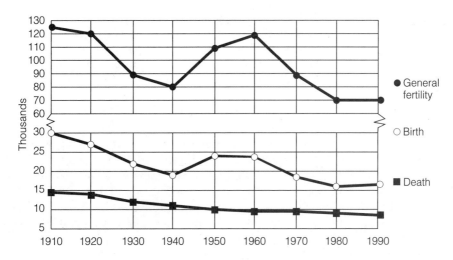

FIGURE 11.1   *General fertility, birth, and death rates in the United States (per 1,000 population).*

United States Bureau of the Census, *Statistical Abstracts of the United States, 1993* (113th ed.), Washington, D.C.

and 1930s, went up after World War II, dropped sharply in the 1960s and 1970s, and were fairly flat in the 1980s. The general fertility rate dropped to 64.9 in 1986—its lowest level in U.S. history (or at least since such records have been kept). In 1991 it moved up to 70. The total fertility rate in 1987 was 1,871 per 1,000 women and in 1991 had moved up to 2,041, which means that women today average about 2 children each. **Replacement level** is defined as the average number of births necessary per woman over her lifetime (say 70 years) for the population eventually to reach zero growth. Although the United States has been below the replacement level, population continues to increase because of large increases in the number of women of reproductive age. The Census Bureau reports that this means that even if the current low fertility rates were to persist for some time, the population of the United States would still continue to grow by natural increase until well into the 21st century.

Understanding why fertility rates fluctuate helps us predict social conditions of the future, but the factors are many and complicated. This is what seems to be happening: More young women are remaining single, and divorce rates remain high (see the discussion in Chapter Seven). Women, especially those in the prime childbearing ages of 20 through 34, are either not having or are putting off having children. The increased freedom of women to work outside the home, the increased use of contraceptives, and changing viewpoints about family roles have contributed to the changing birth rate. Unemployment rates and a sluggish economy are probably also contributing factors.

World birth rates in 1990 showed an average of 3.8 children per woman, compared with about 1.9 in the United States. African countries had the highest rates—an average of 6.4 for the continent and 8.0 in Kenya. Europe had the lowest rates—1.8 for the continent and 1.4 for Denmark. (Some of these comparisons can be seen later in this chapter in Table 11.3, page 345.) High fertility rates may, if mortality rates are low, lead to rapid population growth, and later in this chapter we will examine some of the implications and consequences.

Mortality

Mortality is the number of deaths occurring in a population. It is typically represented by the death rate (or crude death rate), which is the total deaths per 1,000 people in the population. Because death rates vary by age and gender, a more precise measure is the age-gender-specific death rate, which categorizes deaths of males and females in five-year groups—for example, males 15 through 19 and females 15 through 19. A third useful measure is **life expectancy,** which predicts one's expectation of length of life at birth based on risks of death for people born in that year.

Mortality has two major components: life span and longevity. *Life span* refers to the oldest age to which human beings can live. There is some uncertainty as to just what the limits are; reports of people living to 110 and even to 120 years of age are not unheard of, and medical science continues to increase these outer limits. *Longevity* refers to the ability to remain alive from one year to the next—that is, to resist death. Both biological and social factors, of course, are integral to one's longevity. Biological determinants include inherited genetic characteristics, strength of organs, and resistance to disease. Social factors include lifestyle and the amount of stress and conflict it produces, the way society takes care of its members—especially the elderly—and so on. The Abkhasians in Russia and the Vilcabambas in Ecuador seem to live very long lives; it has been suggested that this is due to their simple, unchanging lifestyle, their avoidance of stress, and their tendency to remain physically and mentally active throughout their lives.[2]

The importance of social factors as explanations of longevity and mortality can be seen in other ways. Studies have shown that as occupational prestige goes up, death rates go down. This may be due to greater work hazards in lower-class occupations. However, because the mortality levels of "nonworking" wives seem to parallel those of their husbands, lifestyle is apparently a major factor as well. Studies have also found that deaths from communicable diseases such as tuberculosis and pneumonia are higher among lower-class people, whereas deaths from coronaries, diabetes, and stroke increase with upward social mobility. Yet another study has found that for virtually every major cause of death, white males with at least one year of college had lower risks of death than those with less education. The existence of mortality-rate differences among various racial and ethnic groups is also probably best explained in terms of income, educational, and lifestyle differences.

Death rates in the United States are also shown in Figure 11.1. We see that they have declined, but over the last 50 years they haven't changed much, and over the last decade, hardly at all. Death rates have leveled at between 8.5 and 8.8 from 1975 to the present. Sharp drops in the death rate occurred in the late-19th and early-20th centuries as medical science gained control of many infectious diseases. Life expectancy has risen to a current level of about 73 years for males and 80 for females. Death rates in European countries are very similar to ours, but rates in some other countries (see Table 11.3) are twice ours. This is changing rapidly, however, as health care and disease-control measures spread and countries share their discoveries. So we witness a growth in population caused not by rising fertility, but by declining mortality.

Migration

Migration is defined as a permanent change of residence, with the consequent relocation of one's interests and activities. The number of dimensions—who moves, how often, where, and why—makes migration the most complex of the population processes. Internal migration generally describes migration within a country, whereas international migration refers to migration from one country to another. There are many reasons that people migrate, and these have been summarized into what is called the *push-pull theory*. This theory suggests that one reason people move is that they are pushed out of their original home—by a bad climate, by unpleasant treatment by others, by legal harassment, and so on. The other reason people move is that they are pulled by the attractions of another place. Those who advocate this theory suggest that motivation to improve one's lot is stronger than the desire to escape existing conditions; we could conclude that the *pull* motivation is more influential than is the *push*. It does seem, as Weeks points out, that migration associated with career advancement is a common theme in society today.[3]

Americans are a mobile group: About 43 million people, 17 percent of all people one year old and over, moved during the period between 1991 and 1992. Sociologists have found that some types are more prone to migration than others. Young adults aged 20 to 24 and 25 to 29 have the highest rates. Often these are young married couples with no children or very young children. College graduates are more likely to move than high-school graduates, who in turn are more mobile than those with less education. All of this tends to support the idea that people are being pulled to areas that are more economically attractive for them.

Patterns of internal migration continually change. People used to migrate from farm to city and later from city to suburb. More recently a new trend developed: the rapid growth of small communities near metropolitan areas and increasing migration to rural areas. The first question at the beginning of this chapter dealt with population redistribution. A major trend of the 1980s and 1990s has been a migration flow from "frostbelt" and "rustbelt" to "sunbelt." People are leaving northeastern and midwestern states and moving to the West and South. Eighty-five percent of the population growth in the United States between 1980 and 1990 took place in the South and the West. The five biggest gainers in numbers were California, Florida, Texas, Georgia, and Arizona. North Dakota, Wyoming, Iowa, and West Virginia actually lost population over that decade. Frostbelt states are declining in population, not only because of out-migration, but also because they typically have an older population and lower birth rates. The states that gained the most people between 1991 and 1992 were California (487,000), Texas (307,000), and Florida (221,000). The states gaining the most and least population on a *percentage* basis are shown in Table 11.1. It is clear that population is being redistributed, at least in part, as a result of changing recreation and leisure patterns.

International migration to the United States has gone through several stages. Before World War I, immigrants were admitted freely. Then, as people became concerned about "racial purity," job competition, and political "undesirables," immigration laws became very restrictive. Policies were relaxed slightly in the mid-1960s, but the overall number of immigrants allowed into this country was relatively small and constant throughout the 1970s and 1980s. Then came another change in 1989, and the number of immigrants almost tripled—from 643,000 in 1988 to 1,837,000 in 1991. More immigrants came from Mexico alone (over 900,000) in 1991 than from all countries three years earlier. Figure 11.2 shows the ten countries supplying the most immigrants in 1991. Immigration patterns are clearly affected by economic conditions and standard of living. Political instability and turmoil may also be a factor in some countries.

The consequences of migration are many. For the individual, uprooting oneself and trying to adapt to a new environment involves stress, and the burden of coping is greater in the absence of familiar support groups. The community left behind may lose a valuable segment of its population. The community to which people are migrating has a growing population to assimilate, with the accompanying problems of housing, mass transit, health care, job opportunities, municipal services, and ways to pay for it all. This has led some communities and even states to devise policies to limit migration to their areas.

Other Population Measures

Fertility, mortality, and migration are the major demographic variables, but other measures help describe certain population characteristics. One of the simplest and most direct measures is *population size*. The world population was 5.6 billion in 1993. (July 11, 1987, was designated around the world as "The Day of the Five Billion.") The population of the United States in 1994

Table 11.1 Population Change in the United States 1991–1992 (National Population Increased 1.2 Percent)

States Gaining **Most** Population	Percentage Change	States Gaining **Least** Population	Percentage Change
Nevada	3.5	Connecticut	−0.2
Alaska	2.9	Rhode Island	0
Colorado	2.7	Massachusetts	0
Idaho	2.7	Maine	0.1
Washington	2.5	North Dakota	0.2

United States Bureau of the Census press releases and *State Rankings 1993* (Lawrence, Kan.: Morgan Quitno).

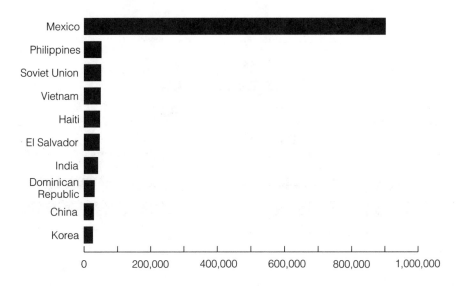

FIGURE 11.2 *Immigrants by country of birth, 1991.*

United States Bureau of the Census, *Statistical Abstracts of the United States, 1993* (113th ed.), Washington, D.C.

was 260 million. This means we are smaller than China (1.18 billion) and India (903 million) but larger than Canada (27 million) and France (57 million).

More detailed profiles based on sex, age, race, religion, and ethnicity are also of interest. **Sex ratio** is defined as the number of males per 100 females. The sex ratio in the United States in 1991 was around 95, or 95 males for every 100 females. Because there are 6 million more women than men in the United States, nearly all states show that same sex ratio, with Massachusetts, New York, Rhode Island, and Pennsylvania having the greatest number of women compared with men. Alaska, Hawaii, Wyoming, Nevada, and California were the only states in 1990 to have more men than women. The *median age* of a population, which designates the midpoint of the age range, is another population measure. The United States median age in 1970 was 28, in 1980 was 30, and in 1991 was 33.1. Our population is getting older; we have fewer teenagers and more elderly people and more people in their thirties and forties. The fastest-growing segments of the population are those aged 35 to 44 and those aged 85 years and older. The number of young people (15 to 24 years old) actually declined between 1980 and 1991.

Several variables, such as age and sex, can be profiled together graphically in a figure called a **population pyramid.** Figures 11.3 and 11.4 illustrate the distinctive age and sex profiles presented by two countries, Mexico and Sweden. Mexico has a high birth rate and a young population, whereas Sweden has a low birth rate and a balanced population in terms of age. These types of profiles could be done on cities—imagine the pyramid you might get in St. Petersburg, Florida (older, retired people and more females), Norfolk, Virginia

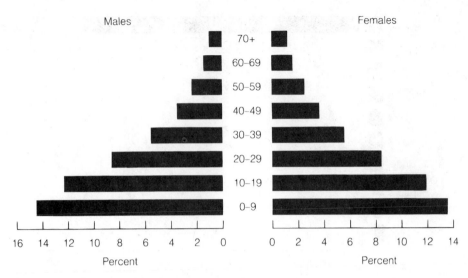

FIGURE 11.3 *Population pyramid for Mexico, 1985.*

Demographic Yearbook, 1985.

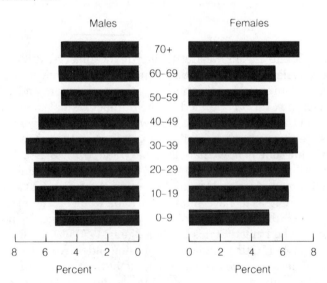

FIGURE 11.4 *Population pyramid for Sweden, 1985.*

Demographic Yearbook, 1985.

(influenced by the number of young males at the large military base), or Davis, California (a university community).

We have examined the major population variables—fertility, mortality, and migration—and other measures that allow demographers to predict future population changes and to provide a statistical portrait of societies. Population data, carefully collected and analyzed, are useful in social planning. Population profiles in terms of age and sex enable planners to determine the number of schools needed, the number and proportion of apartments and homes to build, the types of recreation and child-care facilities required, and so forth. For example, when we hear that a group of city planners is developing a community with no schools (because there are no children), no tennis courts (because tennis is too strenuous for the existing population), and many recreation centers and shuffleboard courts, it is safe to assume that the population data evidenced a need for a planned community for the elderly.

HUMAN ECOLOGY AND THE CITY

Biologists have for some time studied the interrelationships and interdependence of plants and animals—the balance of nature. Living organisms are dependent on other organisms and distribute themselves in space accordingly. Why don't porcupines live in the desert? It may be because they would get sand in their quills, but most certainly because their food supply—the bark of specific types of trees—does not grow there. Likewise, all organisms live in particular places and patterns but not in others. Social scientists noticed that people, too, locate themselves in space in specific patterns, and this gave rise to the study of human ecology.

Flying over the United States, we see that people are not equally distributed but are grouped in clusters. There are fewer clusters in the West and more in the East. Most of the clusters are near rivers, lakes, or oceans. Highways run through the clusters, and various natural resources are nearby. There are fewer clusters where the climate is severe or where the terrain is rugged. People also distribute themselves in space in predictable ways. They group together in cities, where mutual cooperation and a complex division of labor allow people to accomplish much more than they could as isolated individuals. These cities are located to take maximum advantage of natural factors—waterways for transportation, resources for technological needs, and so on.

If we look more closely at the cities, we see that they also have consistent patterns. One city patterning first suggested by Ernest Burgess has five concentric circles or zones (see Figure 11.5). The small inner circle is called the *central business district* and contains the major stores, banks, offices, and government buildings—the downtown area of the city. The next circle is called the *zone in transition* and contains rundown buildings, slums, a high proportion of minority groups, some industry, and higher rates of unemployment. The third ring is called the *workingman's zone* and contains small apartment buildings and older single-dwelling units. The *middle-class residential district* is the fourth zone and contains better private residences and high-class apartments. Finally, the *suburban* or *commuters' zone* contains larger estates and golf courses and makes up the upper-class residential district. Generally speaking, as one moves from the inner zones outward toward the suburbs, population density decreases, home ownership increases, the proportion of foreign-born decreases, social class rises, crime rate decreases, land cost decreases, and family size decreases. The pattern of the city may be affected by how fast the city develops, by the influence of cars, and by natural obstacles—hills, oceans,

FIGURE 11.5 *Three generalizations of the internal structure of cities.*

From C. D. Harris and Edward L. Ullman, *Annals* 242 (November 1945), pp. 7–17.

rivers. So a given city may not fit this design exactly, but many cities are close to this general pattern.

Other consistent city patternings have been suggested and are also depicted in Figure 11.5. Homer Hoyt contends that cities are more accurately described as sectors with a specific type of land use than as concentric zones. He believes that expensive high-rent areas follow patterns of city growth outward from the center like wedges of pie or spokes of a wheel. High-rent areas develop along the paths of major streets, highways, or other major transportation lines. This type of outward growth could be along rivers or rapid-transit systems, for example, and would often cut across the concentric circles suggested by Burgess.

Multiple nuclei describes a city in which a number of specialized centers or downtowns develop, each of which exerts dominance over its particular area. These centers may exist from the beginning of a city's development, as in London, or may develop later as the city grows and expands. Industry and residential areas grow up around each nucleus. Los Angeles, which has been called "a bunch of suburbs in search of a city," is an example of multiple-nuclei development.

More recent theories of city patterning include social area analysis and factorial ecology. These techniques focus on a city's census tracts and obtain information across many variables, such as median rent, median education, percentage of the population married and employed, and the racial and ethnic mix. Tracts are then compared in an attempt to isolate the factors that distinguish among tracts and that explain city patterning. These techniques are usually computerized and are very precise.[4]

How are these patterns explained? People arrange themselves in space according to social and cultural values that they believe to be important. It is apparent from previous paragraphs that economic competition has much to do with spatial arrangement in cities. Land is in greatest demand and, therefore, is most expensive in the center of the city. Banks, large stores, businesses, office buildings, and government buildings can afford the cost of locating there because of their importance, business volume, and/or financial base. Structures whose economic effects are less dramatic—homes, apartment buildings, schools, country clubs—are usually located away from the center of the city where land is cheaper. Other social values are also reflected. Many cities have set aside large green areas in those parts of the city where land is most expensive. Parks near the centers of London and Paris, the Boston Common, and Central Park in New York are but a few of numerous examples illustrating that economic competition is not the sole determining factor in spatial arrangement. Racial and religious minority groups generally reside in separate areas of the city. This separation or segregation sometimes occurs by choice of the minority group, but more often it represents the wishes and values of the majority group. Social-class groupings also separate themselves into different parts of the city, and inevitably areas develop reputations that attract certain types of people and discourage others. As Otis and Beverly Duncan have pointed out, "spatial distances between occupation groups are closely related to their social distance."[5] In a variety of ways, then, we see that spatial arrangement reflects social and cultural values.

It is both important and interesting to note that spatial arrangement *affects* social interaction. People interact with those who live next door or across the street, unless it's a wide and busy street. People living toward the middle of the block have more social contacts than those living on corners. Friendships develop more on the side of the house where the driveway is located, especially if there are adjacent driveways; driveways are natural areas for gathering and talking. A family that doesn't mix or fit in or that is unfriendly often acts as a boundary; neighbors have difficulty interacting past this boundary as they would across a busy street. Such habits become so ingrained that when a new family moves into a house, it may inherit the previous family's reputation as a social boundary.[6]

William Michelson, in his book *Man and His Urban Environment*, outlines a number of other factors that correlate with spatial arrangements. He finds, for example, that physical and social pathologies—crime and social and physical illness—are related to certain spatial factors. The health of women and their children living in self-contained houses was compared with the health of those living in three- and four-story apartments. The sickness rate of the apartment dwellers was 57 percent greater than the sickness rate of those living in houses, as measured by first-consultation rates for any ailment. The differences were attributed to the cramped space and greater isolation of women in apartments removed from the ground. This conclusion is underlined by the fact that the women in apartments who did not have young children (and were therefore able to come and go from their homes more freely) had excellent health. High noise levels are related to the incidence of diseases that involve tension. Studies of overcrowding (many people per room) and high density (many people per acre) have found generally that *density* is more related to pathology than is overcrowding. For example, in one study the researchers found that overcrowding at home was not related to students' achievement in school, but the number of families on the block (density) was. Michelson reminds us that although spatial arrangement may be important, its effects are modified by cultural values. Compare, for example, the way Japanese make use of internal and external space with the way Americans make use of the same space. (See Reading 3.2, page 100.)

Different types of people have different spatial preferences. Michelson reports that people who highly value convenience are likely to prefer more mixed land uses and small lot sizes; people who highly value individualism prefer larger lot sizes. Research indicates that children up to about the age of seven living in high-rise apartments cling much more closely to their parents than do children in single-family homes, who become independent at an earlier age. But past that age the patterns reverse: When they become more mobile, children living in high-rise apartment buildings spend much more time away from home than do children living in single-family houses. Finally, one study reports that although apartment dwellers suffer more from noise disturbance than do people living in homes, their main complaint is the restriction they feel on making noise themselves; this affects their leisure patterns and leads them to such sedentary practices as watching television.[7]

Ecologists use the terms *centralization, decentralization, segregation, invasion,* and *succession* to describe specific types of spatial arrangements. **Centralization** describes the tendency of people to gather around some central or pivotal point in a city. Centralization allows citizens to better fulfill social and economic needs and functions and is represented in most American cities by the central business district. **Decentralization,** on the other hand, refers to the tendency to move away from the central focus. Decentralization probably occurs at least partly because of dissatisfaction with some of the consequences of centralization—traffic, crowds, noise, concrete, and so on—and is seen in the United States in the rush to the suburbs. **Segregation** refers to the clustering together of similar people. These similarities may be along the lines of occupation, race, religion, nationality, ethnicity, or education. **Invasion** refers to the penetration of one group or function into an area dominated by another group or function. Succession refers to the complete displacement or removal of the established group and represents the end product of invasion. The invasion-succession process often produces tension, hostility, and sometimes overt conflict.

A brief look at the ongoing drama of the local faculty dining room illustrates some of these ecological concepts. The room was originally segregated: Faculty ate there, students ate elsewhere. Segregation also existed within the room: Academic faculty generally sat on the south side of the room, and nonteaching staff—secretaries and administrators—sat on the north side of the room. As the college population grew, students were crowded out of other eating areas and began moving into the faculty room. Invasion had begun. Students found that the room made a convenient study area, and they brought stacks of books and papers, which further crowded the faculty. Finally, the faculty gave up and moved elsewhere as the students took over the room. Alas, succession had occurred. . . .

Ecological arrangement in the United States has changed. People who live in large population clusters (50,000 people or more) and in other smaller cities of 2,500 people or more are defined as *urban* by the Census Bureau. It's probably no surprise to learn that, like the rest of the world, we are becoming more and more of an urban nation. In 1860 our population was 20 percent urban; in 1990 we were over 75 percent urban. That 75 percent of the population occupies just 2.5 percent of the nation's land. The number of people living on farms, about 4.6 million in 1991, has been steadily declining over the past 30 years. As nations undergo technological and industrial development, it is probably inevitable that they become more and more urbanized. The transformation from rural to urban and the increased problems of the urbanized society have led to comparisons between the two ways of life. Urban life is seen as secondary, tense, complicated, and anonymous, and rural life is seen as primary, simple, peaceful, and benign. This tendency to idealize rural life may partially explain the current rush to the suburbs. At times in our past, differences between people living in rural areas and those living in cities were pronounced; those differences in religious observance (rural: higher), education (urban: higher), birth rate (rural: higher), and divorce (urban: higher) have been particularly noted. It is likely today, however, that with the effects of migration, technological change, and the mass media, the differences between rural and urban people have become slight.

A related ecological phenomenon is the exodus from the cities to the suburbs, especially by whites. The area near the center of the city (zone in transition or workingman's zone) often offers the least expensive housing in the city. As a result, lower-class and minority-group members migrate into these areas, and as they move in, others who can afford to do so leave for the suburbs. Suburbanites outnumbered city dwellers for the first time in 1970, and this trend has continued. Many people who can afford to move are escaping the central city for the suburbs. The heavy migration of minorities *to* cities, which started in the 1950s, and whites *away* from cities more recently has meant that a number of cities now have a population that is over 50 percent black and

Table 11.2 The World's Ten Largest Urban Areas—Population in 1960, 1985, and Projected for 2025

	1960		1985		2025	
Rank	Urban Area	Population (millions)	Urban Area	Population (millions)	Urban Area	Population (millions)
1	New York	14.2	Tokyo	18.8	Mexico City	36.7
2	London	10.7	Mexico City	17.3	Shanghai	36.1
3	Tokyo	10.7	São Paulo, Brazil	15.9	Beijing	31.9
4	Shanghai	10.7	New York	15.6	São Paulo, Brazil	29.6
5	Beijing	7.3	Shanghai	12.0	Bombay, India	27.0
6	Paris	7.2	Calcutta, India	11.0	Calcutta, India	26.4
7.	Buenos Aires	6.9	Buenos Aires	10.9	Jakarta, Indonesia	23.6
8	Los Angeles	6.6	Rio de Janeiro	10.4	Dacca, Bangladesh	23.5
9	Moscow	6.3	London	10.4	Tokyo	20.7
10	Chicago	6.0	Seoul, South Korea	10.3	Madras, India	20.6

United States Bureau of the Census, *Statistical Abstracts of the United States, 1988* (108th ed.), Washington, D.C., p 801; and the Population Reference Bureau.

Hispanic; Atlanta, Chicago, Detroit, Miami, New Orleans, Newark, Oakland, San Antonio, and Washington, D.C., are examples.

Another ecological trend occurring in the United States in the latter half of the 20th century is the development of *strip cities or megalopolises*. One city and its suburbs merge and grow into an adjacent city and its suburbs. The eventual result is an unbroken series of cities for tens, and even hundreds, of miles. Travelers at night over the East Coast report an unbroken chain of lights from north of Boston to south of Richmond, a 600-mile-long city. If urban or near-urban populations continue to grow, strip cities can be anticipated along the West Coast from San Francisco to San Diego, in the Midwest around the Great Lakes, in the Pacific Northwest, along both Florida coasts, and in many other places throughout the country.

Like the United States, the world is also becoming more urbanized. In many areas, but especially in Asia, people are moving from the countryside to the city in great numbers. They are motivated by the lure of better job opportunities and improved living conditions. Often the reality is quite the opposite—menial jobs (if any), increased poverty, and squalid living conditions. This migration, combined with high population growth (high birth rates), is leading to the development of "mega-cities," as is shown in Table 11.2. Notice several interesting factors in Table 11.2. First, the *huge size* of cities of the future—I can't imagine an urban area (Mexico City is projected at 37 million) with more people than in all of our currently most-populous state (California has 31 million). How can such a city *work*? Can you imagine the traffic? Second, notice the population shifts—the dominant urban centers move from the developed nations of the United States and Europe to the less-developed but high-growth countries of Asia. These patterns of growth and shift bring us to a discussion of some of the problem aspects associated with population and ecology.

EARTH TRENDS—HOW ARE WE DOING?

What's Good?

- Life expectancy on the planet has gone from 46 years in 1950 to 65 years in 1993
- Production of chlorofluorocarbons has been reduced
- Worldwide bicycle sales are triple those of cars
- The number of nuclear weapons has been reduced

What's Not So Good?

- Birds are disappearing—two thirds of all species are on the decline, and 1,000 species are threatened with extinction
- Destructive insects are developing resistance to poisons—now at least 17 species are unaffected by any insecticide on the market
- The sea is yielding about all the food it can—the average yield per person is dropping

- Grain stocks are at their lowest level since the 1970s
- Loss of tropical rain forests is now exceeded by the loss of temperate forests
- There are 19 million refugees worldwide, a new high
- Population continues to grow at alarming rates
- The AIDS virus has now infected 22 million people with no cure in sight
- Americans are 5 percent of the world population but consume 25 percent of the energy
- The average American uses as much energy as 6 Mexicans, 153 Bangladeshis, or 499 Ethiopians
- The U.S. population is growing faster than that of any other industrialized nation

From Worldwatch Institute's "Vital Signs 1994," June 1994; and the Population Reference Bureau.

POPULATION AND ECOLOGY: PROBLEM ASPECTS

Somewhere on this globe every 10 seconds, there is a woman giving birth to a child. She must be found and stopped. —Sam Levinson

As well as providing important information about the characteristics of societies, the study of population and human ecology calls attention to several important problems. One is what many say will become our most serious social problem: the population explosion. The second question at the beginning of this chapter asked if we have a world population problem, and if we do, what caused it: rising birth rates, dropping death rates, or migration patterns? The answer to the first part of the question is easy: *yes*! During the next 24 hours the world will add 260,000 people—it's getting very crowded in many parts of the world. The answer to the second part of the question—what caused it—is more complicated.

For ages the human population remained fairly stable. Such things as disease, wars, and high rates of infant mortality kept births and deaths pretty

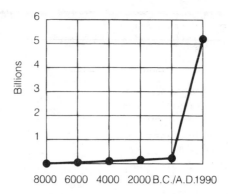

FIGURE 11.6   *World population growth.*

FIGURE 11.7   *World population growth since A.D. 1750.*

much in balance. Then some 200 years ago, a demographic transition occurred as death rates began to fall and population growth suddenly accelerated. A general overview of the growth of the human population over the past 10,000 years is shown in Figure 11.6.

Population increase in the last 250 years is more closely examined in Figure 11.7. The tremendous acceleration in growth since 1900 and especially since 1950 can clearly be seen.

Demographic transition refers to the movement of a population through three stages: (1) high birth rates and high death rates, (2) high birth rates and low death rates, and (3) low birth rates and low death rates. Western industrialized nations are in the third stage, but nations in the underdeveloped regions of the world are generally in the second stage, which means rapid population growth. Table 11.3 shows birth rates, death rates, life expectancy, and population density for several underdeveloped, or Third World, nations and for several industrialized, or developed, nations. Of the underdeveloped countries, Angola, Ethiopia, Nigeria, and Zaire remain on the fringes of the first

Table 11.3 Birth Rates and Death Rates (per 1,000 People), Life Expectancy, and Population Density in Selected Countries

	Birth Rate	Death Rate	Life Expectancy at Birth	Density: People per Square Mile
Underdeveloped Nations				
Angola	46	19	45	20
Egypt	32	9	61	152
Ethiopia	45	14	52	133
Guatemala	36	8	64	240
India	29	11	58	782
Iran	43	8	65	99
Mexico	28	5	73	122
Mozambique	45	17	48	51
Nigeria	44	13	55	270
Zaire	45	14	53	47
Developed Nations				
Austria	12	10	76	249
Canada	15	7	78	6
France	13	9	78	272
Japan	10	7	79	858
Russia	13	11	69	23
Spain	11	9	78	203
Sweden	14	11	78	55
United Kingdom	14	11	77	622
United States	16	9	76	73

United States Bureau of the Census, *Statistical Abstracts of the United States, 1993* (113th ed.,), Washington, D.C.

stage. Their death rates are higher and life expectancy is lower than in other countries. The remaining underdeveloped nations are clearly in stage two, just as the industrialized countries shown are clearly in stage three—low birth and death rates. It is interesting to note that population density is not as closely related to industrialization as one might think. Some of the industrialized countries have high densities and some do not, and likewise with the nonindustrialized countries. If the nonindustrialized nations stay in stage two, however, their population densities will probably increase substantially.

With increasing control of disease and death, population is exploding in many countries of the world. For example, the birth rate in Mexico is nearly twice that of the United States; Mexico's death rate, however, has dropped to even less than ours. The result of this lowered death rate but constant high birth rate is tremendous population growth. World population is growing at a rate of well over 1 million people per week! Every year the world adds over 93 million people—a country the size of Mexico—to the population. The world population is twice what it was 20 years ago. At current growth rates, it is reported that in 600 years or so we will have about one square foot of land per person. (*Then* maybe further reproduction will be impossible.) The world has witnessed headlines of famine and starvation in Bangladesh, in

Ethiopia, and in many other parts of Africa. Food production is not keeping up with population growth in many areas of the world. High oil prices have hurt production of fertilizers, and there is worldwide soil erosion. Population pressure means poverty, disease, starvation, and war.

An influential essay by Thomas Malthus published in 1798 forecast trouble. Malthus stated that people have a natural urge to reproduce and that population would increase faster than would people's ability to produce food. The situation could be helped by preventing births, but Malthus felt that postponing marriage and other types of "moral restraint" were the only acceptable means. Other methods of birth control—contraception, abortion, infanticide—were considered improper means. To Malthus, rational planning was the best hope for controlling the population cycle that would ultimately lead to starvation and poverty. Marx and Engels didn't go along with Malthus, primarily because Malthus felt that poverty was the fault of the poor. Marx and Engels saw poverty as the fault of a poorly organized (capitalist) society. They felt it unnecessary to limit births because advances in science and technology would enable food supplies to keep up with population growth. Modern Malthusians have thrown out the idea of moral restraint as the only acceptable means of birth control and see far more disastrous consequences of population growth than Malthus did. Modern Marxists are more likely to recognize the presence of population-growth problems.[8]

Many population experts are worried about population growth, and organizations ranging from the United Nations to your local Planned Parenthood Association are attempting to inform and enlighten people on the problem and on means to deal with overpopulation. Malthus thought postponing marriage might solve the problem. Today, birth control—the pill and other methods such as abortion and sterilization—is seen as the answer. As we noted earlier in this chapter, birth rates in the United States have declined, reaching their lowest level in history in 1976. They have since increased slightly, and whether this will continue in the future is difficult to guess. The rates will have to stay below the replacement level for some 70 years before population will indeed level off. Many American demographers continue to be cautious, believing that there is a population bomb latent, but not armed, in present society. A spokesperson for the group Zero Population Growth states, "The birth rate is notably fluid and capricious, strongly affected by such psychological factors as the economy or changing attitudes about the role of women." Some countries with high birth rates have encountered difficulty in introducing new population-control measures, especially when they run counter to local customs. Distrust of doctors, ignorance about biological facts or about the consequences of overpopulation, desire for many sons to help farm the land, and numerous other cultural values have frustrated the efforts of people interested in population control. If the problem is ignored or if we give up on it, it may be taken out of our hands; involuntary death control through starvation, war, or selective annihilation may be the final answer. Governmental intervention and national planning already occurring in some countries will have to become worldwide if this problem is to be dealt with effectively.

To further complicate matters, political motivations lead some countries to believe that it is in their best interests to *increase* population. Because of concerns that the nation wouldn't have enough factory workers and army personnel, the Soviet Union tried to increase its birth rate. Mothers were offered financial incentives to have children, extended leave, improved maternity care, and encouragement through pro-baby propaganda. In 1986, West Germany became concerned about declining population, and the government introduced a series of "family-friendly" measures designed to ease the economic and social burden of having children. Other countries, and especially minority groups whose situation is worsened by lack of power at the ballot box, see rising birth rates as a potent weapon. Even in the United States, of all places, this argument has been raised. In his 1987 book, *The Birth Dearth*, Ben Wattenberg warns that the United States and Western nations are not producing babies fast enough! He suggests that our declining numbers may lead to a massive shift in world military, economic, and ideological power and to an overall decline in the influence of Western culture and values. He believes the government should encourage births with cash bonuses, tax deductions, and educational loans. Wattenberg's book has stirred a storm of discussion, with politicians, academics, and demographers lining up on both sides. Some argue that Wattenberg misread the numbers; others suggest that increasing births to combat population growth in non-Western cultures is essentially racist. Others point to the importance of the quality of life and to the ecological strains of trying to accommodate a growing population. Although some experts suggest that our technological sophistication will allow the world to solve overcrowding problems as they happen, others point out that even with a current worldwide surplus of food and fiber, 10 million babies die of malnutrition every year.

On the side of those who believe we must control population, there is encouragement that perhaps the population bomb won't go off after all. Birth rates in some developing nations are beginning to decline. Population research agencies reported in 1985 that the world population growth rate was slowing. This was in large part caused by a dramatic drop in China's birth rate as a consequence of that country's strict family-planning policies (begun in 1979) that limited couples to one child. Mexico has reported a slowing in population growth, indicating that apparently the government's huge family-planning program is working. Still, the growth rates must continue to drop and stay low for perhaps 100 years before world population will level off.

Some people optimistically suggest that population redistribution may be a helpful way of easing pressures in certain areas, and they cite as evidence high population densities in some countries (over 600 people per square mile in Belgium, Holland, Japan, and England and Wales) and low population densities in other countries (fewer than ten people per square mile in Canada and Australia, 23 per square mile in Russia, and 73 per square mile in the United States). It is obvious, they suggest, that some countries are underpopulated and that population redistribution would help. Paul Ehrlich disputes this and argues that overpopulation is a *world* problem, not a problem of individual countries, and that redistribution of population will aggravate rather than solve the problem. More-densely populated countries, such as Holland, are seldom self-sufficient and depend on the relatively underpopulated countries for resources and food. In addition, measurements around the globe indicate that atmospheric dustiness is increasing by perhaps 30 percent per decade, and this poses a serious threat to the climatological balance. Fuels are being burned at rates that will exhaust estimated world reserves in well under a century. Pesticide residues are found everywhere, even in the fat of Antarctic penguins and in the Greenland ice cap. These facts indicate that the *world* has a population problem. Population distribution does present problems, especially in overcrowded urban areas, but in Ehrlich's opinion the first priority is not population redistribution but stopping population growth.[9]

Ecologists and urban sociologists call attention to other serious problems that are as difficult to solve as the population explosion. Specifically, they are worried about the ways we are using and destroying the environment. The automobile has taken over the American city. Most of the land area in our cities must be devoted to car-related functions: streets, highways, parking lots. With the advent of drive-in movies, drive-in restaurants, drive-in banks, drive-in stores, and now even drive-in church services, drive-in marriage ceremonies, and drive-in funeral parlors, it will soon be possible for people to go from birth to death almost without ever leaving their cars. Industries and the car are polluting our environment. Two rivers in Ohio were declared fire hazards. Chemicals developed to control insects have turned out to be more effective than was anticipated and are now destroying other forms of wildlife. Cities are becoming uglier, and urban sprawl is gutting the landscape. Cities are having increasing difficulty disposing of the huge amounts of garbage and refuse that people produce; on the average, Americans produce more than four pounds of pure garbage per person per day. City centers are being abandoned,

A NOVEL SOLUTION

The island of Java in Indonesia has 7 percent of that country's land area but nearly 70 percent of the country's people. The major city, Jakarta, has nine million people and is noisy, dirty, polluted, and crowded with millions of people who are unemployed and poverty-stricken. In an agricultural experiment, Javanese peasants are each given a cow and some land miles away from the crowded city in an area that was once remote jungle. The farmer clears, plows, and plants the land and harvests the crops. A village develops and an agricultural economy begins to thrive. Eventually, the farmer pays for the cow he was given by giving two calves in return. Since 1981 when the experiment started, 130,000 farmers have been given cows and land. In this case, migration—moving people and cows from areas where they were plentiful to areas where they were not—seems to be working.

*Adapted from "Novel Solution for Overpopulation" by Patrice Claude, San Fran-*cisco Chronicle, *November 6, 1990.*

---

and we are creating slums at fantastic rates. More and more people feel that the city is an unfit place to live; only those who are forced to live there do so.[10]

Perhaps we came too quickly into an urban world. It is a fact that living in cities, especially large cities, is a relatively new experience. In 1800, only about 3 percent of the world's population lived in cities of 5,000 people or more. By 1900, the proportion living in cities had grown to 14 percent, and by 1950 it had increased to 30 percent. It is projected that by 2005, half of all the people in the world will live in urban areas. The United States has changed from a sparsely settled rural country to an extremely urbanized nation. Perhaps our problems are to be expected, given our lack of experience with the phenomenon of urbanization.

In his book *The Human Zoo,* zoologist Desmond Morris suggests that the human being is basically a tribal or small-group animal. People are used to interacting on a localized, interpersonal basis. Relatively recently in our history have we, because of population growth and economic advantage, been forced into a supertribal or urban existence. Morris describes this new existence as "a human zoo." We have adapted well to some aspects of the supertribe; the urban environment has stimulated "man's insatiable curiosity, . . . inventiveness, . . . intellectual athleticism." We have adapted less well to other aspects, however, and we therefore see suicide, war, crime, mental illness, and destruction of the environment. The simple tribesman became a citizen, but according to Morris we have not had time to evolve into a new, genetically civilized species. We have adapted somewhat to our new environment through learning and conditioning, but biologically we are still simple tribal animals. Morris concludes, "If [man] is given the chance he may yet contrive to turn his human zoo into a magnificent human game-park. If he is not, it may proliferate into a gigantic lunatic asylum, like one of the hideously cramped menageries of the last century."[11]

THEORY AND RESEARCH: A REVIEW

This chapter was mainly descriptive, and no strong theoretical cast was present. Still, in some instances specific viewpoints could be detected. In the discussion on city patterning, we learned that cities are units and that they develop certain patterns or structures because of functions provided, such as transportation, low rent, and industry; this sounds consistent with *functional analysis*. Analysis of the effects of spatial arrangement on patterns of interaction is moving in the direction of *symbolic interaction*. *Conflict theory* would have much input in the matter of overpopulation—overpopulation consequences, the tactics of governments, and the types of people who are victims. The data are census data, mainly *nonreactive surveys* of national and international birth, death, and migration records. The following questions might arise from this type of data: Do countries in different parts of the world use the same standards? Do they use equal care in collecting their data, and are the data comparable?

SUMMARY

In the previous chapters in this section on social organization we dealt with groups, categories, aggregations, and institutions. These are abstract concepts, but they are valuable for their capacity to describe and analyze the social organization of society. In this chapter we continue to study social organization, but in a somewhat different manner. The focus changes to population and human ecology, and we move from the abstract to the concrete.

Population and human ecology involve the study of numbers of people and their locations in space. The major population variables are *fertility, mortality,* and *migration.* Ecological variables that we examined include study of *city patterns* (slums, suburbs, strip cities), *ecological processes,* and the reasons for and consequences of *spatial patterning.* The study of population and human ecology supplies descriptive data about a society. These data, usually of a statistical nature, develop a picture of a society that often provides an interesting contrast to the picture supplied by the more abstract studies of groups and institutions. Finally, the study of population and ecology has today called attention to several serious problems facing us—*overpopulation, urbanization,* and the *destruction of the environment*—and has encouraged development of the social planning needed to solve these problems.

Los Angeles is the second largest city in the United States and is looking to become number one. But it is difficult to make large cities work effectively, and Los Angeles is no exception; many complex urban problems exist. In the first of two readings that follow, Charles Lockwood and Christopher Leinberger describe the city, its people, its culture, and its problems. The second reading examines China's one-child policy of 1979—how the policy works and what some of its consequences are.

TERMS FOR STUDY

birth rate (330)

centralization (340)

decentralization (340)

demographic transition (344)

demography (328)

fecundity (330)

fertility (330)

general fertility rate (330)

human ecology (328)

invasion (340)

life expectancy (332)

migration (333)

mortality (332)

population pyramid (335)

replacement level (331)

segregation (340)

sex ratio (335)

succession (341)

total fertility rate (330)

Reading 11.1

LOS ANGELES COMES OF AGE

Charles Lockwood and Christopher B. Leinberger

Los Angeles is the second largest city in the United States and has ambitions of becoming the largest. It is also on its way to becoming the first major Third World city in the mainland United States. Los Angeles's economy in 1987 was large enough to rank it as the 11th largest "nation" in terms of gross national product. Can cities this big work effectively? Los Angeles illustrates perhaps better than any other American city the urban problems and issues that cities everywhere must eventually face.

Charles Lockwood and Christopher Leinberger are urban-affairs specialists who write and consult about urban issues and problems.

Los Angeles conjures up a myriad of images: a city dominated by show business, where businessmen take meetings beside swimming pools and don't wear ties to the office; a palm-shaded city where blond young men surf their days away at the beach; a city of transplanted midwesterners who ignore the threat of the next earthquake; an intellectually and culturally vapid city where many residents worship rampant materialism while others pursue various forms of self-realization with equally intense preoccupation; a city with no downtown—a sprawling amoeba of unending stretches of suburban tract houses, connected by more freeways than exist anywhere else in America.

Whatever their suitability for jokes on late-night talk shows, these images of Los Angeles no longer tell anything like the whole story. Within the past decade greater Los Angeles has undergone economic, social, political, and cultural upheavals deeper and broader than those experienced by any other large American city during the same period.

Once a prosperous but provincial regional metropolis, greater Los Angeles has become America's true "second city"—second only to New York in economic power and cultural influence. And Los Angeles rivals New York in attracting ambitious people who want to make it, in almost every professional endeavor and field of business. If present trends continue, and if certain nagging problems don't overwhelm the metropolitan area, Los Angeles might even emerge as the Western Hemisphere's leading city in the early twenty-first century.

Greater Los Angeles, a national trendsetter in popular entertainment, life-styles, and speech, has been wrestling with economic and social issues that are only now beginning to challenge other American metropolitan areas. These issues include how and where Americans will work and live, racial and ethnic integration when Anglos (non-Hispanic whites) are a minority, and which of many groups will assume leadership in the arts and culture. Los Angeles also presents large-scale versions of now widespread urban challenges such as traffic, air quality, waste management, public education, and regional government. To visitors and Angelenos alike, it seems that what many once considered to be an earthly paradise is environmentally befouled and increasingly unlivable.

The changing size and form of Los Angeles set it apart from any other American city, and so does its economy. The 4,100-square-mile metropolitan area, already one of the largest in the world, is rapidly expanding. The direction of the growth is dictated by geography and also politics. The Pacific Ocean forms the city's western boundary. The San Gabriel Mountains, whose peaks reach 10,000 feet, impede but do not completely block growth to the north. To the northwest, Ventura County, where development in the San Fernando Valley will logically extend, has established "slow growth" limits. Greater Los Angeles, then, is expanding southeast, into Orange County, and even more rapidly eastward, into the less expensive and less congested San Bernardino and Riverside counties, more than sixty miles from downtown Los Angeles. . . .

The metropolitan area's great expansion and the revival of downtown reflect the local economy's rapid growth and the area's emergence as a major business center. Last year greater Los Angeles produced $250 billion worth of goods and services, making the 12.6-million-person metropolitan area the world's eleventh largest "nation" in terms of gross national product, ahead of Australia, India, and Switzerland. The Southern California Association of Governments published employment projections in August of 1986 indicating that the area's gross regional product will increase sharply in the coming years. The number of jobs is expected to jump from 5.9 million in 1984 to 8.0 million in 2000.

Some Los Angeles factories can even undersell Third World manufacturers. Last year a UCLA professor interested in labor conditions asked to be taken on a tour of factories in the industrial area near downtown Los Angeles, and his request was granted on condition that he be blindfolded. When the blindfold was removed, the professor found himself standing inside a large factory building with blacked-out windows. Dozens of Hispanic and Asian workers were manufacturing the steel wheels on which automobiles' tires are mounted. The workers' only protection against dust and airborne metal particles was cloth bandannas over their mouths and noses. The greatest shock for the professor was not the sweatshop working conditions or the apparent immigration and pollution violations but the label attached to each wheel at the end of production. It read MADE IN BRAZIL. The goods were headed for Detroit. . . .

. . . Who will own and run the Los Angeles of the future[?] A tangible symbol of Los Angeles's economic might—its gleaming new downtown skyline—has already passed out of local control. According to the *Los Angeles Times,* 75 percent of the large downtown office buildings are foreign-owned or -controlled, up from 25 percent just eight years ago. At their offices in these foreign-owned buildings some Los Angeles banking executives worry that financial institutions based in New York and Tokyo may someday dominate the local market. Already branches of Wall Street firms like First Boston and Morgan Stanley have become Southern California's leading investment banks. Similarly, long-established Los Angeles attorneys fear that an influx of New York law firms now under way will take away some of their business, because the New Yorkers offer the prestige of a national outfit and have already worked with the banks and corporations that have recently opened offices in Los Angeles. By 1987 about forty New York law firms had opened offices in Los Angeles. Most had arrived within the preceding few years; many had come to Los Angeles within the preceding year.

The population of greater Los Angeles is also growing and diversifying. Now 12.6 million, the population is projected to reach 16.4 million by 2000 and 18.3 million by 2010. Greater Los Angeles, therefore, will retain in 2000 its current position as one of the twelve largest cities in the world, according to Peter Hall's *The World Cities*. This fact is impressive when it is considered in the context of urban growth around the globe. In 1950, Hall notes, seven of the world's twelve largest cities were North American or European. By 2000 only two—Los Angeles and New York—will be. Greater New York is not growing rapidly, but it will remain on the list because of its currently enormous population. Greater Los Angeles, however, *is* growing rapidly. Indeed, of the top twelve cities projected for the year 2000, Los Angeles and Tokyo will be the only highly developed post-industrial cities doing so, fast growth being common now only in developing and newly industrializing countries.

As the metropolitan area's population increases, its ethnic composition is becoming less Anglo and more Latin and Asian. The Anglo population will drop from 60 percent to approximately 40 percent in 2010, while remaining stable in absolute terms, according to a recent report by Frank E. Hotchkiss, the director of regional strategic planning for the Southern California Association of Governments. The non-Hispanic black population will rise from 9 percent to 10 percent, increasing by approximately 800,000. The Asian population will grow from 6.2 percent to 9.3 percent, increasing by almost one million. The Hispanic population will rise from 24 percent to 40 percent, increasing by more than four million.

Greater Los Angeles is well on its way to becoming the first major "Third World city" in the mainland United States, with the majority of residents in the metropolitan area of non-European ancestry. Although it is impossible to verify claims about relative ethnic populations in cities all over the world, some Angelenos boast that greater Los Angeles is the largest Mexican metropolitan area outside Mexico, the second largest Chinese metropolitan area outside China, the second largest Japanese metropolitan area outside Japan, the largest Korean metropolitan area outside Korea, the largest Philippine metropolitan area outside the Philippines, and the largest Vietnamese metropolitan area outside Vietnam. . . .

The growth of greater Los Angeles's economy, population, and cultural activity is impressive. But these changes have contributed to serious problems in metropolitan-area government, education, law enforcement, environmental protection, and transportation. Meeting these challenges requires renewed commitment on the part of the region's power structure, which has evolved beyond the typical big-city political, economic, and social hierarchy. Power and influence are now so dispersed that no one circle of interests is dominant.

In the past Los Angeles had its business and social elite, which worked downtown and lived in Pasadena, San Marino, or the Hancock Park neighborhood, several miles west of downtown. The downtown California and Jonathan clubs represent the last vestiges of this system. Until last year, when the city passed an anti-discrimination ordinance, these two clubs still discriminated against nearly everyone except white males. These tradition-bound clubs are becoming less powerful as the Westside, with its highly visible Jewish population, the nouveau-riche Orange County town of Newport Beach, and the Hispanic and Asian communities increase in influence.

Today business and political leaders must be adept at building coalitions if the diverse groups are to cooperate in getting anything done. Dorothy Chandler, the wife of the former *Los Angeles Times* publisher Norman Chandler and the mother of the paper's current publisher, Otis Chandler, was one of the first successful coalition-builders. In the late 1950s and early 1960s she brought together the old families, downtown corporations, show-business leaders, and self-made millionaire businessmen to build the Music Center of Los Angeles County, on Bunker Hill, downtown.

One of the most skillful coalition-builders of recent years is the three-term mayor, Tom Bradley. In a city where blacks constitute only 17 percent of the population, Bradley has had to build coalitions in order to survive. Survive he has, despite what political scientists describe as a "weak mayor" system, to become the most effective mayor in recent Los Angeles history. Bradley's style involves low-key cajoling and constant communication with various groups, which means that he seldom wins sole credit for getting legislation passed or a project completed. But many of the city's finest recent accomplishments would not have happened without his leadership. A good example is the 1984 Olympics, for which Peter Ueberroth received the acclaim. But the Games would never have come to Los Angeles, much less been successful, without the constant support of the mayor.

Problems as Big as the City

Los Angeles has indeed become faster paced, more congested, and meaner in the past ten years. To describe most of today's Angelenos as laid-back would be as inappropriate as calling most New Yorkers polite. Only affluent Angelenos can now enjoy the pleasures that the working and middle classes took for granted ten or fifteen years ago, such as spacious and reasonably priced housing, easy access to nearby beaches and mountains, and exemplary public education.

The Los Angeles Unified School District, which encompasses an even larger area than the city of Los Angeles and is independent of the city government, faces a seemingly impossible challenge. In contrast to the situation in nearly all other big cities, in Los Angeles the school-age population is increasing, at the rate of 14,000 students each year. To accommodate this growth, the district needs to build nine new elementary schools, a junior high school, and a senior high school each year for the next five years, and also do something about the $400 million backlog of deferred maintenance on existing buildings. At the present rates of student increases and school construction, 50,000 students will not find seats in the classrooms in five years.

Educating all these children is increasingly difficult. Of the nearly 600,000 students in the Los Angeles Unified School District, from kindergarten to high school, about 25 percent exhibit only a limited proficiency in English and speak one of more than eighty other languages. More than half the pupils in some classes require remedial English instruction. To compound the district's educational problems, more than half the district's children come from homes whose incomes are under $14,250 a year.

Other social problems affect all Angelenos, whatever their income, as the spate of freeway shootings [in 1987] reminded them. Los Angeles has the nation's third highest crime rate, behind Detroit and Dallas. And if the recent allegations that Los Angeles has emerged as a $1-billion-a-year clearinghouse for cocaine in the western United States are correct, crime rates are likely to go even higher. In the gang-infested South Central ghetto, teenage drug entrepreneurs drive custom $50,000 to $100,000 BMWs and Mercedes, and their gangs' foot soldiers brandish Uzi submachine guns and Soviet-made AK-47 assault rifles. About 70 percent of gang killings involve innocent bystanders.

The list of environmental problems is also daunting. The handsome physical setting of Los Angeles is more threatened than the settings of most of the world's major cities. All the region's residents are affected by the ever-present threat of earthquakes, foul-smelling and chemical-laden tap water, and the potential for water shortages by the year 2000.

An equally serious but often ignored environmental threat is the pollution of Santa Monica Bay, which forms the visually splendid backdrop for the city's beaches from Palos Verdes to Malibu. Raw sewage is occasionally dumped into the ocean by the city's overburdened sewer system, endangering the health of swimmers, and the ocean water is so polluted with cancer-causing DDT and PCBs that much marine life is now deformed or afflicted with tumors and fin rot. Indeed, one meal of local sea bass would exceed Environmental Protection Agency lifetime limits for several toxic chemicals. . . .

The best-known environmental problem, of course, is smog. Los Angeles has the most polluted air in the United States. On two out of three days at least part of the metropolitan area violates federal air-quality standards. On the smoggiest days ozone readings in some areas are often three times more than allowable standards.

Although air quality has continued to improve over the past decade and is far better than it was in the pre-EPA 1960s, it is expected to worsen in the 1990s, simply because of the increases in the population, the number of automobiles, and the amount of industry. Because prevailing winds off the Pacific blow much of the smog into the inland Riverside and San Bernardino counties, where most new "affordable" housing (under $150,000) is being built, air pollution will affect more people in the future. Coastal communities with relatively clean air, like Pacific Palisades, Santa Monica, and Palos Verdes, are almost fully developed, and housing prices in them are among the highest in the metropolitan region.

Besides these difficulties, greater Los Angeles also faces another seemingly insoluble problem: transportation, which affects almost everyone and will eventually limit the region's economic growth. Easy mobility is essential to anyone in Los Angeles and forms an important part of the city's life-style.

But the days of driving simply for fun are over. Los Angeles County has four of North America's five busiest freeways. Here traffic frequently slows to a rush-hour crawl during the middle of the day or late at night. On the other, less-congested freeways, rush hours last from 7:00 to 10:00 A.M. and from 3:00 to 7:00 P.M., unless accidents or rain extend the congestion. Even on "good" mornings more than 300 miles of the 740 miles of freeways in Los Angeles, Orange, and Ventura counties are clogged with commuters going no more than fifteen miles an hour. Only thirty miles of freeways were similarly overburdened in 1963. On some freeways there is enough traffic to fill an additional ten or twenty lanes, if such an expansion were physically and financially possible. Not surprisingly, Los Angeles motorists waste 100,000 hours a day in traffic jams.

If these trends continue, traffic will soon become "intolerable," according to the Southern California Association of Governments. By 2010 the number of miles of freeways that are congested during the afternoon rush hour will more than double. "Roughly 50 percent of the region's travel time will be spent in delay," according to a recent *Los Angeles Times* article. Anthony Downs, of the Brookings Institution, recently said, "The only solution to the traffic problem is to buy a BMW with comfortable seats, a high-quality sound system, and car phone, and enjoy it as best you can." . . .

Because Los Angeles developed in this decentralized way rather than according to the traditional city-suburb pattern, every city in the metropolitan region has always included poor communities, urban employment centers, industrial districts, and cultural institutions. While Angelenos can live, work, play, and shop in one section of the metropolitan area, they cannot isolate themselves from the diversity of urban life, as do the residents of many suburbs. When many Beverly Hills residents drive to the nearest freeway, for instance, they pass through a middle-class Anglo area, the largest Sikh neighborhood in Los Angeles, and a working-class black neighborhood—all within two miles.

Although the jumble of different uses and socio-economic classes in proximity is one reason that Los Angeles is so difficult for out-of-town visitors to comprehend, it gives middle-class and rich Angelenos a greater understanding of the metropolitan region than their counterparts demonstrate in most other cities. . . .

QUESTIONS 11.1

1. What is meant by the statement that Los Angeles is becoming the United States's first major Third World city? Why do you think that is happening?

2. Which ecological patterning described in this chapter best describes Los Angeles? List other concepts from this chapter that can be applied to this reading on Los Angeles.

3. Compare and contrast Los Angeles with the metropolitan area with which you are most familiar. Does yours have the same problems Los Angeles has? Will it? Which of Los Angeles's problems seem to be the most serious? What is the most serious problem in your metropolitan area?

Reading 11.2

THE LITTLE EMPERORS, PART 2

Daniela Deane

China's population in 1993 was about 1.2 billion, about 300 million larger than that of India and about 5 times larger than that of the United States. China has recognized its emerging population problem and in 1979 passed a one child per family policy. Enforcing the policy has been a struggle. It worked better in the crowded cities than it did in the countryside. A growing group of wealthy farmers resisted the policy because they wanted sons to carry on their businesses, and because of their wealth, they were less bothered by the fines levied for violation of the one-child rule. Daniela Deane, a writer now based in Italy, examines some of the other problems and consequences of the policy in this article she wrote for the Los Angeles Times Magazine. *(Also see another part of this article—Reading 7.2—in which Deane describes how the "one child" has become the center of the Chinese family.)*

China's Communist government adopted the one-child policy in 1979 in response to the staggering doubling of the country's population during Mao Tse-tung's rule. Mao, who died in 1976, was convinced that the country's masses were a strategic asset and vigorously encouraged the Chinese to produce even larger families.

But large families are now out for the Chinese—20 percent of the world's population living on just 7 percent of the arable land. "China has to have a population policy," says Huang Baoshan, deputy director of the State Family Planning Commission. With the numbers ever growing, "how can we feed these people, clothe them, house them?"

Thirteen years of strict family planning have created one of the great mysteries of the vast and remote Chinese countryside. Where have all the little girls gone? A Swedish study of sex ratios in China, published in 1990, and based on China's own census data, concluded that several million little girls are "missing"—up to half a million a year in the years 1985 to 1987—since the policy was introduced in late 1979.

In the study, and in demographic research worldwide, sex ratio at birth in humans is shown to be very stable, between 105 and 106 boys for every 100 girls. The imbalance is thought to be nature's way of compensating for the higher rates of miscarriage, stillbirth and infant mortality among boys.

In China, the ratio climbed consistently during the 1980s, and it now rests at more than 110 boys to 100 girls. "The imbalance is evident in some areas of the country," says Stirling Scruggs, director of the United Nations Population Fund in China. "I don't think the reason is widespread infanticide. They're adopting out girls to try for a boy, they're hiding their girls, they're not registering them. Throughout Chinese history, in times of famine, and now as well, people have been forced to make choices between boys and girls, and for many reasons, boys always win out."

With the dismantling of collectives, families must, once again, farm their own small plots and sons are considered necessary to do the work. Additionally, girls traditionally "Marry out" of their families, transferring their filial responsibilities to their in-laws. Boys carry on the family name and are entrusted with the care of their parents as they age. In the absence of a social security system, having a son is the difference between starving and eating when one is old. To combat the problem, some innovative villages have begun issuing so-called "girl insurance," an old-age insurance policy for couples who have given birth to a daughter and are prepared to stop at that.

"People are scared to death to be childless and penniless in their old age," says William Hinton, an American author of seven books chronicling modern China. "So if they don't have a son, they immediately try for another. When the woman is pregnant, they'll have a sex test to see if it's a boy or a girl. They'll abort a girl, or go in hiding with the girl, or pay the fine, or bribe the official or leave home. Anything. It's a game of wits."

Shen Shufen, a sturdy round-faced peasant woman of 33, has two children—an 8-year-old girl and a 3-year-old boy—and lives in Sihe, a dusty, one-road, mud-brick village in the countryside outside Beijing. Her husband is a truck driver. "When we had our girl, we knew we had to have another child somehow. We saved for years to pay the fine. It was hard giving them that money, 3,000 yuan ($550 in U.S. dollars) in one night. That's what my husband makes in three years. I was so happy when our second child was a boy."

The government seems aware of the pressure its policies put on expectant parents, and the painful results, but has not shown any flexibility. For instance, Beijing in 1990 passed a law forbidding doctors to tell a couple the results of ultrasound tests that disclose the sex of their unborn child. The reason: Too many female embryos were being aborted.

And meanwhile, several hundred thousand women—called "guerrilla moms"—go into hiding every year to have their babies. They become part of China's 40-million-strong floating population that wanders the country, mostly in search of work, sleeping under bridges and in front of railway stations. Tens of thousands of female children are simply abandoned in rural hospitals.

And although most experts say female infanticide is not widespread, it does exist. "I found a dead baby girl," says Hinton. "We stopped for lunch at this mountain ravine in Shaanxi province. We saw her lying there, at the bottom of the creek bed."

Death comes in another form, too; neglect. "It's female neglect, more than female infanticide, neglect to the point of death for little girls," says Scruggs of the U.N. Population Fund. "If you have a sick child, and it's a girl," he says, "you might buy only half the dose of medicine she needs to get better."

Hundreds of thousands of unregistered little girls—called "black children"—live on the edge of the law, unable to get food rations, immunizations or places in school. Many reports are grim. The government-run China News Service reported last year that the drowning of baby girls had revived to such an extent in Guangxi province that at least 1 million boys will be unable to find wives in 20 years. And partly because of the gender imbalance, the feudalistic practice of selling women has been revived.

The alarming growth of the flesh trade prompted authorities to enact a law in January that imposes jail sentences of up to 10 years and heavy fines for people caught trafficking. The government also recently began broadcasting a television dramatization to warn women against the practice. The public-service message shows two women, told that they would be given high-paying jobs, being lured to a suburban home. Instead, they are locked in a small, dark room, and soon realize that they have been sold.

Li Wangping is nervous. She keeps looking at the air vents at the bottom of the office door, to see if anyone is walking by or, worse still, standing there listening. She rubs her hands together over and over. She speaks in a whisper. "I'm afraid to get into trouble talking to you," Li confides. She says nothing for a few minutes.

"After my son was born, I desperately wanted another baby," the 42-year-old woman finally begins. "I just wanted to have more children, you understand? Anyway, I got pregnant three times, because I wasn't using any birth control. I didn't want to use any. So, I had to have three abortions, one right after the other. I didn't want to at all. It was terrible killing the babies I wanted so much. But I had to."

By Chinese standards, Li (not her real name) has a lot to lose if she chooses to follow her maternal yearnings. As an office worker at government-owned CITIC, a successful and dynamic conglomerate, she has one of the best jobs in Beijing.

"I had to keep everything secret from the family-planning official at CITIC, from everyone at the office," continues Li, speaking in a meeting room at CITIC headquarters.

"Of course, I'm supposed to be using birth control. I had to lie. It was hard lying, because I felt so bad about everything."

She rubs her hands furiously and moves toward the door, staring continuously at the air slats. "I have to go now. There's more to say, but I'm afraid to tell you. They could find me."

China's family-planning officials wield awesome powers, enforcing the policy through a combination of incentives and deterrents. For those who comply, there are job promotions and small cash awards. For those who resist, they suffer stiff fines and loss of job and status within the country's tightly knit and heavily regulated communities. The State Family Planning Commission is the government ministry entrusted with the tough task of curbing the growth of the world's most populous country, where 28 children are born every minute. It employs about 200,000 full-time officials and uses more than a million volunteers to check the fertility of hundreds of millions of Chinese women.

"Every village or enterprise has at least one family-planning official," says Zhang Xizhi, a birth-control official in Changping county outside Beijing. "Our main job is propaganda work to raise people's consciousness. We educate people and tell them their options for birth control. We go down to every household to talk to people. We encourage them to have only one child, to marry late, to have their child later."

China's population police frequently keep records of the menstrual cycles of women of childbearing age, on the type of birth control they use and the pending applications to have children. If they slip up, street committees—half-governmental, half-civilian organizations that have sprung up since the 1949 Communist takeover—take up the slack. The street committees, made up mostly of retired volunteers, act as the central government's ear to the ground, snooping, spying and reporting on citizens to the authorities.

When a couple wants to have a child—even their first, allotted one—they must apply to the family-planning office in their township or workplace, literally lining up to procreate. "If a woman gets pregnant without permission, she and her husband will get fined, even if it's their first," Zhang says. "It is fair to fine her, because she creates a burden on the whole society by jumping her place in line."

The official Shanghai Legal Daily last year reported on a family-planning committee in central Sichuan province that ordered the flogging of the husbands of 10 pregnant women who refused to have abortions. According to the newspaper, the family-planning workers marched the husbands one by one into an empty room, ordered them to strip and lie on the floor and then beat them with a stick, once for every day their wives were pregnant.

"In some places, yes, things do happen," concedes Huang of the State Family Planning Commission. "Sometimes, family-planning officials do carry it too far."

Large billboards bombard the population with images of happy families with only one child. The government is desperately trying to convince the masses that producing only one child leads to a wealthier, healthier and happier life. But foreigners in China tell a different story, that the people aren't convinced. They tell of being routinely approached—on the markets, on the streets, on the railways—and asked about the contraceptive policies of their countries. Expatriate women in Beijing all tell stories of Chinese women enviously asking them how many sons they have and how many children they plan to have.

QUESTIONS 11.2

1. From the viewpoint of this article, what is the main stumbling block to population control in China?

2. Why is such high value placed on sons in China? Is it similar in the United States? Why or why not?

3. The one-child policy and its implementation may seem harsh. Construct arguments (both sides) on the necessity of a population control policy.

4. Construct and describe a policy (different from the one-child approach) that achieves the same end—population control—that China is trying to achieve.

NOTES

1. The source for this paragraph is *Population* by John Weeks (Belmont, Calif.: Wadsworth, 1978), ch. 4.

2. The major source for these paragraphs on mortality is *Population* by John Weeks (see Note 1), ch. 6, especially pp. 106–108 and 114–119.

3. See *Population* by John Weeks (see Note 1), ch. 7.

4. Original sources referred to in this section include: Ernest Burgess, "The Growth of the City," in Robert Park and Ernest Burgess, *The City* (Chicago: University of Chicago Press, 1925), p. 51; Homer Hoyt, *The Structure and Growth of Residential Neighborhoods in American Cities,* United States Federal Housing Administration (Washington, D.C.: United States Government Printing Office, 1939), ch. 6; and C. D. Harris and Edward L. Ullman, "The Nature of Cities," *The Annals* 242 (November 1945), pp. 7–17.

5. The article by Otis Dudley Duncan and Beverly Duncan, "Residential Distribution and Occupational Stratification," is quoted here from Ralph Thomlinson's book *Urban Structure* (New York: Random House, 1969), pp. 12–13.

6. For a detailed description of these interaction patterns, see William H. Whyte, Jr., *The Organization Man* (New York: Simon & Schuster, 1956), ch. 25.

7. See William Michelson, *Man and His Urban Environment* (Reading, Mass.: Addison-Wesley, 1970). The comments in these two paragraphs are drawn especially from pp. 98–99, 158, 161, and 193–195.

8. The major source for this discussion of Malthus and Marx is *Population* by John Weeks (see Note 1), pp. 16–21.

9. Paul Ehrlich has written a number of articles and several books on population and ecological problems. The ideas I have summarized here come from Paul Ehrlich and John Holdren, "Why Do People Move?" which appeared in *Saturday Review* 5 (September 1970), p. 51.

10. A detailed account of the problems facing cities is provided by Mitchell Gordon in his book *Sick Cities* (Baltimore: Penguin Books, 1965).

11. Desmond Morris, *The Human Zoo* (New York: McGraw-Hill, 1969), especially ch. 1. The quotation is from p. 248.

PART THREE

Social Change and Social Deviance

In Parts One and Two we dealt with the individual as a member of society and with the social organization of society. We saw that social organization could be studied using the following concepts: group, category and aggregation, institution, population, and human ecology. In Part Three we will focus on change and deviance in society. The concepts discussed in this section suggest that all is not as predictable and organized as previous sections may have implied. Uncertainty and instability exist in all societies and are the result of a variety of factors. Social change occurs and affects the ways individuals and groups relate to each other. If social change is rapid and extreme, the organization of society may break down, and social disorganization may result. In some situations, behavior occurs that is spontaneous and unstructured but not necessarily disorganized. This type of behavior is called collective behavior. We learned in Part

One that norms and roles describe what people are expected to do in specific situations and positions. But inevitably, many individuals do not *behave in the ways expected of them. Societies, in turn, devise ways to encourage or even force conformity. This leads us to the final topics of* deviant behavior *and* social control.

CHAPTER TWELVE

Social Change
and Social
Disorganization

A RESEARCH QUESTION

Patterns of violence (for example, lynching of blacks in the deep South) change over time. Are these social changes random, or are they related to other factors? What might such factors be?

[News from Beijing, April 1984] Nomadic herdsmen in northwestern China are trading in their yak-hair tents for a new polyvinyl model, which features optional folding furniture and a wind-powered generator. The new tent, which has more than 70 options, was offered to 500,000 herdsmen in remote Qinghai province as part of a government home-improvement program. Herdsmen interested in the wind-powered generator hope it will make it possible for them to watch TV.

WE HAVE LOOKED at the social organization of society now for the last eight chapters. At this point, society may appear to be a well-oiled machine: groups interacting, highly polished institutions dealing efficiently with society's central issues and conducting business as usual, populations reproducing, and people continuing to distribute themselves in space in very predictable ways. Focusing on the social organization of society is an extremely important perspective, but it unfortunately implies that conditions are constant, that all be-

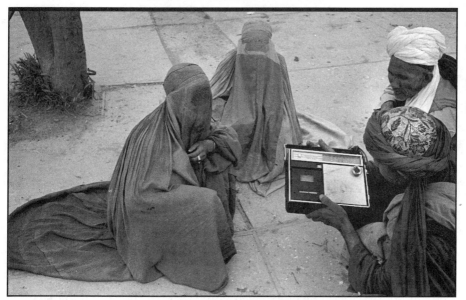

have as they are expected to, and that there are never unpredictable occurrences or failures—either in people or in social conditions.

The fact is that all is *not* constant—life is not the same today as it was yesterday. Right now I am watching the O. J. Simpson preliminary hearing on live television, and two weeks ago I participated, along with the rest of the country, in his bizarre chase around Los Angeles. This sort of coverage would not have been technically possible or even considered until recently. The Concorde travels from the East Coast to Europe in three hours; I can remember when it took more than 12 hours to cross the Atlantic by propeller plane. I used to *dial* my telephone and talk to *people;* now I punch buttons and talk to answering machines. It used to be that when you broke your knee they put on a cast; now they stick you in an MRI machine (and then put on a cast). Test-tube babies are being born to women over the age of 60. Scientists and environmentalists are concerned about the expansion of holes in the ozone layer over the earth's poles. Space stations are being developed, and manned trips to Mars are being discussed. The Berlin Wall and the Soviet Union don't exist anymore, but Myanmar, cellular phones, and fax machines do. You can get an encyclopedia on a computer (CD-ROM) disk. I used to write lectures by hand; now my desktop computer produces transparencies almost without my help. Our students used to stand in long lines to register for classes at the beginning of each semester; now they register by phone through our mainframe computer while sitting at home watching one of 200 channels on cable television. AIDS, a disease that didn't seem to exist a short time ago, looms as the major health issue of today. By 1992, 245,000 AIDS cases and 166,467 deaths from AIDS had been reported in the United States.

The point, of course, is that all societies are dynamic and constantly changing. Some societies change more rapidly than others, and within a society some parts change more rapidly than other parts. Some changes may lead to improvement, to a better life for some or most of the members of a society. Change may also lead to problems because, although some individuals and institutions adjust rapidly, others have difficulty adapting to change. We can anticipate that social change will occasionally lead to periods of social disorganization as old ways erode and collapse and new ways of behaving are developed.

Social change refers to significant alterations in social relationships and cultural ideas.[1] Social and cultural factors are combined in this definition because they are tightly interrelated. Changes in ideas and values quickly lead to changes in social relationships, and vice versa. Scientific and technological developments have led to new roles (astronauts, spacecraft experts) and changed social relationships. People's concerns about pollution have led to new ideas and material inventions, such as smaller cars with fewer harmful emissions. The concern over AIDS mentioned earlier is quickly leading to major changes in other areas; we see changes in laws (is it a crime to knowingly put someone else at risk of infection?), in the nature of health care and how to finance it, in medical research and in tracking the course of a disease, and of course in the area of social and sexual patterns of interaction.

Social change takes place continually and, in modern society, at ever-increasing rates. Change is such a constant factor in our existence that we frequently lose sight of its extent. One way to put change in perspective is to imagine that your own life started ten years earlier than it did or to compare your life experiences with those of your parents. Technological changes probably come to mind first: the growth of jet aviation, voyages to outer space, atomic power, organ transplants, a computerized society. But other changes have occurred as well. Look at the matters in which so much has transpired after only ten years: the role of women; the average family size; educational philosophies and techniques; the control of particular diseases and life expectancy; the likelihood of getting married and the average age at which one marries; demonstrating for or against a cause (apartheid, legalized abortion, the death penalty); deciding whether to go to college, and if so, choosing a course of study; the amount of leisure time available and what to do with it; and so on. And if we can find these considerable differences across ten years, imagine the contrast that would develop over 20 or 30 years.

Change factors are related to each other in complex patterns; often the path of influence is difficult to detect clearly. The development of the contraceptive pill (new idea, invention) led to changes (more freedom) in sexual behavior. Or did it? Perhaps concerns with increased freedom for women and with family size led to the development of contraceptive devices. Keeping in mind the complex nature of social change, perhaps we can outline some of the major factors related to change.

FACTORS RELATED TO SOCIAL CHANGE

Various attempts have been made to relate social change to other factors in the physical and social environment.[2] Some of these hypothesized relationships seem farfetched; others seem more reasonable. For example, social change has been related to the biological characteristics of a nation. Adherents of master-race theories believe that societies fortunate enough to be graced by the presence of a superior race will have greater progress (and more rapid social change) than will "less fortunate" nations. Sociologist Robert Bierstedt points out that such theories—present at all times in all societies—are only a primitive form of ethnocentrism. A more realistic approach might hold that people who differ by race or ethnicity have different social and cultural patterns and that particular mixtures of these types of people in a given society could lead to greater or lesser conflict and, correspondingly, greater or lesser social change. Geographical factors may also be related to change. Some countries are more favored by climate and natural resources than are others, and this may affect their rates of technological advancement and social change. As we discussed in the previous chapter, demographic factors may lead to change as well. When a country faces population growth to the point of overcrowding, social relationships change, both within that society and with surrounding societies that may be less crowded.

Social change may be a result of the influence of individuals. Recall how societies, and even the world, were affected and changed by Adolf Hitler, Napoleon Bonaparte, Albert Einstein, Abraham Lincoln, Karl Marx, and Julius Caesar. The course of events in the United States in the mid-20th century was altered by men such as Franklin D. Roosevelt and Martin Luther King, Jr. It is sometimes argued that significant social changes are initiated only by great and unique men and women; but others maintain that such people are merely products of their society and that if they had not appeared, someone else of the same caliber would have. The truth probably lies somewhere between these views. The cultural and societal conditions might be seen as a stage on which people perform, a few in unique ways having lasting effects, many others in more commonplace and predictable ways. And, certainly, the historical aspect cannot be overlooked as a factor in change. Many great men and women were made by historical conditions—accidents of history. Had they appeared at another time, they probably would have passed unnoticed. Also, historical accidents or unpredictable events—the assassination of a leader or a population-decimating catastrophe such as an earthquake or epidemic—can lead to change.

The research question at the beginning of this chapter inquired into changes in patterns of violence over time. A recent paper looked at the pattern of lynchings of blacks in the deep South between 1882 and 1930. The researchers found that the number of lynchings was related to the price of cotton! In years when the price of cotton was down, the number of lynchings increased. Economic downturns affected all cotton producers, especially the small farmers. In hard times they were probably more frustrated by competition from black laborers, and they favored replacing blacks with unemployed whites. Given the racial caste structure existing in the deep South, white workers would react to other supposed "infractions" and vent their hostility on blacks. Lynchings were one result. Undoubtedly there was anger at the white elite as well, but the small farmers couldn't do anything to them for fear of retaliation.[3]

Ideological and technological factors are related to social change. The appearance of and commitment to ideas of socialism, democracy, Christianity, science, or progress in general have led to major social changes. The industrial revolution of the 18th and 19th centuries was a major force for change in the United States and elsewhere; its dramatic consequences are still with us today. Technological innovations and inventions that have produced tremendous social change include the wheel, the car, the birth-control pill, the atom bomb, the gun, the telephone, the airplane, and the computer. It is easy to point to the dramatic inventions that affect societies, probably because their emergence seems sudden rather than slow and evolutionary, as are many changes. Sometimes the effects of inventions are less obvious. Sociologist William Ogburn recorded numerous examples of the widespread effects of technological innovations.[4] The self-starter on cars aided in the emancipation of women: It allowed women to use cars as easily as do men. The invention of the elevator made possible the construction of tall apartment buildings. Living in these buildings, in turn, changed the family: The rearing of large families was more difficult, and consequently the urban birth rate declined. The invention of the cotton gin was a major factor leading to the Civil War. How? Ogburn suggests that the invention of the cotton gin led to the need for more slaves to cultivate more cotton, and the greater production led to increased trade with England. This in turn led the South, which supported free trade, into conflict with the economic interests of the North, which wanted protective tariffs and restricted

NEW SWEET TALKER

Most Californians before 1987 probably talked to Ramona more than once, even though they had never met her. Ramona was the recorded voice that gave phone numbers when people called information (411) at Pacific Bell.

Pacific Bell dropped Ramona for another woman. In business as in romance, these things happen. The 411 service was evaluated, and a panel of judges rated a new female voice as sounding "more human." Early in 1987, Ramona was replaced by Teri-Ann.

Ironically, neither voice belongs to a Pac Bell operator. A spokesperson said that both Ramona and Teri-Ann were provided by a voice vendor who sells human sounds for computerized recordings.

---

foreign imports, and this conflict led to war. Invention of the six-shooter, barbed wire, and the windmill made possible the settling of the Great Plains. To imply that inventions have such far-reaching effects no doubt seems as much of an oversimplification as does the "great men and women" theory. These examples suggest that we probably cannot single out any one element or invention in a culture and attribute change to it alone.

Social change is introduced into a culture through two processes: invention and diffusion. **Invention** refers to the creation of a new object or idea. Although the end product of the invention (car, airplane, telescope) is usually new, the parts from which it is made are not. The elements have been around, but they are now arranged or put together in a new and significant way. **Diffusion** refers to the spread of objects or ideas from one society to another (Egypt to England) or from one group to another within the same society (upper class to lower class or black to white). Although innovation or invention attracts the most attention, most change is introduced through the process of diffusion. Recall Ralph Linton's "One Hundred Percent American" in Chapter Three.

In sum, social change represents the coming together of a number of events: the conditions in a particular culture at a given time in history and the change agent or catalyst (a particular person or the invention or diffusion of a new technique, object, or idea).

Rates of Change and Planned Change

The various segments of a given society frequently change at different rates. Ogburn called this condition **cultural lag.** A new element may be introduced requiring change in other areas of society, and when these other areas are slow to change, problems result. Faster cars were built before highways could really handle them. Jumbo jets arrived before airports could accommodate the increased passenger load. People's attitudes usually change far more rapidly than laws change, especially in the area of morality laws, and a cultural lag exists.

A society's receptiveness to change is an important factor affecting the rate of social change. If a society has a strictly defined system of stratification or a rigid institutional structure, change will be slower. If a society is isolated from others, change will be slower. Change will be more rapid in a society that emphasizes the values of individualism and self-determination than in a society that emphasizes conformity and reverence for the past and custom. William Graham Sumner believed that there is a fundamental resistance to change in all societies in that people are basically conservative and seek stability or they **"strain for consistency."** Resistance to change is balanced to a certain extent by people's curiosity. They continue to seek new knowledge and better ways of doing things, so change must occur.

Perhaps social change is an evolutionary process beyond our control; we are at the mercy of forces we can do nothing about. Many people believe, however, that we *can* do something more than just describe and accept our social environment. Rather, the environment can actually be manipulated for our benefit. One view of the process of change, as expressed by Sumner, is that social change must be slow, and change in people's attitudes must precede change in legislation. Sumner believed that laws should not move ahead of the customs of the people, or, in his words, "stateways cannot change folkways." An opposite view suggests that new laws can lead the way—they can change people's attitudes and behavior in necessary and beneficial ways. From 1954 on, court decisions and legislative acts have attempted to change the pattern of interaction between the races in the United States. Civil rights groups in the 1960s, feeling that change had been too slow, tried sit-ins, nonviolent demonstrations, and passive resistance to speed up the process. Both sides— those who felt change was beyond control and those who felt that change could be planned and controlled—had fuel for their arguments. The latter group could state that the conditions of minorities in America had improved. The former group, however, could call attention to unanticipated changes that also occurred: increasing violence between the races in the 1960s, and probably a hardening of racial attitudes on the part of many people. At any rate, social scientists today feel that change within limits can be a planned phenomenon. As a result, we see the development of commissions and agencies focusing on a number of matters in which it is felt that planned social change is needed.

THEORIES OF SOCIAL CHANGE

Is there any pattern to social change? Is its direction or speed predictable? These questions have fascinated people for ages, and scientists and philosophers have applied themselves energetically to the task of unraveling the mysteries of social change. Numerous theories have emerged. We will briefly examine five types of theories: cyclic, evolutionary, functional, conflict, and neoevolutionary.[5]

Cyclic Theories

One way of understanding social change is in terms of cycles. Chinese historians gave world history neither a beginning nor an end, but instead saw periods of order and disorder, of prosperity and decline. A 14th-century Arab scholar, Ibn Khaldun, based his theory of change on the clash between nomadic and sedentary peoples. Desert nomads seek the luxuries of the city and therefore continually attack cities and towns. The sedentary city-dwellers are no match for the fierce nomads and are quickly conquered. The conquerers form a new empire, but as they settle in the cities they become comfortable and sedentary like those before them, and they are eventually overrun by a new horde of nomads.

Arnold Toynbee saw civilizations arising out of primitive societies through a process that involved challenge and response. Not all primitive societies become civilizations; those that do, do so because of their response to the challenge of adverse conditions, such as a difficult physical environment, land that has not been settled and tilled, sudden military defeats, and continuing external threats. Two factors are crucial: the severity of the challenge and the development of an elite to manage the response. Most of the society is tied to the past but led by the elite; the civilization can grow by responding successfully to continuing challenges. Eventually civilizations come apart, dividing into several groups, which later end up battling with each other. For Toynbee, civilization's cycle included birth, growth, stagnation, and disintegration. When he looked at the future of Western civilization, Toynbee saw several problems or challenges he felt we had to solve: war, class conflict, population growth, and abundance of leisure time.

Pitirim Sorokin's theory of change had societies fluctuating back and forth like a pendulum through three orientations: ideational, sensate, and idealistic. The **ideational culture** emphasizes feelings and emotions; it is subjective, expressive, and religious. The **sensate culture** emphasizes the senses; it is objective, scientific, materialistic, profane, and instrumental. The **idealistic culture** is an intermediate stage emphasizing logic and rationality. Change from one type to another happens because change is normal and continual in any active system. Sorokin believed that our society was in a stage of decline—that we were an "overripe" sensate culture. He suggested that the future looks bad for us, what with loss of freedom, growth of tyranny, deterioration of the family, and loss of creativity. But good things are ahead: "the magnificent peaks of the new Ideational or Idealistic culture."

Evolutionary Theories

Many scholars have seen social change as linear—a line going in one direction, and usually that direction was progress. Auguste Comte viewed societies as moving through three stages: the theological, or fictitious; the metaphysical, or abstract; and the scientific, or positive. Each stage represents a higher level of existence with the ultimate—the positive stage—resulting in the emergence of priest-sociologists who lead society to a harmonious existence! Comte felt that population increase and density were especially important in influencing change. Herbert Spencer, another evolutionist, likened society to an organism, always growing and increasing in complexity, with its parts dependent on each other. Emile Durkheim saw societies moving from **mechanical solidarity** (small, primitive, homogeneous) to **organic solidarity** (large, complex, specialized).

As all these theories differed, so did the various views of the future: Comte saw increasing progress, with sociologists leading the way, and Spencer saw chances for progress *or* regress.

Functional Theories

Functional analysis sees society as a system of interrelated parts, and it tends to emphasize social equilibrium, stability, and integration of the various pieces. Social change is difficult to accept given this basic perspective. Functional theorists like Talcott Parsons and Neil Smelser suggested that there are tremendous forces resisting change and that these forces may perhaps be overcome slowly and adaptively because of changes outside the system, growth by differentiation, and internal innovations. Generally, however, stability and inertia triumph, and change is viewed as deviant and traumatic.

Conflict Theories

Conflict is universal in the world, and as a force leading to change, it figures in several of the theories we have already mentioned. Current conflict theories spring from the ideas of Karl Marx, who viewed the economic structure to be the foundation of society; changes or contradictions in the economic realm led to changes in other social relations. Capitalist systems were especially liable to produce conflict. The conflict would come from within as the society polarizes into two antagonistic groups: the rulers (bourgeoisie) and the ruled (proletariat). Ultimately the outcome of this struggle between classes would be a revolution and the emergence of a classless society. One of the attractions of Marxism and conflict theory as an explanation for change is that it promises people a role in determining their future. Far from being pawns manipulated by unseen forces, Marx suggests that people can recognize the conditions of their life, react to them, and change them; they can create their own destiny.

Ralf Dahrendorf, like Marx, believed that in every society some members are subject to coercion by others. This means that class conflict is inevitable, and that social change, which emerges out of this conflict, is also inevitable. Dahrendorf believed that human development, creativity, and innovation emerge mainly from conflicts between groups and individuals. The connection between conflict and social change is well documented. Accepting the conflict approach is tempting for those who are frustrated with the slowness of change but lack resources to battle the "system" effectively. Over the last decade, the use of violence and terrorism to produce change has become in-

political power

creasingly common. As Chairman Mao suggested, "Anything can grow out of the barrel of a gun. . . ."

Neoevolutionary Theories

Some theorists have begun to drift back to and rework the evolutionary ideas of yesteryear. Early evolutionary theories explained or analyzed change by focusing on one central factor in a unilinear approach. This broke down when the theory didn't fit the facts of a particular culture and when various theorists came up with different central factors. Modern, or neoevolutionary, theorists still suggest that societies move through a series of evolutionary stages, but now the process is seen as multilinear. By this they mean that change is influenced by varied factors and that although there are similarities or parallels, not all cultures change in the same direction or at the same speed. The newer evolutionary theories also place more importance on cultural borrowing and diffusion than did the earlier ones.

Gerhard Lenski is a contemporary evolutionary theorist. He believes that societies move through a series of forms (hunting and gathering, horticultural, agrarian, industrial) based on their mode of subsistence. Continuity, innovation, and extinction are key elements for Lenski. *Continuity* refers to the persistence of particular elements in society and underlines the fact that, even in a society that appears to be changing rapidly, most of its elements are not changing. In our society, for example, think of the alphabet, the concept of God, the calendar, or driving on the right-hand side of the road. Elements remain because they are useful, because they answer society's needs, or because the cost involved in changing them would be too high. *Innovation* results from inventions and discoveries from within as well as from diffusion from other cultures. Innovation occurs at different rates in different societies because of the amount of information available, the number of people disseminating it, whether it is fundamental (the steam engine is more fundamental than is the can opener), and whether society is receptive to it. *Extinction* refers to the disappearance of cultural elements or of whole societies. The consequence of these three processes is diversity and progress. Societies become more and more diverse as the processes work at different rates in different societies. Progress occurs, not necessarily in the direction of happiness or some higher state of being, but in the sense of continuing technological developments.

The process of change for Lenski goes something like this: Diversity among societies causes them to change at different rates, they develop increasing needs for resources, and this leads to increasing competition among societies. This in turn leads to the survival of some societies and the extinction of others. Because societies with the more advanced technologies are more likely to survive, and because technological changes are basically responsible for other changes, technology becomes the crucial factor in social change. Although the technological advances of the last 150 years have generally raised the levels of freedom, justice, and happiness for people in industrial societies, Lenski is

A TRIP TO GRANDMOTHER'S HOUSE

By Roy Mason and Lane Jennings

During the next decade, the personal computer will change roles—becoming more than just a useful tool. In the following scenario we take a peek at the personal computer of the 1990s at home in its new role of helpmate.

Martha Miller is 85. Since her husband died six years ago, Martha has lived alone in her rural Vermont house. Though the nearest human neighbor is two miles away, and Martha's eyesight no longer allows her to drive to town, she never worries about being unable to cope. Her computer serves as health adviser, housekeeper, secretary, and friend. The system has been equipped with 24-hour biofeedback monitors and a radio-controlled household robot named "Maxine," who acts as companion and roving handyperson.

The house computer wakes Martha every morning at precisely 20 minutes past dawn by drawing the bedroom drapes and, on overcast days, slowly raising the room lights to full brightness. Martha likes to make her own breakfast, and the house computer takes over only if she forgets to turn off the heat under the frying pan when her egg is done or leaves the water running in the sink. This has not happened very often, but if it seemed to be increasing, Maxine would tactfully mention it in conversation.

While Martha eats her breakfast, the house computer plays out an edited version of the broadcast news—omitting ads and sports news (except for Boston Red Sox scores in season) but including multiple versions (recorded at intervals during the night by monitoring different stations) of headline news stories and anything related to Europe, where Martha once served with the State Department.

After breakfast, Maxine rolls into the room and offers to do the dishes—sometimes Martha prefers to do the washing up herself. Next, Martha and Maxine together review the day's schedule. If there are any unusual chores to be done, or a visitor is expected, Maxine may ask for special instructions or remind Martha of what to expect based on the computer's records of similar jobs or visits in the past.

Whenever the weather permits, Martha takes an hour-long walk sometime during the day. Maxine doesn't go along. But the biofeedback monitors attached to Martha's clothes keep the house computer constantly aware of her location and physical condition, and the speaker/phone earrings and necklace Martha wears allow her to converse with the computer—as well as make or receive a phone call—at any time.

Martha enjoys her computer conversations. Maxine has a pleasant voice, and the house computer has been programmed with information about many of the people and places Martha knows well. Moreover, its regular scanning of news broadcasts and on-line information services supplies a constant stream of new material. By monitoring pulse and blood pressure, noting changes in tone of voice and muscle tension, how often Martha laughs, or grows silent, and comparing these with times and frequencies of specific conversation topics, the house computer has gradually built up a personality profile on Martha. Consulting this profile together with her family doctor/medical programmer, Maxine can identify subjects likely to interest Martha, and the doctor obtains data on Martha's mental and physical well-being.

Of course Martha has human visitors and telecommunicates often with family and friends in other parts of the world. But she enjoys her independence because she knows that she is not trapped by it—the companionship and help of her house computer and her robot Maxine keep her from feeling isolated or abandoned.

From Popular Computing, *September 1984. Copyright © 1984 by McGraw-Hill, Inc. Reprinted by permission. All rights reserved.*

not wildly optimistic about the future. His concern with the continuing destruction of the environment, with rapid population growth (especially in
Third World nations), and with increasing international tensions and the
growing potential for nuclear war leads him to conclude that societies *must*
take a more active role in planning their futures. They must learn to control
population and technology rather than continuing to be the victims of uncontrolled growth.[6]

Although the social change theories presented here may appear distinct
unto themselves, there is in fact much overlap among them. Some evolutionary theories seem to incorporate cycle theory, the cycles occasionally look evolutionary, the functionalists borrow from the evolutionists, and conflict seems
to appear everywhere. So what is the best explanation for social change? Does
any one theory fit the facts best? That, of course, depends on which set of
facts you select; explanations for change in nomadic societies must be different from those attempting to explain change in postindustrial society. On the
whole, I find the ideas of the conflict theorists and Lenski's neoevolutionary
approach most interesting. Conflict seems to be a major element in social
change, increasingly so in recent years. Antagonism among classes is commonplace in many societies and acts as a catalyst for change. Lenski's focus
on the importance of technology also makes sense as I write these words on
a computer and read of plans for landing on Mars, of new breakthroughs in
genetic engineering, and of people being released early from prison because
they can be "tethered" to their homes by electronic "handcuffs."

SOCIAL CHANGE AND THE FUTURE: TWO VIEWPOINTS

Every year or so, experts review the state of the world and predict the future.
Alvin Toffler, who analyzed social change in his popular books *Future Shock*
and *The Third Wave*, reports that we are about to be engulfed by a third wave
of change (the first two waves were the agricultural revolution and the industrial revolution). In this third wave, we will see the increasing importance of
electronics and computers: Toffler envisions the home as an electronic cottage
in which people no longer commute to offices but instead work at home, with
results sent to the office by electronics. He predicts space manufacturing—the
development of industries that can take advantage of zero-gravity and vacuum environments. He sees more development of ocean resources; algae and
fish farming will supply oil and protein, and valuable minerals will be mined
from the ocean floor. Toffler also predicts developments in the gene industry
that would produce high-yielding crops that could be grown anywhere; perhaps even human beings could be improved somewhat. More and more
people will raise their own food, weave their own rugs, and make their own
clothing with the new home technologies, according to Toffler.

In their book *Megatrends 2000*, John Naisbitt and Patricia Aburdene analyze
major trends in American society and the world, and they try to predict where
these trends might take us in the future. Their predictions include the following:

1. A global economic boom. The world is becoming a single economy, and factors like increasing worldwide free trade, changing taxes, an Asian consumer boom, and few wars have fueled growth.

2. A renaissance in the arts. As evidence they cite the growth in the number of symphony orchestras and attendance at performances, the increasing number of new book titles, and the booming interest in art and in museums both here and in Europe. New careers and business opportunities seem to be developing in the arts.

3. A global lifestyle. The volume of air travel will continue to increase, foods are transported easily around the world, you can get a Big Mac anywhere (São Paulo, Brazil, has 16 McDonald's), and fashion, film, and entertainment are more international than ever (Disneylands in Japan and France).

4. Rise of the Pacific Rim. Today, Asia has half of the world's population, and by the year 2000 it will have two thirds. Demographically, economically, and culturally, it will continue to replace the Atlantic-Europe connection as a powerful global force. Even now, think of the economic and political influence of Japan, China, Taiwan, Korea, Thailand, and Singapore.

5. Women in leadership. The authors predict that as we continue to pass from an industrial age (male-dominated) to an informational/service age, women will become more and more of an economic force, with the result that the 1990s will become a decade of women in leadership roles.

6. Age of biology. The view of science and technology as having a role in making life "better" is more prevalent. Plants (a super tomato!), animals (more milk from fewer cows), even humans (genetic engineering) can be biologically manipulated in perhaps positive ways (if we can sort out the ethical issues).

7. A religious revival. Naisbitt and Aburdene believe that although there will be a decline in mainline religion, this decline will be countered by increasing interest in "fringe" religions. Catholic, Protestant, and Jewish groups will shrink, but fundamentalist, evangelical, and spiritualist types of religions will increase in what the authors see as a swing from organized religion to spirituality.

8. Triumph of the individual. In a theme carried through from Naisbitt's 1982 book, *Megatrends*, the authors express faith that the power of the individual and individual initiative will continue to be recognized. As one example, they believe that recent inventions (computers, cellular phones, fax machines) have enhanced the power of individuals rather than oppressing them, as it was originally feared they might.

If Naisbitt and Aburdene are correct, these trends are among those that will shape our future.

SOCIAL DISORGANIZATION

Social disorganization refers to the breakdown of norms and roles with the result that customary ways of behaving no longer operate. Suppose the star halfback in the weekend football game is given the ball to run through the opposing team's line, but instead he turns and runs 50 yards the wrong way over his own goal line. The spectators are aghast. Several plays later, a player is running with the ball (in the right direction) only to be tackled by the referee. Again, consternation, because "that's not the way the game is played." The norms are so well known that when they are flagrantly violated and seem to be breaking down, effects on participants are marked. So unusual are these events that sometimes the "norm violators" become famous. In a college game some years ago, a football player had broken loose on a touchdown run. No one had a chance to get him. Suddenly an opposing player, who was out of the game and sitting on the bench, raced onto the field and tackled the ball carrier. The incident made the sports pages throughout the country the next day, and the two players (ball carrier and tackler from the bench) appeared on a nationwide television program a few days later to describe the great event. The event was remarkable because the game just isn't played that way.

But let's get back to the spectators at our peculiar game in which on every play something strange happens: A player runs the wrong way, the wrong person is tackled, eight players come into the game as substitutes but only three leave, and the quarterback runs up into the stands to lead cheers. How long would the spectators tolerate this behavior? It is my guess that very soon they would start leaving the stands: "This is ridiculous—the rules seem to change on every play—you don't know what to expect. . . ." In like manner, extensive breakdown of a society's norms and roles may produce a condition of social disorganization. Extreme social disorganization may lead to personal disorganization, in which the individual becomes upset, disaffected, demoralized, and apathetic; he may want to "leave the game."

Social disorganization is often used to describe living conditions in ghetto and slum areas, which are damaged by the effects of poverty, anonymity, overcrowding, and absence of roots or ties to the community. The person living in these conditions may, like the person watching the peculiar football game, be disgusted at the way the game is being played and decide to leave or fight back. Socially disorganized inner cities have higher rates of crime, mental illness, infant mortality, disease, and family instability. This is not meant to imply that social disorganization causes these to occur, but rather that social disorganization describes a general set of conditions under which these social problems flourish.

Anomie, a particular type of social and personal disorganization, has also been used to describe a generalized condition of "normlessness." French sociologist Emile Durkheim used the concept of anomie to describe a condition in societies in which norms and rules governing people's aspirations and moral conduct had eroded.[7] Solidarity is lost as old group ties break down, and consensus is lost as agreement on values and norms disappears. It is difficult for individuals to know what is expected of them or for them to feel a sense of close identity with the group. Anomie describes a demoralized society in which norms are rapidly changing, uncertain, or conflicting. Durkheim felt that a particular type of suicide occurs in such societies: *anomic suicide,* in which the individual feels lost and disaffected due to an absence of clear-cut rules and standards for behavior. Durkheim believed that abrupt economic changes—sudden inflation or depression, poverty, or wealth—might lead to anomie. This concept has received much attention recently, especially from sociologists in the areas of criminology and delinquency.

As social disorganization occurs, perhaps from rapid social change, new norms replace older ones, role behavior is modified, conflict in values occurs, and institutions assume different forms and functions. Individuals, groups, institutions, and societies must adapt to change, and for some of these it may be difficult, especially if change is rapid or if change is defined as unacceptable and is therefore resisted.

Some sociologists feel that the concept of social disorganization is not very useful. They maintain that the organization of society is constantly in flux and is therefore always changing, and that the result is not disorganization but *reorganization*. They believe that we are really talking about another stage in the social-change process rather than about social disorganization. Further, the critics of the concept argue that there is a tendency to apply the label of social disorganization to those things we dislike, to things that may contradict our own particular values or what we are used to. For example, suppose that in the United States we have high divorce and illegitimacy rates, birth-control pills are freely available, and teenagers commit more crime than any other age group. Does this mean that the institution of the family in America is a victim of social disorganization, or that it is changing to a different form of organization? Is the current situation regarding relations between blacks and whites in America symptomatic of a disorganized society, or of a society undergoing change and reorganization?

My view is that the concept of social disorganization may be helpful in identifying a particular stage that may occur in groups or institutions as they undergo change and reorganization. To produce social disorganization, rapid social change would probably be necessary. Social disorganization involves confusion, disaffection, demoralization, and disorientation for individuals— in short, personal disorganization as well. At the same time, it is important to avoid using the concept of social disorganization to describe only those conditions that we dislike or oppose.

THEORY AND RESEARCH: A REVIEW

Functional analysts and conflict theorists have differing viewpoints on social change, as can be seen in the section on theories of social change. Typically, *conflict theory* is much more applicable to the topic of change. Conflict among groups with different interests, to be expected in any social situation, is an important cause of change. *Functional analysis*, on the other hand, emphasizes stability and order; social change becomes of peripheral interest. The data used in this chapter differed from the survey type of data we have seen in most of the preceding chapters. The "data" of the theorists were analyses of historical records and archives, which led them to their impressions about stages (cyclical, evolutionary, and so on) through which societies pass. Toffler and Naisbitt and Aburdene also analyze this type of data, thereby emphasizing current conditions. For his earlier book, *Megatrends,* Naisbitt used local newspapers, analyzing more than two million articles about events in American cities and towns over a 12-year period. This is another type of *nonreactive survey.*

SUMMARY

In the second section of this book we focused on social organization—the *social fabric* of society. There is a tendency in studying social organization to emphasize the constant, recurring, normal, stable nature of society. In fact, however, all is not as organized, predictable, and fixed as the two previous sections may have implied. Change and deviance are present in all societies, and this section addresses itself to those topics.

This chapter dealt with the concepts of social change and social disorganization. Social change occurs at varying rates in all societies; on this at least there is general agreement. On some other issues there is more debate: Which, if any, factors are related to change? Is change automatic and inexorable, or subject to intervention and control? We have examined cyclic, evolutionary, functional, conflict, and neoevolutionary theories explaining social change. One conclusion is that social disorganization is the condition that sometimes results when norms have broken down and behavior becomes unpredictable. Although there is some disagreement over the usefulness of the concept of social disorganization, the term is often helpful in describing what may happen in periods of rapid and traumatic social change.

In the first of the readings that follow, anthropologist John Yellen describes the !Kung Bushmen of the Kalahari Desert and shows how recent social change is destroying a lifestyle that has existed for thousands of years. The second reading is a study of the social disorganization that occurred in the aftermath of a flood in the mining country of West Virginia.

TERMS FOR STUDY

anomie (379)

conflict theories (374)

cultural lag (371)

cyclic theories (373)

diffusion (371)

evolutionary theories (373)

functional theories (374)

idealistic culture (373)

ideational culture (373)

invention (371)

mechanical solidarity (374)

neoevolutionary theories (375)

organic solidarity (374)

sensate culture (373)

social change (368)

social disorganization (378)

strain for consistency (372)

Reading 12.1

BUSHMEN

John Yellen

Anthropologists led by Richard Lee went to the Kalahari Desert in South Africa in the 1950s and 1960s to study the !Kung Bushmen, one of the last remaining groups in the world still living by hunting and gathering. To their surprise, the anthropologists found the tribe's life to be pleasant and food gathering to be easy, and one of the scientists dubbed the Bushmen "the original affluent society."

Social change is inevitable, however, and anthropologists visiting in the 1970s and 1980s found change among the Bushmen. Drinking vessels made from ostrich eggshells were replaced by pop-top aluminum cans. Before they stalked giraffe and wildebeest; now they gather at the mission station for their daily rations. Before village huts were close together, facing each other and intimate; now they were separated and enclosed by fences demonstrating the newly acquired ethic of privacy. Views about owning property changed. They used to have few possessions and carried them on their backs; now property is collected and stashed in locked trunks. In several decades, a way of life thousands of years old has started coming apart.

John Yellen is an anthropologist at the National Science Foundation. He has visited the Kalahari four times since 1968.

By the mid-1970s, things were different at Dobe. Diane Gelburd, another of the anthropologists out there then, only needed to look around her to see how the Bushman lifestyle had changed from the way Richard recorded it, from how Sahlins described it. But what had changed the people at DBC 12 who believed that property should be commonly held and shared? What had altered their system of values? That same winter Diane decided to find out.

She devised a simple measure of acculturation that used pictures cut from magazines: an airplane, a sewing machine, a gold mine in South Africa. (Almost no one got the gold mine right.) That was the most enjoyable part of the study. They all liked to look at pictures, to guess.

Then she turned from what people knew to what they believed. She wanted to rank them along a scale, from traditional to acculturated. So again she asked questions:

"Will your children be tattooed?"

To women: "If you were having a difficult childbirth and a white doctor were there, would you ask for assistance?"

To men: "If someone asked you for permission to marry your daughter would you demand (the traditional) bride service?"

Another question so stereotyped that in our own society one would be too embarrassed to ask it: "Would you let your child marry someone from another tribe—a Tswana or a Herero—a white person?"

First knowledge, then belief, and finally material culture. She did the less sensitive questions first. "Do you have a field? What do you grow? What kind of animals do you have? How many of what?" Then came the hard part: She needed to see what people actually owned. I tagged along with her one day and remember the whispers inside one dark mud hut. Trunks were unlocked and hurriedly unpacked away from the entrance to shield them from sight. A blanket spread out on a trunk revealed the secret wealth that belied their statements: "Me? I have nothing." In the semidarkness she made her inventory. Then the trunks were hastily repacked and relocked with relief.

She went through the data, looked at those lists of belongings, itemized them in computer printouts. Here's a man who still hunts. The printout shows it. He has a bow and quiver and arrows on which the poison is kept fresh. He has a spear and snares for birds. He has a small steenbok skin bag, a traditional carryall that rests neatly under his arm.

He also has 19 goats and two donkeys, bought from the Herero or Tswana, who now get Dobe Bushmen to help plant their fields and herd their cows. They pay in livestock, hand-me-down clothing, blankets, and sometimes cash. He has three large metal trunks crammed full: One is packed to the top with shoes, shirts, and pants, most well-worn. He has two large linen mosquito nets, 10 tin cups, and a metal file. He has ropes of beads: strand upon strand—over 200 in all, pounds of small colored glass beads made in Czechoslovakia that I had bought in Johannesburg years earlier. He has four large iron pits and a five-gallon plastic jerry can. He has a plow, a gift from the anthropologists. He has a bridle and a bit, light blankets, a large tin basin. He has six pieces of silverware, a mirror and hairbrush, two billycans. His wife and his children together couldn't carry all that. The trunks are too heavy and too large for one person to carry so you would have to have two people for each. What about the plow, those heavy iron pots? Quite a job to carry those through bush, through the thick thorns.

But here is the surprising part. Talk to that man. Read the printout. See what he knows, what he believes. It isn't surprising that he speaks the Herero language and Setswana fluently or that he has worked for the Herero, the anthropologists. Nothing startling there. A budding Dobe capitalist. But then comes the shock. He espouses the traditional values.

"Bushmen share things, John. We share things and depend on each other, help each other out. That's what makes us different from the black people."

But the same person, his back to the door, opens his trunks, unlocks them one by one, lays out the blankets, the beads, then quickly closes each before he opens the next.

Multiply that. Make a whole village of people like that, and you can see the cumulative effect: You can actually measure it. As time goes on, as people come to own more possessions, the huts move farther and farther apart.

In the old days a camp was cosy, intimate and close. You could sit there by one fire and look into the other grass huts, see what the other people were doing, what they were making or eating. You heard their conversations, the arguments and banter.

We ask them why the new pattern?

Says Dau: "It's because of the livestock that we put our huts this way. They can eat the grass from the roofs and the sides of our houses. So we have to build fences to keep them away and to do that, you must have room between the huts."

I look up from the fire, glance around the camp, say nothing. No fences there. Not a single one around any of the huts, although I concede that one day they probably will build them. But why construct a lot of separate small fences, one around each hut? Why not clump the huts together the way they did in the old days and make a single large fence around the lot? Certainly a more efficient approach. Why worry about fences now in any case? The only exposed grass is on the roofs, protected by straight mud walls and nothing short of an elephant or giraffe could eat it.

Xashe's answer is different. Another brief reply. An attempt to dispose of the subject politely but quickly. "It's fire, John. That's what we're worried about. If we put our houses too close together, if one catches fire, the others will burn as well. We don't want one fire to burn all our houses down. That's why we build them so far apart."

But why worry about fire now? What about in the old days when the huts were so close, cheek by jowl? Why is it that when the huts were really vulnerable, when they were built entirely of dried grass, you didn't worry about fires then?

You read Diane's interviews and look at those lists of how much people own. You see those shielded mud huts with doors spaced, so far apart. You also listen to the people you like and trust. People who always have been honest with you. You hear their explanations and realize the evasions are not for you but for themselves. You see things they can't. But nothing can be done. It would be ludicrous to tell these brothers: "Don't you see, my friends, the lack of concordance between your values and the changing reality of your world?"

Now, years after the DBC study, I sit with data spread out before me and it is so clear. Richard's camp in 1963: just grass huts, a hearth in front of each. Huts and hearths in a circle, nothing more. 1968: more of the same. The following year though the first *kraal* appears, just a small thorn enclosure, some acacia bushes cut and dragged haphazardly together for their first few goats. It's set apart way out behind the circle of huts. On one goes, from plot to plot, following the pattern from year to year. The huts change from grass to mud. They become larger, more solidly built. Goats, a few at first, then more of them. So you build a fence around your house to keep them away from the grass roofs. The *kraals* grow larger, move in closer to be incorporated finally into the circle of huts itself. The huts become spaced farther and farther apart, seemingly repelled over time, one from the next, people, families move farther apart.

The bones tell the same story. 1947: All the bones from wild animals, game caught in snares or shot with poisoned arrows—game taken from the bush. By 1964 a few goat bones, a cow bone or two, but not many. Less than 20 percent of the total. Look then at the early 1970s and watch the line on the graph climb slowly upwards—by 1976 over 80 percent from domesticated stock.

But what explains the shattering of this society? Why is this hunting and gathering way of life, so resilient in the face of uncertainty, falling apart? It hasn't been a direct force—a war, the ravages of disease. It is the internal conflicts, the tensions, the inconsistencies, the impossibility of reconciling such different views of the world.

At Dobe it is happening to them all together. All of the huts have moved farther apart in lockstep, which makes it harder for them to see how incompatible the old system is with the new. But Rakudu, a Bushman who lived at the Mahopa waterhole eight miles down the valley from Dobe, was a step ahead of the rest. He experienced, before the rest of them, their collective fate.

When I was at the Cobra Camp in 1969, Rakudu lived down near Mahopa, off on his own, a mile or so away from the pastoral Herero villages. He had two hats and a very deep bass voice, both so strange, so out of place in a Bushman. He was a comical sort of man with the hats and that voice and a large Adam's apple that bobbed up and down.

The one hat must have been a leftover from the German-Herero wars because no one in Botswana wore a hat like that—a real pith helmet with a solid top and a rounded brim. It had been cared for over the years because, although soiled and faded, it still retained the original strap that tucks beneath the chin. The second hat was also unique—a World War I aviator's hat, one of those leather sacks that fits tightly over the head and buckles under the chin. Only the goggles were missing.

I should have seen then how out of place the ownership of two hats was in that hunter-gatherer world. Give two hats like that to any of the others and one would have been given away on the spot. A month or two later, the other would become a gift as well. Moving goods as gifts and favors along that chain of human ties. That was the way to maintain those links, to keep them strong.

When I went to Rakudu's village and realized what he was up to, I could see that he was one of a kind. The mud-walled huts in his village made it look like a Herero village— not a grass hut in sight. And when I came, Rakudu pulled out a hand-carved wood and leather chair and set it in the shade. This village was different from any of the Bushman camps I had seen. Mud huts set out in a circle, real clay storage bins to hold the corn— not platforms in a tree—and *kraals* for lots of goats and donkeys. He had a large field, too, several years before the first one appeared at Dobe.

Why shouldn't Bushmen do it—build their own villages, model their subsistence after the Herero? To plant a field, to tend goats, to build mud-walled houses like that was not hard to do. Work for the Herero a while and get an axe, accumulate the nucleus of a herd, buy or borrow the seeds. That year the rains were long and heavy. The sand held the water and the crickets and the birds didn't come. So the harvest was good, and I could sit there in the carved chair and look at Rakudu's herd of goats and their young ones and admire him for his industry, for what he had done.

Only a year later I saw him and his eldest son just outside the Cobra Camp. I went over and sat in the sand and listened to the negotiations for the marriage Rakudu was trying to arrange. His son's most recent wife had run away, and Rakudu was discussing a union between his son and Dau the Elder's oldest daughter who was just approaching marriageable age. They talked about names and Dau the Elder explained why the marriage couldn't take place. It was clear that the objection was trivial, that he was making an excuse. Even I could see that his explanation was a face-saving gesture to make the refusal easier for all of them.

Later I asked Dau the Elder why he did it. It seemed like a good deal to me. "Rakudu has all that wealth, those goats and field. I'd think that you would be anxious to be linked with a family like that. Look at all you have to gain. Is the son difficult? Did he beat his last wife?"

"She left because she was embarrassed. The wife before her ran away for the same reason and so did the younger brother's wife," he said. "Both brothers treated their wives well. The problem wasn't that. It was when the wives' relatives came. That's when it became so hard for the women because Rakudu and his sons are such stingy men. They wouldn't give anything away, wouldn't share anything with them. Rakudu has a big herd just like the Herero, and he wouldn't kill goats for them to eat."

Not the way Bushmen should act toward relatives, not by the traditional value system at least. Sharing, the most deeply held Bushman belief, and that man with the two hats wouldn't go along. Herero are different. You can't expect them to act properly, to show what is only common decency; you must take them as they are. But someone like Rakudu, a Bushman, should know better than that. So the wives walked out and left for good.

But Rakudu understood what was happening, how he was trapped—and he tried to respond. If you can't kill too many goats from the herd that has become essential to you, perhaps you can find something else of value to give away. Rakudu thought he had an answer.

He raised tobacco in one section of his field. Tobacco, a plant not really adapted to a place like the northern Kalahari, has to be weeded, watered by hand, and paid special care. Rakudu did that and for one year at least harvested a tobacco crop.

Bushmen crave tobacco and Rakudu hoped he had found a solution—that they would accept tobacco in place of goats, in place of mealie meal. A good try. Perhaps the only one open to him. But, as it turned out, not good enough. Rakudu's son could not find a wife.

Ironic that a culture can die yet not a single person perish. A sense of identity, of a shared set of rules, of participation in a single destiny binds individuals together into a tribe or cultural group. Let that survive long enough, let the participants pass this sense through enough generations, one to the next, create enough debris, and they will find their way into the archeological record, into the study of cultures remembered only by their traces left on the land.

Rakudu bought out. He, his wife, and his two sons sold their goats for cash, took the money and walked west, across the border scar that the South Africans had cut, through the smooth fence wire and down the hard calcrete road beyond. They became wards of the Afrikaaners, were lost to their own culture, let their fate pass into hands other than their own. At Chum kwe, the mission station across the border 34 miles to the west, they were given numbers and the right to stand in line with the others and have mealie meal and other of life's physical essentials handed out to them. As wards of the state, that became their right. When the problems, the contradictions of your life are insoluble, a paternalistic hand provides one easy out.

Dau stayed at Dobe. Drive there today and you can find this mud-walled hut just by the waterhole. But he understands: He has married off his daughter, his first-born girl to a wealthy Chum kwe man who drives a tractor—an old man, more than twice her age, and by traditional Bushmen standards not an appropriate match. Given the chance, one by one, the others will all do the same.

QUESTIONS 12.1

1. List and illustrate the concepts (invention, diffusion, and so on) mentioned in this chapter with examples from this reading.

2. Which theory of social change best fits the Bushmen's situation? Why? How would conflict theorists and func-

tional theorists describe what happened?

3. How do the anthropologists seem to have done their research? List the different techniques they used.

Reading 12.2

EVERYTHING IN ITS PATH

Kai Erikson

On February 26, 1972, a makeshift mining-company dam in the hills of West Virginia gave way, and a flood of water swept down Buffalo Creek. Communities along Buffalo Creek were destroyed and 125 people were killed. For the survivors, the disaster turned what had been a tightly knit community into a collection of disorganized people with no roots, no neighborhood, and little of the feeling they formerly held for each other. Sociologist Kai Erikson described the effects of the disaster in Everything in Its Path. *In this excerpt from his book, Erikson describes the trauma of loss of communality. By communality, Erikson means a state of mind shared among a particular gathering of people.*

In most of the urban areas of America, each individual is seen as a separate being, with careful boundaries drawn around the space he or she occupies as a discrete personage. Everyone is presumed to have an individual name, an individual mind, an individual voice, and, above all, an individual sense of self—so much so that persons found deficient in any of those qualities are urged to take some kind of remedial action such as undergoing psychotherapy, participating in a consciousness-raising group, or reading one of a hundred different manuals on self-actualization. This way of looking at things, however, has hardly any meaning at all in most of Appalachia. There, boundaries are drawn around whole groups of people, not around separate individuals with egos to protect and potentialities to realize; and a person's mental health is measured less by his capacity to express his inner self than by his capacity to submerge that self into a larger communal whole.

I am going to propose, then, that most of the traumatic symptoms experienced by the Buffalo Creek survivors are a reaction to the loss of communality as well as a reaction to the disaster itself, that the fear and apathy and demoralization one encounters along the entire length of the hollow are derived from the shock of being ripped out of a meaningful community setting as well as the shock of meeting that cruel black water.

- It is almost like a ghost town now.
- It has changed from the community of paradise to Death Valley.
- Some reason or other, it's not the same. Seems like it's frozen.
- I have found that most of the people are depressed, unhappy, mournful, sick. When you go up Buffalo Creek the only remains you see is an occasional house here and there. The people who are living in the trailers have a depressed and worried look on their faces. You don't see children out playing and running as before. Buffalo Creek looks like a deserted, forsaken place.

- What I miss most is the friendliness and closeness of the people of Buffalo Creek. The people are changed from what they were before the disaster. Practically everyone seems despondent and undecided, as if they were waiting for something and did not know what. They can't reconcile themselves to the fact that things will never be the same again.

- It's kind of sad around there now. There's not much happiness. You don't have any friends around, people around, like we had before. Some of them are in the trailer camps. Some of them bought homes and moved away. Some of them just left and didn't come back. It's like teeth in an old folk's mouth down there now.

- People don't know what they want or where they want to go. It is almost as though they don't care what happens anymore.

- My husband and myself used to enjoy working and improving on our home, but we don't have the heart to do anything anymore. It's just a dark cloud hanging over our head. I just can't explain how we feel.

- I don't know. I just got to the point where I just more or less don't care. I don't have no ambition to do the things I used to do. I used to try to keep things up, but anymore I just don't. It seems I just do enough to get by, to make it last one more day. It seems like I just lost everything at once, like the bottom just dropped out of everything.

The clinical name for this state of mind, of course, is depression, and one can hardly escape the conclusion that it is, at least in part, a reaction to the ambiguities of post-disaster life in the hollow. Most of the survivors never realized the extent to which they relied on the rest of the community to reflect back a sense of meaning to them, never understood the extent to which they depended on others to supply them with a point of reference. When survivors say they feel "adrift," "displaced," "uprooted," "lost," they mean that they do not seem to belong to anything and that there are no longer any familiar social landmarks to help them fix their position in time and space. They are depressed, yes, but it is a depression born of the feeling that they are suspended point-lessly in the middle of nowhere. "It is like being all alone in the middle of a desert," said one elderly woman who lives with her retired husband in a cluster of homes. As she talked, the voices of the new neighbors could be heard in the background; but they were not *her* neighbors, not *her* people, and the rhythms of their lives did not provide her with any kind of orientation.

This failure of personal morale is accompanied by a deep suspicion that moral stan-dards are beginning to collapse all over the hollow, and in some ways, at least, it would appear that they are. As so frequently happens in human life, the forms of misbehavior people find cropping up in their midst are exactly those about which they are most sensitive. The use of alcohol, always problematic in mountain society, has evidently increased, and there are rumors spreading throughout the trailer camps that drugs have found their way into the area. The theft rate has gone up too, and this has always been viewed in Appalachia as a sure index of social disorganization. The cruelest cut of all, however, is that once close and devoted families are having trouble staying within the pale they once observed so carefully. Adolescent boys and girls appear to be slip-ping away from parental control and are becoming involved in nameless delinquencies, while there are reports from several of the trailer camps that younger wives and husbands are meeting one another in circumstances that violate all the local codes. A home is a moral sphere as well as a physical dwelling, of course, and it would seem that the boundaries of moral space began to collapse as the walls of physical space were washed down the creek. The problem is a complex one. People simply do not have enough to do, especially teenagers, and "fooling around" becomes one of the few available forms of recreation. People have old memories and old guilts to cope with, especially the seasoned adults, and drinking becomes a way to accomplish that end. And, for everyone, skirting the edges of once-forbidden territory is a way to bring new excitement and a perverse but lively kind of meaning into lives that are otherwise without it.

A retired miner in his sixties speaking of himself:

- I did acquire a very bad drinking problem after the flood which I'm doing my level best now to get away from. I was trying to drink, I guess, to forget a lot of things and get them moved out of my mind, and I just had to stop because I was leading the wrong way. I don't know what the answer is, but I know that's not it. I don't want to drink. I never was taught that. I've drunk a right smart in my life, but that's not the answer.

And a woman in her late twenties who had recently moved out of the largest of the trailer camps:

- There was all kinds of mean stuff going on up there. I guess it still does, to hear the talk. I haven't been back up there since we left. Men is going with other men's wives. And drinking parties. They'd play horseshoes right out by my trailer, and they'd play by streetlight until four or five in the morning. I'd get up in the morning and I'd pick up beer cans until I got sick. The flood done something to people, that's what it is. It's changed people. Good people has got bad. They don't care anymore. "We're going to live it up now because we might be gone tomorrow," that's the way they look at it. They call that camp "Peyton Place," did you know that? Peyton Place. I was scared to death up there. I don't even like to go by it.

- Many marriages have broken up that seemed secure before the flood. My husband and I can agree on only one thing: we won't go back to Lorado. When the time comes to buy us a house, we both agree that we will face a major problem in our marriage. I hope we can agree on where to live. If not, then we may have to come to a parting of the ways after twenty-six years of marriage.

- My husband and I, we was happy before the flood. We got along real good, other than just a few quarrels that never amounted to nothing. But after the flood we had fights, and it was constantly we were quarreling about something or other. We had fights. He would hit me and he would choke me and he would slap me around.

- My children are changed. I sit and try to talk to them, tell them they are a family and should love each other and treat each other like brothers and sisters. But most of the time they treat each other like enemies. They're always on the firing line at each other. It's always screaming and yelling.

- My grandchildren. It used to be we was the loveliest people you ever seen. We was, together. Now my grandchildren won't hardly give me a look. I don't know what's wrong. They seem like they are moody or something. My grandson there, used to be he loved me better than anything, and now he won't even look at me. He don't want to be around me. One of the granddaughters, too, is about the same. She has spells that way. I don't know. I can't understand it.

- I have good new neighbors, but it's not the same. The neighbors I had before the flood shared our happiness when our babies were born, they shared our troubles and our sorrows. Here is the change. My husband has been sick going on three weeks. My old neighbors would ask about him or go see him or send him a get-well card. But he only got one card, and it was from someone away from here. The day the flood came, the people of Buffalo Creek started running, and they are still running inside their minds. They don't have time to stand and talk.

One result of all this is that the community, what remains of it, seems to have lost its most significant quality—the power it generated in people to care for one another in times of need, to console one another in times of distress, to protect one another in times of danger. Looking back, it does seem that the general community was stronger than the sum of its parts. When the people of the hollow were sheltered together in the embrace of a secure community, they were capable of extraordinary acts of generosity; but when they tried to relate to one another as individuals, separate entities, they found that they could no longer mobilize whatever resources are required for caring and nurturing.

- It used to be that everyone knew everyone. When you were hitchhiking, you just put out your thumb and the first car along would pick you up. But it's not like that now. They just don't care about you now. They got problems of their own, I guess.

- The changes I see are in the people. They seem to be so indifferent toward their fellow man. I guess it's because they had to watch a whole lifetime go down the drain.

- I'm getting old, too, and I can't get no help. Nobody'll help you do nothing. You have to pay somebody, and they'll come and start a project for you, but then they'll walk off and leave you. It's just too much.

In general, then, the loss of communality on Buffalo Creek has meant that people are alone and without very much in the way of emotional shelter. In the first place, the community no longer surrounds people with a layer of insulation to protect them from a world of danger. There is no one to warn you if disaster strikes, no one to rescue you if you get caught up in it, no one to care for you if you are hurt, no one to mourn you if the worst comes to pass. In the second place—and this may be more important in the long run—the community can no longer enlist its members in a conspiracy to make a perilous world seem safe. Among the benefits of human communality is the fact that it allows people to camouflage what might otherwise be an overwhelming set of realities, and the question one should ask about Buffalo Creek is whether the people who live there are paralyzed by imaginary fears or paralyzed by the prospects of looking reality in the eye without the help of a communally shared filter.

QUESTIONS 12.2

1. Apply the terms *social disorganization* and *anomie* to the previous selection.

2. Analyze and comment on the following statement: "It was not social disorganization but reorganization that the

people of Buffalo Creek were experiencing."

3. Forecast the future of the Buffalo Creek area using the social-change theories discussed in this chapter.

NOTES

1. See the chapter on social change by Alvin Boskoff in Howard Becker and Alvin Boskoff, *Modern Sociological Theory* (New York: Dryden Press, 1957).

2. This discussion is drawn in part from Robert Bierstedt, *The Social Order*, 4th ed. (New York: McGraw-Hill, 1974), ch. 20.

3. "The Killing Fields of the Deep South: The Market for Cotton and the Lynching of Blacks. 1882–1930," by E. M. Beck and Stewart Tolnay, *American Sociological Review* 55 (August 1990), pp. 526–539.

4. Ogburn's works include *Social Change* (New York: Viking Press, 1922, 1950); *The Social Effects of Aviation* (Boston: Houghton

Mifflin, 1946); *Machines and Tomorrow's World*, rev. ed., Public Affairs Pamphlets, no. 25, 1946; and Ogburn and Nimkoff, *Sociology*, 2d ed. (Boston: Houghton Mifflin, 1950).

5. The major source for this section is *Perspectives on Social Change*, 2d ed., by Robert Lauer (Boston: Allyn and Bacon, 1977), especially chs. 2, 3, 6, 9, and 12, and pp. 47 and 89.

6. See *Human Societies*, 4th ed., by Gerhard Lenski and Jean Lenski (New York: McGraw-Hill, 1982), especially chs. 2–4 and 14; also see Lauer (see Note 5), chs. 2 and 12.

7. Emile Durkheim, *Suicide*, translated by John A. Spaulding and George Simpson (New York: Free Press, 1951).

CHAPTER THIRTEEN

Collective Behavior

TWO RESEARCH QUESTIONS

1. Does the information spread in a rumor become more and more distorted as the rumor is passed on, or does the information become more accurate?

2. Why do people wear T-shirts with messages and pictures on them? Are there any meaningful patterns among the types of T-shirt messages?

COLLECTIVE BEHAVIOR IS group behavior that is spontaneous, unstructured, and unstable. It may be either sporadic and short-term or more continuous and long-lasting.[1] Collective behavior is often difficult to predict because it is not rooted in the usual cultural or social norms. Spontaneous and unstructured behavior is difficult to observe or record objectively and is, therefore, difficult to study. Ethically and practically, the researcher cannot yell "Fire!" in a crowded theater, start a downtown riot, or produce a natural disaster and then observe how people behave. Although some artificial or laboratory-created studies of rumor and panic have been conducted, most studies of collective behavior by social scientists are after-the-fact analyses and discussions with people who happened to be involved.

These studies have revealed that collective behavior may follow reasonably

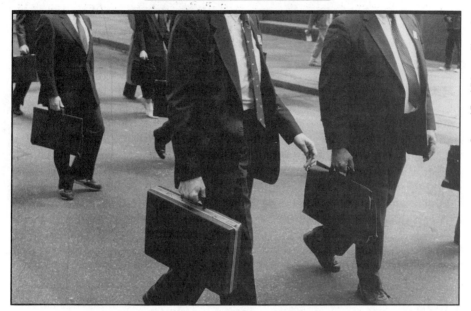

Leonard Freed/Magnum Photos Inc.

consistent patterns. For example, Neil Smelser[2] has suggested that specific conditions must be present for collective behavior to occur:

1. *Structural Conduciveness:* Given the setting or structure of a specific group or society, a particular type of collective behavior, such as panic, riot, craze, or lynching, *could* happen; although the behavior would not necessarily be encouraged, it would still be possible because of historical or other structural reasons.

2. *Structural Strain:* Conditions in society place strain on people; a general feeling of deprivation or conflict is produced by such things as economic failure, hostility among races, social classes, or religions, or sudden changes in the existing order.

3. *Generalized Belief:* A set of feelings, beliefs, or rationalizations must be present to explain the cause of the strain, to create a common culture prepared for action.

4. *Precipitating Factors:* A specific event provides a concrete reason for taking action.

5. *Mobilization for Action:* The participants become organized and act.

6. *Social Control:* Factors that occur after the event has started may prevent or inhibit the effect of the above conditions from continuing and may change its course.

Look at Reading 13.1, which describes an episode of collective behavior, and use Smelser's conditions to analyze it: (1) Conduciveness: American society has a tradition of violence, of taking the law into one's own hands to deal with situations in which it is felt that the law is moving too slowly; this is not legally encouraged, but it happens nevertheless; (2) Strain: Perhaps strain developed because of poor crops, bad economic conditions, or racial enmity between blacks and whites; (3) Beliefs: Perhaps blacks were seen as the cause of the situation—"they are stepping out of their place and citizens had to respond to put the situation right"; (4) Precipitating Factors: Over some drink and discussion, the idea emerged that something had to be done, and a lynching seemed the appropriate response; (5) Mobilization: A group distracted the sheriff to another part of the county and, with collars up and hats pulled down, gathered in front of the jail; (6) Social Control: The lynching was stopped by an unanticipated type of informal social control; what was it?

Keep this general model in mind as we look at some types of collective behavior. We will examine crowds (audiences, mobs, and riots); panic and mass hysteria; rumors; fads, crazes, and fashions; disasters; publics, public opinion, and propaganda; and social movements.

CROWDS

An outline of the major collective behavior concepts would start with the crowd. A **crowd** is a temporary collection of people in close physical contact reacting together to a common stimulus. For example, the passengers on the flight from New York to San Francisco whose pilot suddenly decides he would like to go to the North Pole might be transformed from an aggregation into a crowd (or even a mob). Crowds have certain characteristics in common. **Milling** usually occurs as a crowd is being formed. In one sense milling refers to the excited, restless physical movement of the individuals involved. In a more important sense, milling refers to a process of communication that leads to a definition of the situation and possible collective action. Not long ago, a classroom I was in suddenly started shaking with the first tremors of an earthquake. Almost at once the people began turning, shifting, and looking at each other, at the ceiling, and at the instructor. They were seeking some explanation for the highly unusual experience, and whether spoken aloud or not, the questions on their faces were clear: "What is it?" "Did you feel it?" "What should we do?" Buzzing became louder talking, and someone shouted "Earthquake!!!" The students began to get up and move toward the doors. Many continued to watch the ceiling. . . . Milling may involve the long buildup of a lynch mob or the sudden reaction in a dark and crowded theater when someone shouts "Fire!!!" Milling helps ensure the development of a common mood for crowd members.

When they are part of a crowd, people tend to be *suggestible*. They are less critical and will readily do things that they would not ordinarily do alone. This is true in part because, as members of a crowd, they are *anonymous*. The prevailing feeling is that the crowd is responsible, not the individual. Once one becomes a member of an active crowd, it is extremely difficult to step back, to get perspective, and to evaluate objectively what one is doing. Crowd members have a narrowed focus, a kind of tunnel vision. The physical presence of a crowd is a powerful force; people almost must separate themselves from the crowd physically before they can critically examine their own behavior. There is also a *sense of urgency* about crowds. Crowds are oriented toward a specific focus or task: "We've got to do *this,* and we've got to do it *now!*" Some form of leadership usually appears in the crowd, but as the mood of the crowd changes, the leadership may shift quickly from one individual or group to another.

There are many different types of crowds. Some are passive, such as those watching a building burn or those at the scene of an accident. Some are active, such as a race riot or a lynch mob. Some crowds have a number of loosely defined goals. Other crowds are focused on a specific goal. Turner and Killian distinguish between crowds that direct their action toward some external object—harassing a speaker until he leaves the platform or lynching a person—and expressive crowds that direct their focus on the crowd itself—cheering at a football game or speaking in tongues at a church service.

Controlling the behavior of a crowd is difficult because of the mass of people involved and the spontaneous nature of their behavior. Some methods of dealing with a potentially riotous crowd have been suggested, however: Remove or isolate the individuals involved in the precipitating incident. Reduce the feelings of anonymity and invincibility of the individuals; force them to focus on themselves and on the consequences of their actions. Interrupt patterns of communication during the milling process by breaking the crowd into small units. Remove the crowd's leaders if it can be done without use of force. Finally, attempt to distract the attention of the crowd by creating a diversion or a new point of interest, especially if this can be accomplished by some-one who is considered to be in sympathy with the crowd.

Audiences, Mobs, and Riots

Audiences and mobs are specific types of crowds. An **audience** at a concert, football game, lecture, religious service, or burning building may usually be likened to a passive crowd. Emotional contagion is possible in such situations, and individuals in a group are responsive in ways that they would not be as individuals. Audiences at performances of rock-and-roll stars and in Pentecostal church services may become very expressive in a variety of possibly unpredictable ways. Comedians expect audiences to laugh at their jokes. But most comedians have "bits" that they deliver when audiences are unpredictable and don't laugh. (Johnny Carson got audience sympathy by jokingly mentioning his "war injuries.") Much of audience behavior is predictable, or at least predictably unpredictable. The football fan knows he is going to cheer at the game, the comedian's rejection bits are well prepared, and rock concerts are adequately staffed with police to protect the musicians and with paramedics to minister to injured fans. At the same time, audiences demonstrate collective behavior characteristics in that their behavior is frequently spontaneous, and members are suggestible and anonymous.

A **mob** is a focused, acting crowd. It is emotionally aroused, intent on taking aggressive action. A lynch mob would be an example of such a crowd. A waterfight that started among several fraternities in Berkeley on a warm spring day and spontaneously turned into a panty raid would be another example. In that instance, thousands of eager young men marched with determination through the campus community, methodically stealing panties from numerous sorority houses. Afterward, many of the participants expressed amazement that they had been involved in such behavior.

Mob behavior also occurs when unusual events, such as a severe winter storm in New Jersey and power outages in New York City, have been followed by looting and vandalism. The lights went out in New York City at about 9:35 in the evening on July 13, 1977. The looting and property damage that followed cost businesses $135–$150 million. Robert Curvin and Bruce Porter studied the mob behavior that occurred during the blackout, and they discovered some interesting patterns. Different types of people looted at different stages of the blackout. Stage-one looters started shortly after the lights went out and were generally criminal types; over 80 percent of them had previous arrest records. They moved in quickly and selected valuable merchandise that they could sell quickly and profitably. Here's a typical example: A man gathered several friends and a 38-foot moving van and filled the truck twice with such goods as Pampers, baby food, color TVs, leather coats, and sneakers. Stage-two looters started a little later and were primarily teenagers out looking for fun and excitement. Stage two also included unemployed, poorly educated ghetto poor under 35. Their looting tended to be random and unorganized. Stage-three looting started several hours later and lasted into the next day. Looters at this stage were likely to be working-class people who got caught up in the street activity and felt social pressure to get involved. Better-off employed people with little or no experience in crime were common. Unfortunately for them, police were now becoming more effective, and more arrests were taking place. As one participant (a salesman with a family income of $375 a week) said, "I wasn't thinking about a color TV or anything that the professional dudes were after, like cars and things. I just wanted to snatch something." A store was broken into, items were thrown out onto the street, he gathered up a pile of women's clothes, and then a police car pulled up and he was arrested. Curvin and Porter suggest that because the looters arrested during the blackout had very high unemployment rates, the main cause of the theft and destruction was the blackout opportunity combined with the national economic decline, high unemployment, high prices for essentials, and the worsened living conditions of the poor.[3]

A **riot** describes the situation in which mob behavior has become increasingly widespread and destructive. Riots may involve a number of mobs acting independently. Throughout our history, the United States has had riots over the issue of race: New York, 1863; Chicago, 1919; and Detroit, 1943, are just three examples. In the 1960s, urban riots occurred with increasing frequency. The issues were social class and poverty as well as race. The Watts riot in Los Angeles in the summer of 1965 lasted six days. Nearly 4,000 people were arrested, and property damage was estimated at over $40 million. Thirty-four people were killed and over 1,000 people injured. Most of the dead and injured were blacks. In the Detroit riot in the summer of 1967, more than 40 people died, and property damage was estimated at $50 million. The relatively minor incident that set off the week-long confrontation was the police closing of a "blind-pig," an after-hours tavern. A crowd collected as the police carted off the tavern's patrons, a stone was thrown, a shoe store was set on fire, and the riot was on. There have been riots in prisons and, increasingly, riots related to sporting events. Teams win a championship, their fans celebrate, and small-scale riots occur. In June 1985, a fight between British and Italian soccer fans in Brussels escalated into a riot and left at least 38 dead and 437 injured. In 1969, a disagreement at a soccer match led to a riot, which in turn led to a short war between Honduras and El Salvador. Recent sports riots appear to be accompanied by a hot evening, large crowds, and widespread consumption of alcohol.

England had its worst riots in a century during July 1981. Gangs of white, black, and Asian youths burned cars, stoned police, and smashed shops. In the first week, 300 policemen were injured, millions of dollars' worth of property was destroyed, and more than 1,000 people were arrested in disturbances in London, Liverpool, Manchester, and other cities. In the analysis that followed, there was much disagreement over causes, but the following conditions were mentioned most often: high rates of unemployment (especially among black youth), racism, and young people's resentment of police and authority.

On April 29, 1992, four Los Angeles police officers who had been charged with assault in the beating of Rodney King were acquitted by a jury in Simi Valley, California. The verdict, which surprised, angered, and saddened many, was announced at 3:10 P.M. West Coast time. Around four P.M., five young men stole some bottles from a Korean-owned liquor store near the intersection of Florence and Normandie avenues in South Central Los Angeles. A little after five P.M., motorists in the area reported being attacked and having their car windows broken out. Police responded in numbers, felt overwhelmed by the crowd, and left the area. Around six P.M., looting, especially of liquor stores, began. At 6:46 P.M. and watched by hundreds of thousands of Americans on television, truck driver Reginald Denny was pulled from his truck, assaulted, and nearly killed. Thus began the 1992 Los Angeles riot, which, by the time it ended on May 4, resulted in 53 deaths, 2,325 reported injuries, over 5,000 arrests, more than 600 buildings completely destroyed by fire, and approximately $735 million in total damages. The causes are still being studied—clearly the Simi Valley verdict was a flash point, but it played against a backdrop of the 1990–1992 recession that worsened already bad economic conditions. There was also continuing anger toward a criminal justice system that appeared to be racially biased; there was resentment toward Korean merchants who had been buying out black store owners in South Central; perhaps even the presence and use of alcohol in the area was a factor—the health department in Los Angeles reported that four times as many blacks died in 1990 from alcohol-related causes as from cocaine use and that two of three accidental drug deaths in the county were alcohol-related. South Central has an amazing number of liquor outlets, and looters and arsonists made liquor stores a prime target. It is interesting that the apparent causes of the L.A. riot in 1992 are similar to those suggested for the riots in England a decade earlier (see previous paragraph).

Social scientists have studied riots extensively since the activities of the 1960s. Ralph Conant has examined a number of urban conflicts, and he suggests four phases that describe the possible life history of a riot. Conant believes that not all civil disturbances go through all four stages; in fact, most do not even reach Phase Three but die out earlier. Typically, the temper or mood of the participants is ambivalent and unstable throughout the riot:

> *Phase One—The precipitating incident:* Some gesture, act, or event is taken by the community as concrete evidence of injustice. The incident—use of force by police or eviction of a tenant, for example—is inflammatory because it is seen as typical of the antagonist. It is seen as ample excuse for striking back, and it tends to draw together a large number of people. In a community in which there is much unrest, the precipitating incident may be a minor event to others, but it is very important to the members of the community.

> *Phase Two—Confrontation:* More and more people gather. Individuals suggest targets for violence and attempt to focus crowd reaction, a process called **keynoting.** Other individuals attempt to calm the crowd and encourage moderation. Police officials may appear and may attempt to disrupt the keynoting by dispersing the crowd. This may have the opposite effect and raise a hostile keynoter to an even more prominent position. Conant states that the outcome of Phase Two is of crucial importance. The temper of the crowd may calm, or it could escalate explo-

sively. The responses of civil authorities, of local leaders, and of news media (whether they over-report and sensationalize or show restraint) are all critical.

Phase Three—Roman holiday: Success of hostile keynoting and an explosion of activity mark the beginning of Phase Three. The crowd, dominated now by younger people, begins moving. Conant describes their mood as an angry, gleeful intoxication. Buildings and cars are attacked. Targets are not necessarily random but more likely are selected and patterned, representing objects that rioters resent or fear. Police are taunted and attacked. Later in Phase Three, excitement subsides and systematic looting begins. According to Conant, behavior in the carnival-like atmosphere of the Roman holiday is explained at least in part by the amazing contradictions and ambivalences present in the urban ghetto: stores owned by hated white people and the beloved contents of the stores; despised police and their admired weapons. This ambivalence extends to the riot participants, who show great changes in behavior. They can be violent one day and calm the next day. These ambivalences toward violence and in mood make it unlikely that most riots will reach Phase Four.

Phase Four—Siege: Polarization of the two sides signals the end of effective communication. Curfew is declared. Warlike acts occur. Snipers attack, buildings are firebombed, and a general state of siege exists.[4]

PANIC AND MASS HYSTERIA

Panic represents a particular type of reaction in a crowd situation. Sociologically, **panic** is defined as nonadaptive or nonrational flight resulting from extreme fear and loss of self-control. Usually there is a severe threat, a limited number of ways out, and a feeling of being trapped. Flight in itself does not necessarily mean panic. Some flight is rational and sensible, as when people leave in an orderly manner an area that is threatened by hurricane or flood. Panic refers specifically to nonadaptive flight—people stampeding through a burning building and attempting to fight their way out a door that is already hopelessly blocked. Panic differs somewhat from other forms of collective behavior in that although it is frequently a result of the crowd situation (spontaneous and contagious), it is essentially an individualistic and competitive reaction. During a panic, each person is desperately trying to obtain an objective on his or her own, and because many others are doing the same thing, there is a good chance that some will not make it.[5]

The Iroquois Theater fire in Chicago provides an extreme example of panic and its possible consequences. During a performance on a December afternoon in 1903, draperies on the stage caught fire. Somebody yelled "Fire," and most of the audience panicked. Actors and musicians attempted to calm the crowd, but did not succeed. There were many exit doors, but some were poorly marked and some were difficult to open. People were crushed against doors and on stairways; others jumped to their deaths from fire escapes. The fire was not very serious, and the fire department arrived quickly—the panic lasted only eight minutes. The death toll, however, was 602.

A SNAKE WITH BAD BREATH

A venomous snake has been haunting the people of the Pokharan desert area of Rajasthan in India. The snake, known as *peevana* or *piana*, is about 20 inches long. It steals into huts at dawn and delicately crawls onto the chest of someone asleep. It then opens its mouth wide and holds it in front of the sleeping person's nose, and the victim inhales the snake's poisonous breath. The snake appears to be sadistic as well; it lets its victims know of their fate by whipping them in the face with its tail as it slips away. Newspapers in Rajasthan recently reported that five people fell victim to *peevana* in one village in a single day.

The *peevana* myth has endured for centuries. According to the local maharaja's gazette, the ruler once offered 1,000 rupees to anyone who could produce the reptile, dead or alive. No one has yet been able to do so.

Mass hysteria is in some ways similar to panic and describes the situation in which a particular behavior, fear, or belief sweeps through a large number of people—a crowd, a city, or a nation. Examples of mass hysteria might include fainting at a rock concert, the fear that flying saucers are after us, or the belief that particular women are practicing witchcraft.

RUMORS

Rumor is a type of collective behavior that may or may not be crowd-oriented. Much **rumor** is merely a form of person-to-person communication. Rumor is defined as unconfirmed, although not necessarily false, communication. It is often related to some issue of public concern. Rumors change constantly as they spread. They tend to grow shorter, more concise, and more easily told. Some attractive details of the rumor become magnified. New details are manufactured to complete the story or to make it internally consistent. For example, a rumor that Beatle Paul McCartney had been killed in an automobile accident swept the country in the fall of 1969. Fans of the rock-music group frantically sought further proof of Paul's demise by playing Beatle records backwards and by finding clues on album covers and in photographs of the group. Even several personal appearances by the supposedly dead Beatle did little to quash the rumor. In fact, some are still suspicious today of that fellow passing himself off as Paul McCartney. . . .

People pass rumors for many reasons. The rumor may fit with what we want to believe, with what we know to be true: "Teachers are absentminded" or "Sex education is Communist-inspired." We may pass rumors on to increase our status in the eyes of others. If we were the first to get the news, we want others to know it. Or it's such a tremendous story that others are bound to think more of us when we tell it to them. Did you hear about the flying saucers over Minneapolis? About those giant birds perching on trees in the

Southwest? About Bigfoot wandering through downtown Yakima? You *didn't?* Well, let me tell you. . . .

Rumor may have a variety of functions. It may be gossip and serve to enliven the Thursday-afternoon bridge club, or it may provide a core of communication during a crowd's milling period and thus be the stimulus for crowd action.

As is implied in the first question asked at the beginning of this chapter, there is a common assumption that rumors become less accurate as they are passed on. Taylor Buckner makes the point that in some situations rumors become *more* accurate. The important variables to consider are rumor set and interaction setting. Buckner suggests three rumor sets: critical, uncritical, and transmission. *Critical set* occurs when people have knowledge about the rumor's subject matter, have personal experience with it, or are habitually suspicious of rumors anyway. These people tend to eliminate inaccurate and irrelevant parts of rumors they hear and pass on only the most important parts. An *uncritical set* may exist when people have no knowledge about a situation, when passing the rumor fills a need, or during crisis situations when little information is available. People with an uncritical set may modify the rumor so it sounds better, change it to fit their own needs, devise totally new rumors to fit the situation, or in some other way contribute to the inaccuracy of the rumor. *Transmission set* means that the content of the rumor is irrelevant to the person. The person's only interest is to pass the rumor on. People who merely transmit a rumor forget or reword parts but usually do not purposely distort or correct the rumor.

Group characteristics—the interaction setting—also affect a rumor's accuracy. Buckner suggests that a rumor is heard more often and evaluated more carefully in a close, more primary group (college dorm, sorority, army unit, social club, clique) than in a more diffuse, or secondary, group. It also follows that the more interested and involved a group is with a rumor, the more the rumor is told and evaluated. The rumor that then-Beatle Paul McCartney was dead was undoubtedly more actively discussed among rock-music and Beatles fans than it was in the boardroom at General Motors or among people passing on rumors about flying saucers. Buckner concludes that the multiple interaction likely to occur in a close, highly involved group combined with a critical set leads to increased accuracy in the rumor. However, if the critical set is missing, greater inaccuracy results.[6]

FADS, CRAZES, AND FASHIONS

Fads, crazes, and fashions are forms of collective behavior that are usually more widespread and long-lasting than is crowd behavior. **Fads** and **crazes** refer to the relatively short-term obsessions that members of society or members of specific groups have toward specific mannerisms, objects, clothes, or ways of speaking. Among the more recent fads (in California at least) are roller blades, running shoes, Lycra tights, sweat suits, and video games. Past examples include pet rocks, the hula hoop, swallowing goldfish, and stuffing phone booths. Fads and crazes may interest many and may burn brightly for a while, but they usually die out quickly. **Fashions** are similar to fads and crazes but are more widespread and last longer. Examples of fashions might include miniskirts, two-piece bathing suits, long hair on men, wide neckties, Mustang style in cars, ranch style in homes, and rock and folk rock in popular music.

The second research question at the beginning of this chapter asked a fashion question: Why do people wear distinctive T-shirts? In a mass society, one way of establishing your identity—who you are, what you've done, where you've been—is through the clothes you wear. T-shirts fit these conditions well. They are generally acceptable, easily available, and inexpensive. Collecting and wearing T-shirts has become something of an international clothing fad. In an attempt to learn more about T-shirts and the search for identity, the authors of a recent paper asked a number of college students to describe their favorite T-shirts and why they liked them. Four categories of T-shirts emerged: idiosyncratic, commercialized, elitist, and utilitarian. *Idiosyncratic* T-shirts announce uniqueness and individualism. Often these shirts are designed and made by the wearer. These shirts may show off-beat humor or may be souvenirs from places others haven't heard of. These shirts make the individual "stand out"; they demonstrate his or her uniqueness. *Commercialized* T-shirts were the majority. "Corona Beer," "Hard Rock Cafe," "Gold's Gym," and "Bart Simpson" are easily recognized, are available to everyone, and are often good conversation starters with strangers. *Elitist* shirts tell others that the wearer is a member of a group that most others can't get into. They might announce an exotic and expensive place (Club Med—Phuket, Thailand), an athletic event (Pikes Peak Invitational), an accomplishment (I Survived Kilimanjaro), or a school or organization (Dartmouth '92 or Phi Beta Kappa). *Utilitarian* shirts are the plain type and just another article of clothing. T-shirts, then, are both fad and fashion. They have become a typical and popular style of clothing, and they are used in part to establish one's identity.[7]

Turner and Killian report an interesting peculiarity of fads and fashions: Fashions follow the social class structure, but fads *do not* necessarily follow the social class structure; that is, fashions usually start at upper-class levels and seep downward. Fads, on the other hand, may appear anywhere in the class structure and may be adopted more quickly by members of lower social classes than by members of the upper classes. Because following fads and fashions brings prestige to people, fashion supports the *status quo* (the current prestige system), whereas fads may upset the *status quo* by granting prestige to people who otherwise have low status.[8]

DISASTERS

Studies of disasters provide us with data about collective behavior. A **disaster** is defined as a situation in which there is a basic disruption of the social context within which individuals and groups function—a radical departure from the pattern of normal expectations. Such a situation almost by definition results in collective behavior.[9]

Disaster research reveals three general time periods: immediately before, during, and after the disaster. *Predisaster* studies have noted that making people aware of impending disaster is extremely difficult unless the group concerned has already experienced a disaster—a flood, a bombing, or an earthquake. Otherwise, people do not take the warning seriously. They look around, note that others don't seem bothered, and define the situation as normal or at least not serious. During Hurricane Camille in Mississippi, some 20 people were killed in an apartment house where they had decided to ride out the storm and have a "hurricane party." People tend to interpret disaster cues as normal or familiar events.

Studies of behavior *during* a disaster indicate that, contrary to popular belief, panic and loss of control are rare. Flight is a frequent response, but it usually represents an adaptive rational reaction to the situation rather than panic. Studies of behavior *after* a disaster indicate that people tend to underestimate the scope and destructiveness of the disaster. Again, people define events in terms of the familiar and normal. Often those who have clearly defined tasks in case of disaster—civil-defense workers, nurses—experience severe role conflict. The conflict is between helping people in general, as they are trained to do, or finding and helping family and loved ones. Frequently the latter choice is made. Also, people tend to flock *to*, rather than away from, a disaster area. Within 24 hours after the atom bomb was dropped on Hiroshima, thousands of refugees streamed *into* the city. Also, airliners in trouble at major airports attract crowds of people to the scene. One such incident at New York's Kennedy Airport brought so much automobile traffic that emergency vehicles had difficulty getting through. Communication lines often become clogged with calls coming *into* disaster areas. Material convergence occurs as all sorts and varieties of material are shipped into the disaster area in an effort to help the victims. Within 48 hours after a tornado in Arkansas, truckloads full of material began arriving. Among the mass of material that people sent were button shoes, derby hats, a tuxedo, and a carton of falsies. It took 500 workers two weeks to sort it all out. After Hurricane Camille, when supplies began pouring into Mississippi, authorities had to make television appeals to stop the shipments. The same thing happened after an earthquake in Guatemala. People in an area heavily hit by the quake who usually eat corn and beans were sent canned goods they could not open and would not eat. They were also sent jars of peanut butter, boxes of cornflakes, and cans of anchovy paste. Other useless items that piled up in warehouses after the quake included canned cherries, high-heeled shoes, see-through shirts, and wigs. A Red Cross official reported that the same thing happened after an earthquake in Nicaragua and after a hurricane in Honduras; they concluded that the public simply responds, partly out of a desire to be helpful in a bad situation, partly because they just want to empty their closets. Finally, the studies report that in many cases disaster as an immediate threat produces solidarity among those experiencing it. People tend to grow closer together as a result of their attempts to fight and survive the situation.

PUBLICS, PUBLIC OPINION, AND PROPAGANDA

Publics and public opinion represent another aspect of collective behavior. A **public** is a number of people who have an interest in, and difference of opinion about, a common issue. A public engages in communication and discussion (more often indirectly than face-to-face), and, contrary to the crowd, a public is dispersed or scattered rather than in close physical contact. **Public opinion** refers to opinions held by a public on a given issue. **Propaganda** refers to attempts to influence and change the public's viewpoint on an issue.

In a complicated mass society, a great number of issues appear, and each issue has its concerned interest group or public. Membership in these publics is transitory and constantly changing. Some members of a public are vitally interested, others only marginally so. Members of a public may communicate by special magazines, newspapers, television, and letter writing. Professional football and baseball have their publics; those people interested in gun control or their opposites interested in the free right to bear arms are each a public. The issues of socialized medicine, legalization of prostitution, and elimination of the death penalty all have publics. Some publics are widespread; if the issue is the type of job the president is doing, the whole nation may become the public. Other publics may be much smaller.

Public opinion is registered in a variety of ways. It is registered when an incumbent running for office is defeated. It is recorded in the letters-to-the-editor column of a newspaper or magazine that prints a story with which the general public agrees or disagrees. More recently, public opinion polls have become a popular way of determining what the general public is thinking. Polls are avidly followed by politicians as well as by people on the street. Polls have even been criticized because it is feared that some people look at the polls before deciding how to vote or what to think on a given issue.

Public opinion polling has many pitfalls, and occasionally even the best pollsters go wrong. Probably the most famous miss was the *Literary Digest* poll of 1936, which predicted that Landon (Republican) would defeat Roosevelt (Democrat) for president. The *Literary Digest* obtained a biased sample by mailing ballots to people whose names, for the most part, were selected from telephone directories or from lists of automobile owners. Only 40 percent of all homes had phones in 1936, and only 55 percent of all families had cars, so the magazine was sampling people who were economically well-off. Another error was introduced by *mailing* ballots; people of higher income and education are more likely to return mailed questionnaires than are people with lower incomes and education. Ballots were mailed early (September) so that last-minute changes were missed. The upshot was that *Literary Digest* asked a group of reasonably well-to-do people (who generally tend to vote Republican) for whom they intended to vote. Their prediction was so grossly wrong that it is given as the reason that the magazine went out of business.[10]

Because it is usually impossible in a public opinion poll to talk to all members of the public, a sample is taken. Frequently the success of the poll hinges on the accuracy of the sampling technique. The timing of the poll is also important: Opinions change rapidly, and if the latest political poll is three or four weeks before the election, pollsters can expect to have missed many last-minute changes. A more basic fault in polling is that some people are very interested and know a lot about the issue, and others are only slightly interested and know little or nothing. Yet in most polls, the responses of these two types are given the same weight. Other difficulties in sampling public opinion occur when people give pollsters answers they think the pollsters want, or when a person is a member of several conflicting publics at once, as most of us are. What if a member of the Catholic Church who is a single mother with more kids than she can comfortably support is asked her opinion of government-supported artificial birth control? She may find herself a member of conflicting publics. Finally, as far as action is concerned, it is often much more important to know how a few powerful people or opinion leaders feel about an issue than it is to know what many relatively powerless people think.

In conclusion, let's examine how the public, public opinion, and propaganda might relate to a given issue, say capital punishment. Members of the public interested in this issue would include police officers, prison officials, sociologists, psychologists, social workers, and future criminals. To obtain a measure of public opinion, the pollsters might randomly select 1,000 people and ask them what they think about keeping the death penalty. The sample includes interested members of the general public and disinterested bystanders, and all opinions are counted equally. What if, for example, we find that most of those who know most about capital punishment are opposed to it, and most of those who know least about it are in favor of it? What do we do? Suppose we find (as Gallup did in 1991) that 76 percent of those polled want to keep the death penalty? The public in its various forms has spoken. Because the margin is not decisive, however, the other side decides to hire a public-relations firm to persuade the public a little. Propaganda begins to appear. For example, a public-interest story in a popular magazine tells how the family of an executed man is getting along. A newspaper story appears describing a famous case in which a condemned man was cleared of the crime he was supposed to have committed—but just too late. . . . An ex-warden of San Quentin is interviewed on a nationwide television news broadcast, and he relates in detail his opposition to the death penalty. A reporter, witness to an execution in the electric chair, describes it in vivid detail in a Sunday newspaper supplement. The effectiveness of the propaganda will depend on its ability to reach the audience, on the sophistication of the audience, and on the receptiveness of the audience to the new viewpoint. Let's assume that the propaganda has been cleverly conceived, and a new sampling indicates that public opinion has changed; now only 40 percent want to keep the death penalty. Still, nothing happens until power is applied. A lobbyist in Washington, D.C., working for the Anti-Death Penalty League, puts pressure on legislators. Thirty thousand anti–capital-punishment letters (all remarkably similar) flood in. Finally a new law is passed, more because of the League, the lobbyist, and the letters than because of public opinion. We can see from this imaginary and oversimplified example that boundaries of what constitutes a public are obscure, polls may be misleading, public opinion may change rapidly, propaganda is frequently very useful at building or changing public opinion, and finally, public opinion must be backed by political power if meaningful action is to result.

SOCIAL MOVEMENTS

In the chapter on social change (Chapter Twelve), we mentioned that many people believe that through conscious effort the environment can be manipulated and changed. One way this can happen is through a form of collective behavior called a social movement. A **social movement** is defined as a group of people acting with some continuity to promote or to resist a change in their society or group. Social movements are lasting rather than temporary, they have a distinct perspective or viewpoint, they are oriented toward a specific goal or goals, and members have a sense of solidarity or *esprit de corps*.[11] Social movements seem to be especially popular in modern mass society. In mass society there is mass confusion; that is, there are a variety of viewpoints on every issue. The United States, like other mass societies, has a multiplicity of groups from numerous backgrounds, and each group has its own values, its own way of looking at things. Individual discontent, anxiety, and frustration about the condition of the world and about one's own opportunities are all conducive to social movements. Mass communication exposes us to all the confusing aspects of the mass society and puts us immediately in touch with other people who may also be frustrated, anxious, and looking for a way to change things. The combination of these elements—complex societies with a variety of peoples and viewpoints, numbers of people who feel discontented or shortchanged, and a system of mass communication to tie them together—provides fertile soil for social movements.

Current examples of social movements in the United States would include pro- and antiabortion groups, the Moral Majority, antinuclear power groups, people interested in gay rights, and groups on both sides of the gun-control issue. Our recent history would also include social movements centered on civil rights, women's rights, peace, anticommunism, religion, and many other issues. Some of these are **revolutionary movements,** in that they demand a complete change in the social order. Others are **reform movements,** in that they seek modification of certain aspects of society. Social movements of either type, however, provide an important impetus for social change.

Social movements frequently make use of public demonstrations to bring attention to an issue. The pro-life and pro-choice movements and the gay-rights movement are examples. Farmers have descended on Washington, D.C., in several massive "tractorcades" to bring attention to their economic problems. Sometimes, certain pressures lead groups that otherwise would not do so to use public protest as a means for change. The farmers may be one example of this, and "Anatomy of a Public Protest Action" on the following page describes another.

Often one social movement will develop to counter the effects of another. A current example focuses on the legalization of abortion. Until 1966, abortion was illegal throughout the United States, although a few states would allow it if the mother's life was in danger. A social movement developed among people who, for a variety of reasons—rights of women, population control, objection to laws that discriminated against the poor—felt that abortion should be legalized. Pamphlets to citizens, letters to senators, and articles in newspapers and magazines made people aware of the issue. In 1967, Colorado passed a liberalized abortion law, and soon California, New York, Hawaii, and other states followed. By early 1972, 16 states and the District of Columbia had changed their laws. Then, in January 1973, the U.S. Supreme Court overruled all state laws that prohibited or restricted women from obtaining abortions during the first three months of pregnancy. Seldom is social change, especially legal social change, so rapid, and seldom are the effects of a social movement so dramatic. There was an immediate outcry from those who objected to the Supreme Court's action. Groups such as Right to Life began forming, and a social movement aiming toward restrictive abortion legislation was born.

Collective-behavior experts feel that social movements have a life history. One description of the life span of movements sees them going through four stages.[12] The *preliminary stage* involves individual excitement, discontent, and unorganized and unformulated restlessness. Mechanisms of suggestion, imitation, and propaganda are important, and an "agitator" type of leader appears. The *popular stage* involves crowd or collective excitement or unrest. The unrest is now open and widespread, and *esprit de corps* and an ideology or viewpoint become important mechanisms. The leader at this stage is a "prophet" who puts forth a general ideology, or a "reformer" who focuses on specific evils and develops a clearly defined program. Eric Hoffer believes that the early stages of social movements are dominated by "true believers." These are people of fanatical faith for whom mass movements have an almost irresistible appeal. The true believer is ready to sacrifice his or her life for a cause, any cause.[13] The *formal stage* of the social movement involves the formula-

ANATOMY OF A PUBLIC PROTEST ACTION

Independent truck drivers are generally viewed as middle-class and politically conservative. They are not the type to get involved in challenging, confronting, mass protest actions. But on December 3, 1973, a group of owner-operator truckers blocked the entrance to the Delaware Memorial Bridge for several hours. Two days later they blockaded the Ohio Turnpike and highways in several other states. The national guard and highway patrol broke up the blockades with force. Truck windshields were broken and trucks towed off highways. Some truckers were beaten and arrested.

Charles Bisanz studied the truckers and found them to be independent and self-reliant. They have a belief in success through individual accomplishment, and they view trucking as a way of life. Social networks are developed by getting together at truck stops and by communicating through citizens' band radios. In December 1973, truckers were facing a severe economic squeeze—diesel fuel had suddenly become expensive, but the increases couldn't be passed back to the consumer because hauling rates were fixed by the ICC. Truckers felt they had no clout, that Congress didn't care about them. They also felt they had no exit; trucking was all they knew, and there were no other jobs available to them except perhaps unskilled work in factories. These feelings led to the blockade in early December. Widespread publicity of the blockades led to increased interaction among truckers and further confirmation of their problems. A nationwide shutdown took place from December 13 to 15, and again from January 31 to February 9, 1974. After eight days of the second shutdown, a rate increase was granted, and the independent truckers were given recognition by the government. Bisanz concludes by suggesting that when an individual (or group) is faced with frustration and a threat to a way of life, the more cut off he perceives himself to be from established channels of institutional response and the fewer exits he perceives, the more likely he will engage in public protest actions as a strategy for change.

Adapted from "The Anatomy of a Mass Public Protest Action: A Shutdown by Independent Truck Drivers," by Charles Bisanz, Human Organization *36 (Spring 1977), pp. 62–66. Reprinted with permission of* Human Organization.

---

tion of issues and the formation of publics. The movement becomes "respectable" to gain wider appeal. Issues are discussed and debated. The leaders are "statesmen." The *institutional stage* involves the legalization of the movement; the movement becomes part of the society. The leaders are of the "administrator-executive" type.

Remember that this is a generalized description; some movements follow this pattern, some do not. The time involved in any one stage is variable. The unsuccessful social movement disappears before it can progress through all stages. The successful social movement that attains its objectives must adapt and broaden its goals, or it too will die out.

Finally, the *charismatic leader* figures prominently in both sociological and popular literature. As we discussed in Chapter Four, **charisma** refers to a particular quality an individual has that sets that person apart from other people.

Charismatic leaders are treated as though they are superhuman and capable of exceptional acts. The charismatic movement follows the charismatic leader, who is a symbol for the movement and is above criticism. Some movements are so closely identified with the leader that they have no goal beyond that of following the leader. Other movements, although they follow such a leader, have well-defined goals that allow the movement to continue even if the charismatic leader is lost. Examples of charismatic leaders might include Gandhi, Joan of Arc, Martin Luther King, Jr., Hitler, and Charles de Gaulle.

THEORY AND RESEARCH: A REVIEW

Conflict-theory and symbolic-interactionist perspectives are useful in this chapter. *Symbolic interactionists* will be interested in looking at patterns of communication and the person-to-person aspects of collective behavior. There is a contagion in collective behavior; waves of feeling sweep through a group of people and the group becomes a crowd or a mob. How does this happen? Are there consistent patterns? Is there a leader? If so, what are his or her special traits? How are feelings transmitted? What are the communication processes in rumor? Are recurrent patterns present? *Conflict theorists* might have some ideas about the underlying causes of riot behavior. They might suggest, noting that participants in urban and sports riots often seem to be of the working class or the lower class, that perhaps the class structure is a good place to look for the sources of the unrest and alienation that lead to such events. Revolution and reform social movements attempting to bring about social change would be of central interest to conflict theorists.

The study of the New York blackout was a *reactive survey* using interviews. Data obtained from polls are also a product of reactive surveys. As examples in this chapter have illustrated, careful *sampling technique* is essential. Disaster research may also involve surveys, but some research teams collect information by arriving at disaster scenes as quickly as possible and then using *nonreactive observations* of people's behavior: reviewing the scene, observing disaster agencies at work, watching the victims' behavior, listening to the media reports, and so on.

SUMMARY

In the previous chapter in this section on change and deviance in society we dealt with social change and social disorganization. In this chapter we examined another form of activity that is outside the organized and the ordinary: collective behavior. *Collective behavior* refers to spontaneous and somewhat unstructured actions by groups of people. A number of concepts and terms central to the study of collective behavior were discussed. We examined types of collectivities: crowds, audiences, and mobs. We described types of behaviors: panic and mass hysteria, rumors, fads, crazes, and fashions. Special situations involving collective behavior were noted: riots and disasters. Our discussion of collective behavior concluded with an analysis of publics, public opinion, propaganda, and social movements.

A mob is a focused, acting crowd. A mob is difficult to deal with once set in motion, but the first reading that follows relates how one of the characters in Harper Lee's novel *To Kill a Mockingbird* distracted a lynch mob. The second reading describes the beginning of the 1992 Los Angeles riot. A disaster situation is defined as one in which there is a radical departure from the pattern of normal expectations. In the third reading in this section, D. L. Webber describes what happened when Darwin, Australia, was hit by a cyclone.

TERMS FOR STUDY

audience (396)

charisma (408)

collective behavior (392)

craze (402)

crowd (394)

disaster (403)

fad (402)

fashion (402)

keynoting (398)

mass hysteria (400)

milling (394)

mob (396)

panic (399)

propaganda (404)

public (404)

public opinion (404)

reform movement (407)

revolutionary movement (407)

riot (397)

rumor (400)

social movement (406)

Reading 13.1

MAYCOMB JAIL

Harper Lee

A mob is a focused, acting crowd. It is emotionally aroused and intent on taking action. A mob is a powerful force, and it is difficult to change its direction. When the direction is changed (as it is in this fictional account), the change can be understood in collective-behavior terms. Here, Scout and Jem Finch help their father, Atticus, protect a black man from a lynch mob in Harper Lee's novel To Kill a Mockingbird.

The Maycomb Jail was the most venerable and hideous of the county's buildings. Atticus said it was like something Cousin Joshua St. Clair might have designed. It was certainly someone's dream. Starkly out of place in a town of square-faced stores and steep-roofed houses, the Maycomb jail was a miniature Gothic joke one cell wide and two cells high, complete with tiny battlements and flying buttresses. Its fantasy was heightened by its red brick facade and the thick steel bars at its ecclesiastical windows. It stood on no lonely hill, but was wedged between Tyndal's Hardware Store and *The Maycomb Tribune* office. The jail was Maycomb's only conversation piece: its detractors said it looked like a Victorian privy; its supporters said it gave the town a good solid respectable look, and no stranger would ever suspect that it was full of niggers.

As we walked up the sidewalk, we saw a solitary light burning in the distance. "That's funny," said Jem, "jail doesn't have an outside light."

"Looks like it's over the door," said Dill.

A long extension cord ran between the bars of a second-floor window and down the side of the building. In the light from its bare bulb, Atticus was sitting propped against the front door. He was sitting in one of his office chairs, and he was reading, oblivious of the nightbugs dancing over his head.

I made to run, but Jem caught me. "Don't go to him," he said, "he might not like it. He's all right, let's go home. I just wanted to see where he was."

We were taking a short cut across the square when four dusty cars came in from the Meridian highway, moving slowly in a line. They went around the square, passed the bank building, and stopped in front of the jail.

Nobody got out. We saw Atticus look up from his newspaper. He closed it, folded it deliberately, dropped it in his lap, and pushed his hat to the back of his head. He seemed to be expecting them.

"Come on," whispered Jem. We streaked across the square, across the street, until we were in the shelter of the Jitney Jungle door. Jem peeked up the sidewalk. "We can get closer," he said. We ran to Tyndal's Hardware door—near enough, at the same time discreet.

In ones and twos, men got out of the cars. Shadows became substance as lights revealed solid shapes moving toward the jail door. Atticus remained where he was. The men hid him from view.

"He in there, Mr. Finch?" a man said.

"He is," we heard Atticus answer, "and he's asleep. Don't wake him up."

In obedience to my father, there followed what I later realized was a sickeningly comic aspect of an unfunny situation: the men talked in near-whispers.

"You know what we want," another man said. "Get aside from the door, Mr. Finch."

"You can turn around and go home again, Walter," Atticus said pleasantly. "Heck Tate's around somewhere."

"The hell he is," said another man. "Heck's bunch's so deep in the woods they won't get out till mornin'."

"Indeed? Why so?"

"Called 'em off on a snipe hunt," was the succinct answer. "Didn't you think a' that, Mr. Finch?"

"Thought about it, but didn't believe it. Well then," my father's voice was still the same, "that changes things, doesn't it?"

"It do," another deep voice said. Its owner was a shadow.

"Do you really think so?"

This was the second time I heard Atticus ask that question in two days, and it meant somebody's man would get jumped. This was too good to miss. I broke away from Jem and ran as fast as I could to Atticus.

Jem shrieked and tried to catch me, but I had a lead on him and Dill. I pushed my way through dark smelly bodies and burst into the circle of light.

"H-ey, Atticus!"

I thought he would have a fine surprise, but his face killed my joy. A flash of plain fear was going out of his eyes, but returned when Dill and Jem wriggled into the light.

There was a smell of stale whiskey and pigpen about, and when I glanced around I discovered that these men were strangers. They were not the people I saw last night. Hot embarrassment shot through me: I had leaped triumphantly into a ring of people I had never seen before.

Atticus got up from his chair, but he was moving slowly, like an old man. He put the newspaper down very carefully, adjusting its creases with lingering fingers. They were trembling a little.

"Go home, Jem," he said. "Take Scout and Dill home."

We were accustomed to prompt, if not always cheerful acquiescence to Atticus's instructions, but from the way he stood Jem was not thinking of budging.

"Go home, I said."

Jem shook his head. As Atticus's fists went to his hips, so did Jem's and as they faced each other I could see little resemblance between them: Jem's soft brown hair and eyes, his oval face and snug-fitting ears were our mother's, contrasting oddly with Atticus's graying black hair and square-cut features, but they were somehow alike. Mutual defiance made them alike.

"Son, I said go home."

Jem shook his head.

"I'll send him home," a burly man said, and grabbed Jem roughly by the collar. He yanked Jem nearly off his feet.

"Don't you touch him!" I kicked the man swiftly. Barefooted, I was surprised to see him fall back in real pain. I intended to kick his shin, but aimed too high.

"That'll do, Scout," Atticus put his hand on my shoulder. "Don't kick folks. No—" he said, as I was pleading justification.

"Ain't nobody gonna do Jem that way," I said.

"All right, Mr. Finch, get 'em outa here," someone growled. "You got fifteen seconds to get 'em outa here."

In the midst of this strange assembly, Atticus stood trying to make Jem mind him. "I ain't going," was his steady answer to Atticus's threats, requests, and finally, "Please Jem, take them home."

I was getting a bit tired of that, but felt Jem had his own reasons for doing as he did, in view of his prospects once Atticus did get him home. I looked around the crowd. It was a summer's night, but the men were dressed, most of them, in overalls and denim shirts buttoned up to the collars. I thought they must be cold-natured, as their sleeves were unrolled and buttoned at the cuffs. Some wore hats pulled firmly down over their ears. They were sullen-looking, sleepy-eyed men who seemed unused to late hours. I sought once more for a familiar face, and at the center of the semi-circle I found one.

"Hey, Mr. Cunningham."

The man did not hear me, it seemed.

"Hey, Mr. Cunningham. How's your entailment gettin' along?"

Mr. Walter Cunningham's legal affairs were well known to me; Atticus had once described them at length. The big man blinked and hooked his thumbs in his overall straps. He seemed uncomfortable; he cleared his throat and looked away. My friendly overture had fallen flat.

Mr. Cunningham wore no hat, and the top half of his forehead was white in contrast to his sunscorched face, which led me to believe that he wore one most days. He shifted his feet, clad in heavy work shoes.

"Don't you remember me, Mr. Cunningham? I'm Jean Louise Finch. You brought us some hickory nuts one time, remember?" I began to sense the futility one feels when unacknowledged by a chance acquaintance.

"I go to school with Walter," I began again. "He's your boy, ain't he? Ain't he, sir?"

Mr. Cunningham was moved to a faint nod. He did know me, after all.

"He's in my grade," I said, "and he does right well. He's a good boy," I added, "a real nice boy. We brought him home for dinner one time. Maybe he told you about me, I beat him up one time but he was real nice about it. Tell him hey for me, won't you?"

Atticus had said it was the polite thing to talk to people about what they were interested in, not about what you were interested in. Mr. Cunningham displayed no interest in his son, so I tackled his entailment once more in a last-ditch effort to make him feel at home.

"Entailments are bad," I was advising him, when I slowly awoke to the fact that I was addressing the entire aggregation. The men were all looking at me, some had their mouths half-open. Atticus had stopped poking at Jem: they were standing together beside Dill. Their attention amounted to fascination. Atticus's mouth, even, was half-open, an attitude he had once described as uncouth. Our eyes met and he shut it.

"Well, Atticus, I was just sayin' to Mr. Cunningham that entailments are bad an' all that, but you said not to worry, it takes a long time sometimes . . . that you all'd ride it out together. . . ." I was slowly drying up, wondering what idiocy I had committed. Entailments seemed all right enough for living-room talk.

I began to feel sweat gathering at the edges of my hair; I could stand anything but a bunch of people looking at me. They were quite still.

"What's the matter?" I asked.

Atticus said nothing. I looked around and up at Mr. Cunningham, whose face was equally impassive. Then he did a peculiar thing. He squatted down and took me by both shoulders.

"I'll tell him you said hey, little lady," he said.

Then he straightened up and waved a big paw. "Let's clear out," he called. "Let's get going, boys."

As they had come, in ones and twos the men shuffled back to their ramshackle cars. Doors slammed, engines coughed, and they were gone.

I turned to Atticus, but Atticus had gone to the jail and was leaning against it with his face to the wall. I went to him and pulled his sleeve. "Can we go home now?" He nodded, produced his handkerchief, gave his face a going-over and blew his nose violently.

"Mr. Finch?"

A soft husky voice came from the darkness above: "They gone?"

Atticus stepped back and looked up. "They've gone," he said. "Get some sleep, Tom. They won't bother you any more."

From a different direction, another voice cut crisply through the night: "You're damn tootin' they won't. Had you covered all the time, Atticus."

Mr. Underwood and a double-barreled shotgun were leaning out his window above *The Maycomb Tribune* office.

It was long past my bedtime and I was growing quite tired; it seemed that Atticus and Mr. Underwood would talk for the rest of the night, Mr. Underwood out the window and Atticus up at him. Finally Atticus returned, switched off the light above the jail door, and picked up his chair.

"Can I carry it for you, Mr. Finch?" asked Dill. He had not said a word the whole time.

"Why, thank you, son."

Walking toward the office, Dill and I fell into step behind Atticus and Jem. Dill was encumbered by the chair, and his pace was slower. Atticus and Jem were well ahead of us, and I assumed that Atticus was giving him hell for not going home, but I was wrong. As they passed under a streetlight, Atticus reached out and massaged Jem's hair, his one gesture of affection.

QUESTIONS 13.1

1. Outline how mob and crowd characteristics are illustrated in this selection.

2. Describe in collective-behavior terminology why and/or how the lynch mob was broken up.

Reading 13.2

LOOKING FOR JUSTICE IN L.A.

Peter J. Boyer

The riot in South Central Los Angeles that started on April 29, 1992, left 53 dead, over 600 buildings destroyed by fire, and $735 million in damage. Five thousand people were arrested, one of whom was Damian Williams. Peter Boyer wrote "Looking for Justice in L.A." for The New Yorker *in which he examines the aftermath of the L.A. riots and the impending trial of the four defendants in the beating of Reginald Denny. This excerpt from the beginning of his article describes how the riot got started.*

Last April 29th, while Georgiana Williams was sitting in a hair-dresser's chair at Clinton Young's Beauty Shop, she heard on the radio that there was a disturbance at the intersection of Florence Avenue and Normandie Avenue, a block from her house, on Seventy-first Street, in the section of Los Angeles known as South Central. When her hair was done, she hurried home, but the streets were so clogged that it was hard for her to make her way through. She was stunned by what she saw: "People running and hollering and crying and screaming, truckloads of people, vans of people," she recalls. "So I said, 'What in the world's going on?' And they told me something had happened up on Florence." Her youngest son, Damian, was at home, with a group of his friends. They told her that her oldest son, Mark Jackson, had been arrested that afternoon along with two other young men from the neighborhood, for fighting with a police officer. She recalls that she went into her room to pray, then went to bed. Although she was at ground zero, Georgiana Williams, who watches only the Christian Trinity Broadcasting Network, did not see any of the innumerable replays of the violence of that day, nor did she know that her son Mark's arrest had ignited the worst riot of the century and that her son Damian would become its most notorious figure.

Georgiana Williams is a soulful woman with an easy smile and a strong will—strong enough to have carried her from Mississippi, where as a girl she worked the fields of a plantation with twelve brothers and sisters, to California, where she got a job as a nurse. She bought the house on Seventy-first Street twenty-two years ago, for eight hundred and fifteen dollars. In other circumstances, Miss Williams, as she is called in the neighborhood, would be considered one of those quiet heroines of the American inner city, a hardworking, churchgoing mother who brought up her four children mostly alone, and brought up some others who weren't hers as if they had been. Now, unexpectedly—and although there is nothing in her manner to suggest it—Georgiana Williams has become the mother of the L.A. riot.

In fact, the narrative of the early hours of the riot sounds like an entry in the Williams family chronicle. While Mark was in his mother's back yard that afternoon, installing brakes in a neighbor's car, and Damian was hanging out with friends at a vacant house next door, a neighbor told them that the jury had just acquitted the four white policemen charged in the beating of the black motorist Rodney King. "We was all saying that was jacked up, it was fucked up, we was all mad about what happened," Mark Jackson says. "We just sat down on the porch and said, 'Aw, man, that was crazy. That was messed up, what they did.' Everybody just kept on talkin' about it." All around the intersection of Florence and Normandie, the mood was an eerie combination of holiday and natural disaster: people pouring outdoors, jamming the sidewalks, shouting "Rodney King!"

"If the police hadn't started harassing us, we'd have all walked back to the house and the riot would never have started," Jackson says. According to police reports, rocks were thrown at two squad cars driving up Normandie; one officer chased down a teen-age boy from the neighborhood, Shondell Daniels. When Daniels was apprehended, the crowd got close and vocal. "Everybody was just whooping and hollering and going off," Jackson says. "They say I had the biggest mouth in the crowd. They say I was trying to provoke people to start a fight." When officers tried to arrest Jackson, he resisted, and after a scuffle he was handcuffed and brought to a squad car, where his head slammed into one of the car's windows. By now, several more squad cars had been called to the scene. The officers formed a "skirmish line"—standing their ground, nightsticks in hand. Shots of the scene on neighborhood home videos show that the police were fairly restrained, but the crowd pressed in, shouting insults at the officers. Damian Williams was there, too, and at one point, in a gesture of disapproval of the police presence, he pulled down his pants and underwear and mooned the police. By the time Jackson and two others were taken away, the crowd had grown to more than a hundred and fifty, and the police numbered about twenty-five. At that point, the officer in charge ordered his people to leave the scene. Bart Bartholomew, a photographer working for the New York *Times,* was among the first to feel their absence. He headed back to his car, which was parked on Seventy-first, near the Williams house, but as he walked he acquired a crowd. He reached his Volvo and somehow got inside; someone pulled open the passenger door and stole his camera equipment, while others jumped up and down on the hood. As Bartholomew frantically started his car and drove away, a rock smashed through the window on the driver's side, shattering glass over his face.

With the police gone, there was suddenly a power vacuum in South Central. The crowd, Damian included, moved a block south, to Florence and Normandie, and soon Tom's Liquor Store, an Asian-owned business, was hit with rocks and overrun by looters. Liquor fuelled the anger and the urge to act on it, and passing cars were soon being pelted with rocks and bottles. One motorist was dragged out of his car and beaten, and then another. The crowd, at first indiscriminate in its attacks, began targeting whites. "Leave the Mexicans alone," one voice yelled from the street. "We're only getting the Buddha-heads and white boys!" The white driver of a medical-supply delivery truck was stopped, pulled out, beaten, and struck with an oxygenator stolen from his load. A mile or so away, Reginald Denny, hoping to avoid the afternoon rush hour, exited from the Harbor Freeway onto Florence Avenue, en route to Inglewood, where he was to deliver a load of sand and gravel. Before he had gone far on Florence, he noticed the medical-supply truck stopped in the right lane, with its rear loading door open, and people helping themselves to its cargo. Denny thought of turning around and heading back to the freeway, but he was driving an eighteen-wheeler, a forty-ton lummox of a truck without power steering, and he was afraid he couldn't make a U-turn, so he eased into the left lane and crept toward Normandie. At the intersection, people walked in front of his truck, and he stopped. Denny, who had been listening to music on the radio, didn't know about the Rodney King verdict, or that it had anything to do with the object that shattered the passenger window of his truck, which is the last thing he remembers about that day.

QUESTIONS 13.2

1. Which of Conant's riot phases are illustrated in the previous reading?

2. Using collective-behavior concepts, come up with some ways the tense situation in L.A. might have been defused and the following riot and destruction avoided.

3. Construct an argument that supports the idea that the incident described in this reading is best described as (1) a mob, or (2) a riot, or (3) a social movement.

Reading 13.3

DARWIN CYCLONE: AN EXPLORATION OF DISASTER BEHAVIOR

D. L. Webber

A disaster is defined as a situation in which there is a basic disruption of the social context within which individuals and groups function, a radical departure from the pattern of normal expectations. Studies of disasters provide us with data about a particular type of collective behavior. In this paper, D. L. Webber describes what was called Australia's greatest natural disaster, the Darwin cyclone of 1974.

D. L. Webber is a psychologist. He and his family were in Darwin during the cyclone and remained to help with the emergency.

On Christmas Eve 1974 Darwin was being battered by the fringe winds of Cyclone Tracy. By daylight on Christmas Day the cyclone had passed through Darwin, wrecking the city. Cyclonic winds exceeding 217 km per hour at their peak had brought about the greatest disaster in Australian history. Forty-nine people were dead and 16 others

From "Darwin Cyclone: An Exploration of Disaster Behavior," by D. L. Webber, *Australian Journal of Social Issues*, February 1976. Reprinted by permission of the author.

were missing, presumed dead, on the harbour. More than 140 people were admitted to hospital, mostly on Christmas Day, and over 500 people were treated at Outpatients for injuries sustained during the cyclone. Countless others were treated at suburban emergency health centres for lacerations and injuries. As the city struggled to its feet on Christmas Day it was without power, water, sewerage, or any communication contacts to the suburbs. Approximately $500 million damage had been done to buildings and facilities. Massive destruction and damage to housing left much of the population homeless. Roads were blocked, debris was everywhere, and it continued to rain.

Some weeks before 24 December 1974 there had been a cyclone alert that had ended as an anticlimax. There had in fact been no major cyclone strike in Darwin since 1937. There was no doubt about the current situation however. The Bureau of Meteorology was on full alert, regular warnings were broadcast, emergency advice given, and later people were cautioned to take cover as destructive winds would begin in a few hours. Many heeded this advice. Some even booked into hotels or moved to larger buildings.

It was Christmas Eve, an earlier cyclone had passed away, and many people had no experience of cyclones. Despite the advice of authorities and the example of those who took precautions, a substantial portion of the population either went to parties or got undressed as usual and simply went to bed. For very many people there was a complete misperception of the situation.

Cyclone Strike

By 11 P.M. Christmas Eve, roofs were being lost from buildings and trees were being uprooted. By 2 A.M. Christmas Day the power was off and radio stations had failed. People were now experiencing the nerve-shattering sound of sheet iron grinding and scratching as it was hurled about by the wind. The thudding and exploding sound of beams falling, walls collapsing and heavy airborne objects smashing through houses was everywhere. Many houses simply disintegrated about the occupants. Faced with an immediate threat to life people sought shelter wherever they could find it. Some hid in baths, others lay on floors or under beds. As their houses were swept away some were able to shelter in cars parked between house piers. Most could only lie where they were, hang on and hope.

The sense of isolation during the cyclone was extreme. It was a common feeling that one was at the centre of the destruction and that other areas could not be so badly off. Only later did people realize that the destruction was almost city wide. As might be expected fear was a fairly universal and justifiable experience. Dissociation also occurred. Some people, whose houses were reasonably intact, wandered about inside in a dissociated state looking out of windows.

Post Cyclone

With the approach of dawn the winds abated and people realized they could come out of hiding. Most were simply relieved that the ordeal was over and that they were alive. Immediately neighbours and friends were checked and given aid. Shelter had to be found for children and others. As the only major buildings in most suburbs were the schools, an exodus of people began to these buildings. Some were able to grab a few personal items, others had sense enough to bring a little food. Many just arrived in night attire or in what clothing they could quickly gather. A major social and community crisis existed.

In the meantime key personnel, maintenance crews, tradesmen, and the like reported for duty as soon as they could. The hospital and police station had continued to work throughout. Water and emergency power had to be restored to such places. At dawn plant and machinery at the works depot were operating and emergency procedures that had been adopted were being implemented. Front end loaders and graders began clearing major roads; the evacuation of the sick continued.

News of the disaster had by this stage reached the outside world. The government had been notified and the National Disaster Organization moved into action. Following a review of the situation, a full scale evacuation of the city was ordered. Over the next few days more than 25,000 people were moved to southern centres by air while nearly 10,000 others left Darwin by road. The greatest natural disaster in Australia had been followed by the greatest mass movement of civilian personnel in its history. During this process, food, water, clothing, medicines, and other basic needs had to be met for thousands now forced to accommodate themselves in schools and elsewhere. Behind the scenes local professionals, technicians, tradesmen, and others strove to tackle these problems.

Social and Individual Breakdown

The trauma experienced by family units in their efforts to survive, the mass destruction of homes, the termination of jobs for many in the private sector and the breakdown of public facilities were too much for many people. After they reached safety shock set in. Hundreds milled about looking stunned and unable to act. With others panic had already taken over. They dropped all and simply drove off down the North-South Highway. By 7 A.M. Wednesday there was a stream of traffic heading out of the city. In their haste to leave, many people left wardrobes full of clothes, personal possessions, valuables and pets. Often not even readily accessible tins of food were packed. The stream of evacuees moving down the highway was to increase over the next day. Fortunately for them relief parties were ready and waiting at towns and hamlets on the route. Free fuel, food, and assistance was provided at these points.

For the thousands seeking refuge at the schools, stress, anxiety, and lack of communications began to tell. There had been no radio contact since 1:40 A.M. Wednesday. The ABC could not resume local broadcasting for 36 hours. No official communication existed to the suburbs and the people. Collective anxiety was expressed about where the cyclone might be or whether it would come back. Stories of destruction in other parts of town and the numbers killed were quickly passed on and exaggerated; the situation was ripe for the spread of rumours. Many common experiences were misinterpreted and became the basis of these rumours. One such rumour concerned a supposed outbreak of gastroenteritis in the population. This had its origin in the large number of people suffering gastric upset immediately after the cyclone as a result of stress.

As in all sudden emergencies an immediate and far-reaching change in personal goals and the status of people occurred. New priorities were established. Skills were revalued in terms of immediate needs, and role and status changes were called for on the part of many. Suddenly plumbers, plant operators, cooks, nurses, and the like became more essential than petty bureaucrats. A town that had thrived as a basically bureaucratic structure with well-ordered clerical hierarchies could not make the transition to a new order. Hundreds were disorientated, panic-stricken, and without any apparent social place in the city. The stage was set for mass panic and exit.

Other signs of social breakdown began to show. Looting began in the early hours of Wednesday 25th and was to continue as a problem throughout. Scapegoats were looked for. Even the Bureau of Meteorology was being blamed for the situation. As the stress continued prejudices became heightened, fighting and violence broke out, and petty crime increased.

For those with a clearly defined role or the ability to adopt a new role the task of meeting immediate needs was immense. Such people worked with only a few hours' sleep for several days. As the emergency continued and the central disaster organization worked non-stop to deal with the situation, the effects of a breakdown in cooperative behaviour and its replacement by competitive behaviour on the part of large numbers of citizens became all too apparent.

Following the decision to evacuate by air as many people as possible from the city to southern centres, much of the competitive behaviour now became focused on this aspect as the great bulk of the population tried to escape. Women fought to get on planes, others grabbed nearby children so that they might get priority. Those women who were staying behind or helping directly in the emergency had other women approach them with a view to using their children for evacuation priority. As social dislocation continued and anxiety and panic mounted, evacuation centres were sometimes mobbed and gates to planes rushed. On occasion order had to be restored by police resorting to a show of arms. The height of the disruption came with family units breaking up, men feigning injuries to get evacuated, and some even dressing up as women.

Of the many who were evacuated and the few who remained, those who stayed seemed to suffer less emotional trauma and adjusted quicker to their circumstances. Those who were evacuated had the additional stress associated with the evacuation to cope with. At the airport children and adults often milled for hours awaiting aircraft seats. Confusion, crowding, and stress heightened anxiety. Sleeping on wet floors or standing in rain for hours, anxious mothers, or separation from families did little for most children. Under continued stress, mothers often could not attend effectively to the needs of their children.

Discussion and Conclusion

The basic behaviour patterns that emerged in response to Cyclone Tracy were shock, apathy, psychosomatic reactions, anxiety, and panic. Sleeplessness, loss of appetite, and heightened sensitivity to wind and sound also occurred. Exhaustion, depression, and grief reactions were common later. Detachment and the sense of attack on personal image through home and property loss were noticeable. Loss of public communication and rumours affected the course of events for many days. The wholesale substitution of competitive behaviour for cooperative behaviour threatened the entire social fabric. The aftermath of the cyclone continues and will do so for a long time.

Recognition of these emotional factors and the implementation of programmes to cope with them have been a feature of situations elsewhere. *Unfortunately no such programmes were implemented at Darwin.* On the whole the crisis was approached as essentially a physical crisis by authorities and public alike, with only belated recognition that there were social overtones. A small group of personnel provided mental health resources at the local hospital. This was much too small a group to handle Australia's largest natural disaster. Although general welfare aid in the form of food, clothing, and bedding, together with personnel to organize and implement relief programmes were rushed to the city, there was no infusion of mental health personnel and no opportunity for the one or two that were on hand to either influence or advise administration. Australia's greatest peace time emotional crisis had passed by without even preliminary recognition that it was indeed an emotional and social as well as physical crisis.

To the writer it appears axiomatic that where basic social units such as family structures are collapsing in response to a disaster, efforts should be directed towards reestablishing such units, and developing a sense of social cohesion. Care should be exercised not to accelerate such breakdown by unnecessarily separating families, or isolating children. Evacuation of the injured, unfit, and other appropriate elements of the population was essential as a first step in dealing with the disaster. The continued freighting out of disturbed and upset people many days after the event, instead of providing local support and treatment first, is questionable.

QUESTIONS 13.3

1. In what ways does the behavior in Darwin fit the general characteristics of disaster discussed in the text? In what ways does it differ?

2. List the collective-behavior concepts involved in the study of disaster. Give examples of each from the events surrounding the Darwin cyclone.

NOTES

1. Treatment of concepts in this chapter in general follows the description given by Ralph Turner and Lewis Killian in their book, *Collective Behavior,* 2d ed. (Englewood Cliffs, N.J.: Prentice-Hall, 1972).

2. Neil Smelser, *Theory of Collective Behavior* (New York: Free Press, 1962).

3. See *Blackout Looting: New York City, July 13, 1977,* by Robert Curvin and Bruce Porter (New York: Gardner Press, 1979) and their article, "Blackout Looting," in *Society,* May/June 1979, pp. 68–76.

4. Ralph Conant, "Rioting, Insurrection, and Civil Disobedience," *The American Scholar* 37 (Summer 1968).

5. This discussion of panic is drawn from Duane Schultz, *Panic Behavior* (New York: Random House, 1964), and from Turner and Killian, *Collective Behavior* (see Note 1), chs. 5 and 6.

6. H. Taylor Buckner, "A Theory of Rumor Transmission," in *Readings in Collective Behavior,* 2d ed., edited by Robert Evans (Chicago: Rand McNally, 1975), pp. 86–102. Also see Warren A. Peterson and Noel P. Gist's article, "Rumor and Public Opinion," *American Journal of Sociology* 57 (September 1951), pp. 159–167, for a summary and analysis of some theories on rumor.

7. "Identity Announcement in Mass Society: The T-Shirt," by Donna Darden and Steven

Worden, *Sociological Spectrum* 11 (January–March 1991), pp. 67–79.

8. Turner and Killian, *Collective Behavior* (see Note 1), pp. 152–153.

9. This section on disaster is taken from Charles Fritz's chapter on disaster, which appears in Robert Merton and Robert Nisbet's book, *Contemporary Social Problems* (New York: Harcourt Brace Jovanovich, 1961), ch. 14.

10. This discussion is summarized from Julian Simon's very complete discussion in his book *Basic Research Methods in Social Science* (New York: Random House, 1969), ch. 8.

11. Turner and Killian, *Collective Behavior* (see Note 1), p. 246.

12. This description of the life history of a social movement is from Rex Hopper, "The Revolutionary Process: A Frame of Reference for the Study of Revolutionary Movements," *Social Forces* 28 (March 1950), pp. 270–279.

13. See Eric Hoffer, *The True Believer* (New York: New American Library, 1951).

CHAPTER FOURTEEN

Deviance and Social Control

THREE RESEARCH QUESTIONS

1. Suppose a person's favored method of committing suicide is not available. Will the person choose another method, or not commit suicide?

2. When capital punishment (an execution) takes place and is publicized, does the homicide rate drop (deterrent effect), rise (brutalization effect), or do nothing?

3. Suppose that a group of inmates who are on death row and are about to be executed are instead released! Will they go on to commit more violent crime on the outside or become law-abiding citizens?

EXPERTS TELL US that no two people have the same fingerprints. People differ from each other in their appearance and behavior as well. Variations in behavior occur even though we have norms and roles that specify what *should* happen and what people *should* do in almost any situation. Social differentiation is not only tolerated but expected in social interaction. We expect people to be different from ourselves in behavior and appearance, and we would be very surprised if they were not. At a particular point, however, when differences from a group norm become great enough, social differentiation becomes social deviance.

DEFINING DEVIANCE

There are a number of ways to define deviance. Sometimes deviance and normality are defined statistically. Those around the average are considered normal, and those at the extremes are seen as deviant. If the average height of American males is 5'10", then males from 5'7" to 6'2" are probably viewed as normal, and a fellow 6'8" or 5'2" is considered deviant. Because most people have an I.Q. of between 95 and 115, that I.Q. level is "normal," and a person with an I.Q. of either 145 or 55 is deviant. Because most Americans are heterosexual and fewer are homosexual, heterosexuality is considered normal and homosexuality deviant, if viewed statistically. Sociologists, however, generally do _not_ look at deviance as a statistical phenomenon.

Another way to define deviance is to liken it to a disease. The disease
medical, model sees deviance as something that a person has that we wi
or she could get over, like a cold or the measles. People wonder, "V
wrong with Susan? Why does she continue to screw up? She
crazy. . . ."

The person behaves differently because something is wrong w
The logical next step: If we study carefully, perhaps we can
wrong with the person and then correct it. The medical a
fully with this view of deviance: Find the nature of the
a cure.

425

This view of deviance is a popular one; it's easily understood and it makes sense. This has meant that for years many scholars, scientists, and graduate students alike have scurried about looking for *the* cause, the explanation for the disease. As you can imagine, a number of answers have been found. Several categories of explanations come to mind.

Biological Explanations

In 1875, Cesare Lombroso stated that deviant behavior was inherited and that criminals were throwbacks (closer in the evolutionary chain) to apes. Lombroso came to this conclusion by taking a series of body measurements of institutionalized criminals and noncriminals and comparing them with those of primitive humans. (The criminals' measurements were closer to apeman's than to the noncriminals'.) Other scientists had difficulty obtaining the same results when they repeated his work, but Lombroso's research was influential. It explained a complicated activity—crime—in a reasonably simple and straightforward way. Other studies examined criminal families (for example, the Jukes and the Kallikaks) and identical and fraternal twins to try to prove that crime was inherited. Attempts were made to relate physique and body type (fat, skinny, or muscular) to being criminal and even to the type of crime committed. Peculiar chromosome patterns, especially the XYY pattern, were thought to be more prevalent among prison populations and therefore a potential "disease" factor. More recently, researchers have examined brain and body chemistry and diet to see how they are related to behavior.

Psychological Explanations

The ideas of Sigmund Freud (see Chapter Two) had a great influence on those seeking the causes of deviant behavior. Researchers stressed how early childhood experiences and guilt can affect the personality. The "unconscious" became an important concept. It was seen as that part of the mind in which unpleasant, perhaps antisocial, memories are repressed or stored. Psychologists believed that these memories, even though repressed, would seek expression and would often make their presence known in other ways, such as ams, slips of the tongue, and other difficult-to-explain behaviors. Or it was d that an individual becomes delinquent or criminal because his or go never developed adequate strength to deal with the antisocial owerful id. A whole new vocabulary became available that fo- ogy or heredity, but on the *mind*, and it proved to be a conve- ining antisocial behavior. One outgrowth of this research personality tests, which could be given to trouble- nts, or people suspected of committing crimes in etect the problem ahead of time.

or as behavior contrary to generally ac- e family, group, organizational, or social norms, often established by custom and sometimes supported by law. Whereas the medical approach to deviant behavior infers weakness in the individual, be it biological, psychological, or whatever, sociologists believe that deviant behavior is *normal* behavior—normal in the sense that, because of individual differences, diversity is to be expected in all societies. Sociologists recognize that biological and psychological factors play a part in the development of the individual (as we outlined in Chapter Two), but they emphasize that deviant behavior, like nondeviant behavior, is learned—developed through the socialization process.

One problem with defining deviance this way, however, is that it is relative; what is deviant to one audience may be normal to another. For example, take the member of an urban delinquent gang; he is conforming to gang norms when he defends his turf, carries a weapon, and saves face by fighting rather than backing down to a threat. His "deviant" behavior is learned and is as conforming in its own way as is the behavior of the middle-class college student who conforms to the fashions, behaviors, and attitudes of those around her. There will always be differences of opinion about norms; likewise, there will be differences over what is deviant and what is normal behavior.

Again, sociologists believe that deviant behavior and crime are learned; they are consequences of social conditions. What the sociologist does, then, is try to discover the situations in society that are most likely to lead to crime and deviant behavior. Poverty, lack of legitimate opportunity to move up the social-class ladder, racism, sexism, and other social factors place tremendous pressures on people and may very well push some of them into delinquency and crime. Conflict between social classes, which results in one class overpowering another, can produce pressure leading to crime. These are *structural* conditions—they are a part of the social structure.

For example, Robert Merton saw that our society encourages people to strive for goals, but limited resources means that the goals aren't necessarily available. A discrepancy may exist between aspiration and expectation. How does a person respond to this sort of situation? Merton suggests five patterns, or modes, of adaptation based on goals (what you want) and means (how you get them). The *conformist* follows the rules. For a college student the goal is a degree, and the accepted means are to register, go to class, read books, and pass tests; and that is what the conformist does. But suppose you don't have the time or resources to use the accepted means. This may produce pressure to become an *innovator*. The innovator is one who wants the same goal but uses unacceptable means to get it; the student innovator cheats on tests, buys term papers written by others, copies other people's notes, and so on. The *ritualist*, on the other hand, sticks to the accepted means and does what he or she is supposed to (studies, goes to class), but he or she has lost sight of the goal. The student ritualist wanders around taking course after course, changing majors frequently, becoming a professional student, and never graduating. The *retreatist* abandons both means and goals, gives up, and drops out of the fight totally. The student retreatist has given up the goal of a degree and the means for obtaining it and has dropped out of school. Merton's final mode of adapting, the *rebel*, abandons accepted means and goals like the retreatist; but rather than giving up, he or she substitutes new goals and means. The student rebel might try to set up an alternative college in which various goals are sought using different means (open classes without tests, grades, or assignments). To theorists following Merton, the innovator response was especially interesting as a way of explaining property crime committed by people feeling blocked from achieving desired goals. Merton's ideas influenced many, including several of the gang theorists we discussed earlier (Chapter Five), who also based their explanations of delinquent behavior on structural conditions.

Other sociological approaches address the specific ways in which crime is learned. Edwin Sutherland's theory of **differential association** says that behavior, normal and deviant alike, is learned during interaction with others. Sutherland felt that the prostitute, the professional thief, and the white-collar criminal learn to behave through association and communication with people who portray deviant behavior in a favorable light. They are taught the positive rewards of that behavior—that crime does pay.

Before leaving the sociological theories of deviant behavior, we should briefly look at another approach that has attracted wide interest. This approach focuses not so much on the deviant act or deviant person, but rather on the process of defining the act or person as deviant. Who makes the rules? Who enforces them? On whom and why? These questions emerge from what is generally called **labeling theory.**

Labeling Theory

A 6'8" basketball player for the University of Arkansas, sporting collar-length hair and a long drooping mustache, fouled out of three straight games. Before his next one, against Baylor, he decided to trim his hair and shave his mustache. In the Baylor game he was called for only two fouls. Two nights later, against Texas Tech, the neat, trim player played forty minutes and was still around at the end of a 93–91 double-overtime victory that knocked Tech out of the Southwest Conference lead.

The haircut? The shave?

"I think they helped," the player stated. "I hate to say it, but I guess the hairstyle did affect the refs."

Labeling theory sees deviant behavior as a product of group definitions. Becoming a deviant involves a labeling process; one is a deviant because a particular group labels him or her as such.[1] As with the basketball player, the focus changes from an examination of why the individual does peculiar things to an examination of why these things are called deviant. In 1919, lawmakers succeeded in passing the Eighteenth Amendment to the United States Constitution, which made the sale and manufacture of intoxicating beverages a crime. In 1933, the Twenty-First Amendment repealed the Eighteenth and decreed that drinkers were no longer criminal. Which was the majority view? Probably the latter, but the important point is that both amendments were results of the actions of rule makers who defined the same act as deviant and criminal at one time and permissible at another time. In other words, labeling groups determine who and what is deviant; *deviation is socially defined.*

Why are some acts labeled as deviant and others not? Sometimes the rule makers are motivated by moral or religious beliefs. They are strongly convinced that acts like drinking alcohol, smoking marijuana, or legalizing abortion are morally wrong and then make tremendous efforts to see their beliefs written into law. You can probably think of examples in American history, some quite recent, in which people have successfully crusaded for a cause that ultimately became law.

Perhaps rule makers are motivated by profit. If new clean-air standards are going to mean retooling a company's whole assembly line and spending $5 million on research and development, then you can be fairly sure that the company may spend about $3 million fighting to delay that law. Or turn it around. Imagine that I just invented an ingenious little catalytic converter that makes exhaust air *cleaner* than normal air, and I am lobbying for a law to get my invention placed on all cars. Would you believe me if I told you that I am motivated solely by concern for the environment?

Sometimes the rule makers are interested in supporting the *status quo* and in protecting vested interests. Often, young politicians recently elected to Congress arrive in Washington with the idea of changing the entrenched seniority rules in the House and Senate. But the club remains.

So, all sorts of motivations and justifications are used by rule makers to make rules. The result is a set of personal judgments about what is "right" and "wrong," which become justification for attempts to control people's behavior. Now rule enforcers take over. They seek out and punish those who violate the rules. Howard Becker refers to rule creators and rule enforcers—people whose profession centers on defining and regulating deviant behavior—as moral entrepreneurs. Labeling theory looks at it this way. Some people *make* rules and others *enforce* them, and out of this process particular individuals are *labeled* deviant. It is a *social* process.

"IT'S A BOY! . . . I THINK . . ."

For about one in every three thousand births (about one thousand babies a year in the United States) the determination of the newborn baby's sex may be difficult. These are babies who for a variety of reasons have what doctors call ambiguous genitalia. When the external sex organs do not clearly identify the infant's sex, what the doctor does next sets the course for the infant's future.

For the doctor to say, "I don't know yet," or "I'm not sure," will be very traumatic to the parents and ultimately to the baby. The father and mother should have no doubts that they have a boy or a girl. A clear-cut sex announcement is needed, because of the great importance that "the sex of rearing" plays in forming the child's own sense of sexual identity.

What does the doctor do? One obstetrician who has dealt with many such cases tells us that the size of the visible genitals is the key. "If it's small, say it's female. If in doubt, say it's female. If there is any uncertainty about the baby's eventual ability to perform sexually, it would be a serious mistake to call it a male." The doctor believes that a "male" who is incapable of functioning biologically as a male might be better off functioning as a female, even if by other standards (XY chromosomes) he might be judged male.

From The Sacramento Bee, *February 15, 1978.*

To whom does the deviant label get applied? Well, not to everyone equally, of course. It depends in part on the *seriousness* of the act. On very serious behaviors (homicide, armed robbery) about which there is general agreement, the deviant label is quickly and formally applied. If the act is less serious (belching in public, dressing very strangely), the deviant label, if applied at all, will be applied less stringently. It also depends on the *vulnerability* of the person being labeled. Lower-class people and minorities are more likely to be labeled as deviant than are others. It depends as well on *visibility*. If the act is obvious and occurs in full view of others, it is more likely to get a reaction.

These and other factors interact in complicated ways. Thus, in the view of the labeling theorists, calling people deviant may be an arbitrary and capricious act—an act that tells us more about the person doing the labeling than about the person being labeled. Take the long-haired basketball player once more. As you recall, he cut his hair and no more fouls were called after the Baylor game. Tell me, which of the following would apply? Was he: (1) a mild-mannered, peaceful, and orderly player victimized by referees who have a bias against grubby long-haired-hippie types, or (2) a brutal "hatchet man" whose behavior became less visible to the rule enforcers (referees) when he cut his hair? We'll never know.

An example of the labeling process in action has been provided by Stanford psychologist David Rosenhan. As you may recall from our discussion of **Situation 6** in Chapter One, Rosenhan and seven colleagues each went to the admissions office of a mental hospital and told the same story: They were hearing unclear voices that seemed to be saying words like *empty, hollow,* and *thud.* That was enough for the admitting psychiatrists at 12 of the hospitals; Rosenhan and his friends were diagnosed as schizophrenic and admitted. The label was applied, and it stuck. The diagnoses were not questioned by other staff members, and the pseudopatients were incarcerated for between 7 and 52 days before being released. Although they behaved calmly and normally in the hospital, they were continually given pills (a total of nearly 2,100) to "help" them. None was considered cured upon release—"in remission" was the final diagnosis. When a label is firmly affixed, it is very difficult to escape it. While in the hospitals, the pseudopatients took careful notes about what was happening; the hospital staff either ignored this or saw it as evidence of insane compulsiveness. The pseudopatients' normal behavior was either overlooked entirely or profoundly misinterpreted. Often they would approach hospital staff with reasonable questions. In response, nurses and attendants would move hurriedly away, eyes averted. Psychiatrists also ignored the questioners. This sort of depersonalization is a predominant characteristic in labeling situations. Rosenhan is discouraged about the inability to correctly diagnose mental illness; but more to the point of our study of labeling is his comment that once a label is applied to people it sticks, a mark of inadequacy forever.[2]

PRIMARY AND SECONDARY DEVIANCE

In an approach to deviance that emphasizes society's *reaction* to behavior as the labeling approach does, the distinction between primary and secondary deviance is of central importance. **Primary deviance** is not too serious and does not affect the individual's self-concept. People can rationalize primary deviant acts as being within the bounds of reasonable, acceptable, and normal behavior for people like them. It fits in with or does not seriously detract from the person's typical status and role. The doctor who splits fees, the person who drinks too much, and the eccentric college professor are all deviant to be sure; but these types of deviance are considered primary because they do not go far beyond the boundaries of the conventional statuses and roles they occupy.

A process also occurs whereby society reacts to continued primary deviance by expressing moral indignation, rejection, and penalties. **Secondary deviance** refers to deviance that represents a defense, attack, or adjustment to the problems created by the societal reaction to the primary deviance. The individual becomes labeled and stigmatized. The person begins to see himself or herself as others do, and the self-concept begins to change; in short, he or she wears the label and the self-fulfilling prophecy has done its work. It is not long before the clever lawyer who drinks a little too much becomes the drunk who occasionally practices law. As one's self-concept changes in secondary deviance, the person may become involved in a deviant subculture and gain further knowledge and skill regarding the behavior, including techniques for avoiding detection. The important factors in secondary deviance are the redefinition of self that follows society's reaction and the consequences of that redefinition for future behavior.[3]

A paradoxical and important outcome develops from this sequence: The societal reaction has the intent of stopping primary deviance, but it may force one into secondary deviance. This in turn often leads to an *increase* in deviant involvement and behavior. Here are some examples: The basketball player fouls out of a couple of games. The local paper labels him a rough and aggressive player, he begins to see himself differently, and suddenly he fouls out of *every* game. The juvenile is picked up on a minor charge, such as a runaway or curfew violation. She is locked up in a detention home and treated like the other delinquents. She learns from them, starts to see herself differently, and when she is released she is ready for a full-blown career in delinquency.

DEVIANT BEHAVIOR IN PERSPECTIVE

Deviant behavior is often seen as a malignant element in society to be eradicated at all costs. To say that something is deviant is to imply that something is wrong and needs to be fixed. This is an incorrect view for several reasons. First, it is important to remember that deviant behavior is basically the product of people's definitions rather than something natural or inherent. Knowing this allows us to shift our focus from the person's behavior to the question, "Why is his behavior being called deviant?" Second, Durkheim, Erikson, and others have pointed out that deviant behavior is not necessarily harmful to group life. In fact, it often plays an important part in keeping the social order intact. Deviant behavior enables groups to define boundaries; it preserves stability within the group by pointing out the contrast between what is inside the group and what is outside. Erikson suggests that without the battle between "normal" and "deviant," the community would lack a sense of identity and cohesion, a sense of what makes it a special place in the larger world.[4]

Finally, deviant behavior appears to be one of the processes involved in social change. Often, what is deviant today is accepted tomorrow and expected the next day. Maybe the people who seem to be listening to a different drummer just heard the beat before the rest of us. *Some* forms of deviant behavior represent an early adaptation to changing conditions. There are many examples of this: Deviant clothing styles of the past are the fashion of today. Yesterday's pornography is accepted literature today. Obviously there are limits; some behaviors should be considered deviant no matter what. The point is, however, that deviant behavior does not necessarily represent the evil its title implies; often it serves necessary and important functions in society, including those of defining the group's boundaries and helping to bring about social change.

CATEGORIES OF DEVIANCE

It is often difficult for people to reach agreement on what is deviant and what is normal. At the same time, there are some behaviors and characteristics about which there is more general agreement. In the following paragraphs we will briefly examine some of these categories of deviance.

Individuals who lack intelligence to the extent that they are unable to perform the normal tasks expected of people their age are considered *mentally deficient*. Current practice defines five levels of mental deficiency or retardation: *borderline* (I.Q. range of 83–68), *mild* (67–52), *moderate* (51–36), *severe* (35–20), and *profound* (below 20). The extent of mental retardation in the United States is estimated at approximately 3 percent of the population. More than half of the retarded receive special educational services, and a great majority of those in the 50–80 I.Q. range make a satisfactory adjustment in the community as adults. Those with I.Q. levels between 25 and 50 are viewed as trainable in that they can learn certain skills necessary for living in the home, neighborhood, or sheltered workshop. They will, however, need some care and supervision throughout life. About 5 percent of all retarded fall in the profoundly retarded category and require complete care and supervision throughout life. Mental deficiency, especially severe retardation, may be related to genetic factors, birth trauma, or diseases of the mother prior to birth. Many instances of retardation, particularly at the higher I.Q. levels (50–80), seem to be related to factors in the social environment.[5]

Mental illnesses are commonly classified in two categories: neuroses and psychoses. **Neuroses** are the mildest and most common; **psychoses** are more severe and less common. The neurotic is anxious, nervous, and compulsive, but generally able to function adequately in society. The psychotic loses touch with reality. The psychotic may have hallucinations, incoherent speech and thought, or delusions of persecution; or he or she may withdraw into a dream world. Psychoses fall into two categories: organic and functional. **Organic disorders** are those caused by brain injury, hereditary factors, or physiological deterioration (as a result of aging or the effects of alcoholism or syphilis). **Functional disorders** are those without organic cause; they are based on environmental factors. Functional disorders represent a reaction to such things as stress or rejection, or they may possibly be a chosen alternative to an unlivable and intolerable world. Examples of functional disorders include schizophrenia and manic-depressive psychoses. Social scientists are generally more interested in the functional psychoses because of their relationship to social factors. It is important to note that these categories (organic and functional) are neither clear-cut nor mutually exclusive. Some research has indicated, for example, that hereditary factors affect schizophrenia and that psychoses resulting from aging may be as much related to social isolation (functional) as to physiological deterioration (organic).[6]

It is difficult to estimate the exact number of neurotics in the United States. Some say we are all a little neurotic; other estimates of the number of neurotics in the population run from a high of 40 percent to a low of 5 percent. It has been estimated that there are over two million psychotics in the United States. Half of all hospital beds in the country are used for the care of mental patients, and at current rates one person in ten will be hospitalized for mental illness sometime during his or her life. It is easy to see why mental illness has been called America's *major* health problem.

Drinking alcoholic beverages is relatively common in the United States, but excessive drinking, or *alcoholism*, is viewed as deviant behavior. A 1992 Gallup poll reported that 64 percent of Americans over 18 drink. Drinking and alcoholism are not identical. Some groups (for instance, those for whom drinking has mainly a ritualistic significance) may have relatively high alcohol-consumption rates but little alcoholism. The major factors here are the group associations and cultural factors associated with drinking.

There are several types of drinkers. *Social drinkers* drink when the occasion suggests or demands it, and they are relatively indifferent to alcohol—they can take it or leave it. Some social drinkers drink regularly, others infrequently. *Heavy drinkers* drink more frequently and consume greater quantities when they drink. They occasionally become intoxicated. *Acute alcoholics* have much trouble controlling their use of alcohol. They may go on weekend binges or drunks. They have sober periods, but they rely more and more on alcohol. *Chronic alcoholics* drink constantly. They "live to drink, drink to live." It is difficult if not impossible for them to hold jobs, and their health will inevitably be affected. It is estimated that 18 million Americans have serious drinking problems. The social costs are enormous. It is estimated that alcoholism costs society $117 billion a year in everything from medical bills to lost work days. Cirrhosis of the liver kills at least 14,000 people a year; drunk drivers were responsible for about half of the 49,000 driving deaths in 1988; alcoholism is involved in approximately two thirds of 4,000 drowning deaths and probably 30 percent of the 30,000 suicides. It is believed that alcohol is a major factor in 40 percent of family-court cases and accounts for between 25 and 50 percent of violence between spouses and a third of child-molestation incidents. Of an estimated 14.1 million arrests in 1992, approximately 3.2 million were for offenses directly related to alcohol (drunkenness, drunk driving, and disorderly conduct).[7]

A 1992 Gallup poll found that 24 percent of respondents said that drinking had been a cause of trouble in their family. There has been a tremendous increase in the availability and use of treatment centers throughout the country. The efforts of a number of people and organizations, including MADD and SADD (Mothers and Students Against Drunk Driving), are having some success in changing harmful behaviors. Indeed, Americans may be moderating their use of alcohol; consumption of alcohol has declined by 5 percent since 1981. Sociologists believe that alcoholism is primarily a learned behavior because drinking patterns tend to be associated with occupational groups, social classes, religious categories, and nationalities. Genetic links to alcoholism continue to be hypothesized by some experts, and current research is attempting to clarify the roles played by environmental and genetic factors.

The use of *narcotics* is a central issue in American society. Drug use is viewed as deviant in a society dominated by the values of the Protestant ethic: hard work, self-control, and self-discipline. But use of drugs is widespread. Many youths are using marijuana, amphetamines, and LSD and other hallucinogens. Their elders are using tranquilizers, pep pills, and sleeping pills. Use of anti-anxiety drugs is widespread. The most used, Valium, ranked eighth of all drugs sold in 1986, and a similar drug, Xanax, ranked 12th. The other drug of the 1980s and 1990s seems to be cocaine, usually a white powder that is sniffed, and crack cocaine, which is smoked. Crack is popular because it can be cut into small portions that make it easier to carry, because it's cheaper, and because it can get into the system faster (by smoking). Cocaine generally produces a euphoric state, followed by a variety of different feelings, including depression. Increased amounts may need to be taken to obtain the desired effect, and it is often used in combination with alcohol. Some of the current violent urban youth-gang behavior has been connected with the selling and distribution of crack cocaine.

In the mid- and late 1960s, there was great concern about drugs. This led to new laws, increased enforcement, numerous arrests, and harsh punishment. In the mid- and late 1970s, public opinion moderated, things changed, and there were fewer arrests, although it's unlikely that people's drug-taking behavior changed. Then in the 1980s the picture changed again and arrests increased. The usual enforcement approach has been to work on supply (trying to stop the production and importation of drugs) rather than on demand (working on the reasons that people want drugs in the first place). This approach hasn't worked well in the United States. In the early 1990s, law-enforcement officials saw an increasing and alarming connection between narcotics and violent crime, especially in big cities like Los Angeles, Miami, and New York. The United Nations estimates that the international narcotics trade grosses $300 billion a year. The U.S. Justice Department figures America's own drug economy at $100 billion annually, almost 2 percent of the gross national product.

In the late 1980s and again in the 1990s use of drugs became a scandal in professional sports. Players in baseball, basketball, and football were suspended and sent to drug rehabilitation programs or kicked out of the sport completely. Several athletes in the 1988 and 1992 Olympics, including a gold medalist, were disqualified for steroid use.

Suicide is viewed as deviant behavior in the United States, as it is in most societies dominated by Christian and/or Jewish religions. Reactions vary, however. Some countries have called suicide a crime and have buried suicides in special cemeteries. In some Asian societies suicide is looked upon with less disfavor. Ceremonial self-destruction, known as *hara-kiri*, has long been a custom in Japan. Several states in the United States have classified attempted suicide as a crime, but it is unlikely that this has any deterrent effect, except possibly to ensure the success of the attempt. It is estimated that 31,000 Americans committed suicide in 1990, a rate of about 12 per 100,000 people. Probably at least five times as many people attempt suicide as actually commit it.[8]

Patterns of deviant behavior often vary substantially from one country to another, and this is certainly true of suicide, as can be seen in Table 14.1. However, suicide statistics may be somewhat inaccurate. Countries vary in how they report suicides, and in most places it is probably underreported. In the United States, the coroner usually determines whether or not a death is reported as suicide. A coroner who is uncertain about the cause is more likely to label the death "due to natural causes." In addition, relatives, motivated by personal or religious beliefs, may press for a "death by natural causes" decision from the coroner.

Suicides in America are more likely to be male; white; over 45; single, divorced, or widowed; and Protestant. Although men commit suicide more often than women do, more women than men attempt suicide. Men more often use guns to commit suicide; women tend to use poison. Some occupational groups—military officers, police officers, psychiatrists—have high suicide

Table 14.1 Suicide Rates (per 100,000 People) in Selected Countries, 1987–1990

Hungary	40	China	18	England	8
Finland	29			Italy	8
Denmark	24	Australia	13	Spain	7
Austria	24	Canada	13	Greece	4
Switzerland	22	United States	12	Mexico	2
France	21			Kuwait	1
Sweden	19	Ireland	8	Egypt	1
Czechoslovakia	18	Israel	8		

Demographic Yearbook, 1991

rates. And, interestingly, suicide is a leading cause of death for young adults.

In the late 19th century Emile Durkheim studied suicide in Europe and described in detail three types of suicide: egoistic, altruistic, and anomic.[9] **Egoistic suicide** occurs when interpersonal relationships are secondary, distant, and not group-oriented. In such situations, the individual lacks group attachments, and when personal problems appear, the individual, in the absence of emotional support from others, resorts to suicide. The high suicide rate of single people might be an example of egoistic suicide. **Altruistic suicide,** on the other hand, is the result of strong group attachments. The individual commits suicide to benefit the group and to follow group norms. Suicides of Japanese soldiers and airmen during World War II and the self-immolation of Buddhist monks in Vietnam are described as altruistic suicides. **Anomic suicide** occurs in situations in which norms are confused or are breaking down. Economic depressions or rapid social changes that lead to disequilibrium in society and a state of "normlessness" may result in anomic suicides.

The first research question asked at the beginning of this chapter concerned suicide. A recent paper discussed some interesting data on how people commit suicide. There clearly are cultural or national variations. For example, suicide by gunshot is favored in Australia and the United States, the Golden Gate Bridge is used in San Francisco, the most common technique in Holland for a long time was drowning, and the typical method in India involved drinking a cheap and easily available insecticide. It has been claimed that jumping out of tall buildings and suicide by gunshot depend directly on how many tall buildings and guns are available. But what would happen if the favored technique of committing suicide were not available? Would people choose another technique (displacement), or would they not commit suicide?

Between 1963 and 1975 the number of suicides in England dropped suddenly (35 percent) at a time when suicides were on the increase in nearby European countries. It was difficult to figure out why. It turns out that one of the favored techniques for suicide in England was gas poisoning. The remarkable decline in suicide occurred because of the removal of carbon monoxide from the domestic gas supply. Improved production methods and natural gas from the North Sea meant that the gas was less toxic and no longer lethal. The method some people wanted to use didn't work, and they did not switch

TERRORISM

Terrorism has been defined as the premeditated, deliberate, systematic murder, mayhem, and threatening of the innocent to create fear and intimidation in order to gain a political or tactical advantage. Terrorism differs from common crime in that it is *politically* motivated. Examples of terrorism exist throughout history. A religious sect called the Sicarii were active terrorists in the Roman Empire from A.D. 66–73. The 19th century was marked by many political crimes throughout the world. However, only in the 1970s and 1980s has the world really become aware of terrorism. In part this is because technological changes have made terrorism seemingly more serious. These changes include the advent of the jet airliner and the accompanying increase in travel, of modern weapons of mass destruction, and of a global media network that quickly broadcasts the events around the world. All of these work to provide small terrorist groups with remarkable leverage. The U.S. State Department reports that 928 people were killed by terrorists in 1985. There were 800 terrorist incidents in 1986, and 840 in 1987. During the early 1980s nearly half of all terrorist incidents occurred in western Europe, by 1987 that proportion had declined to 20 percent. The Middle East and Latin America had about 25 percent each and North America was relatively terrorism-free, having about 3 percent of the incidents. Americans, although relatively safe at home, are at greater risk when traveling abroad; in 1987, nine of 28 hostages known to be held in Moslem West Beirut were Americans. Terrorism takes many forms—assassinations, hijackings of airplanes and cruise ships, bombings in train stations and airports, hostage dramas—and appears in many places— Northern Ireland, western Europe, Latin America, Asia, the Middle East. The bombing in a Bologna railway station in 1980 killed 83 civilians. A ransom of $60 million was paid during the 1970s by a corporation in Argentina for the release of three executives. The future of terrorism is difficult to assess. Some believe that given the current international political climate, terrorism as a low-intensity conflict (a type of unconventional war) will continue to be a popular means to achieve political goals.

Tom Kando, sociologist

---

to another technique. Displacement didn't happen, and the authors of the paper suggest that this might be a lesson in suicide prevention—an argument for a suicide barrier on the Golden Gate Bridge, for restricting the number of guns in circulation, and so on.[10]

DEVIANT BEHAVIOR AND CRIMINAL BEHAVIOR

A distinction should be made between deviant behavior and criminal behavior. As we have mentioned, deviant behavior, like beauty, is in the eye of the beholder. Behaviors are not naturally deviant; they are defined as deviant by groups. Similarly, reactions to deviance vary greatly. Some deviant behaviors are ignored, some deviant behaviors are tolerated, and some deviant behaviors elicit severely critical reactions from others. As behaviors fall further outside the range of what is defined as normal, societies feel that they must for-

BLACK ON BLACK CRIME

Blacks represent 12 percent of our population, but they make up 55 percent of all arrests for homicide, over 60 percent of the arrests for robbery, and close to 40 percent of the arrests for aggravated assault (1992). Blacks also run a much greater risk of being victims of these crimes. Much violent crime, especially homicide, is intraracial (within the same race). In 1992, 94 percent of blacks killed were killed by blacks; 83 percent of whites were killed by whites. Why the high incidence of black on black crimes? There are several factors. Homicide in the United States usually happens between friends, relatives, and acquaintances. Homicide rates are higher in big cities and among economically disadvantaged people. According to William J. Wilson, there is a growing split between the black middle and lower classes. As one group thrives and moves upward, economic conditions for the "underclass" worsen. For many blacks, absence of equal educational and economic opportunity, along with the absence in the inner city of a stabilizing black professional middle class, has led to higher unemployment, increased teenage pregnancy and single parenthood, and an increase in crime and violent behavior. Further, organized crime has traditionally supplied a path of upward mobility for those who were blocked from more respectable pursuits. Increased involvement of blacks in organized criminal activity likely increases the incidence of black on black crime. Finally, today in the inner city violence is much more a part of American culture— "murder is in style."

See "Black on Black Crime," Howard Palley and Dana Robinson, Society 25 (July/August 1988), pp. 59–62.

mally proscribe the behaviors. Rule makers decide that particular acts are a threat to the organization and structure of society and must therefore be prohibited. Laws are passed, and these forms of deviance become illegal or criminal if performed.

Conflict theorists tell us to be alert here and ask crucial questions. Who are the rule makers? Who passes the laws? The answer is middle- and upper-class representatives of the power structure. Who enforces the laws? Police, who, if they are not members of the middle class, are agents of and are working for the middle class. The theme continues for courts and prison systems. Conflict theorists suggest that we should not be surprised when arrest figures show that those without power and wealth commit most crimes; the system is designed to work that way. According to this view, the whole criminal-justice system from law to prison is an instrument the ruling class uses to perpetuate existing patterns of power and privilege.[11]

A law violation is considered **criminal behavior,** and there are several categories of crimes. More serious acts are called **felonies,** punishable by a year or more in a state prison or by death. Less serious crimes are called **misdemeanors.** The FBI has developed a list of eight major crimes: homicide, rape, robbery, aggravated assault, burglary, larceny-theft, motor-vehicle theft, and arson. Trends in crime are studied by noting the number of crimes reported

Table 14.2 Reported Crime in the United States, 1992

Crime	Number	Type
Homicide	23,760	
Forcible rape	109,060	
Robbery	672,480	1,932,270 Violent (13%)
Aggravated assault	1,126,970	
Burglary	2,979,900	
Larceny-theft	7,915,200	12,505,900 Property (87%)
Motor-vehicle theft	1,610,800	
Arson*		
Total	14,438,170	

*Total not available.
Uniform Crime Reports, 1992.

to police throughout the country. Table 14.2 shows the number of major crimes reported in the United States in 1992.

Aside from the eight major crimes, a number of other acts are illegal: prostitution, gambling, fraud, embezzlement, vandalism, traffic offenses, and so on. Some of the forms of deviant behavior previously discussed in this chapter may fall into criminal categories. If alcoholism means drunk driving, it becomes a crime. The sale, possession, and/or use of particular drugs become a crime.

One might assume that if an act is defined as criminal, there must be general agreement about it. Such a serious move by a society must mean that people are in accord with each other and that group differences and vested interests have been set aside. This, of course, is not necessarily so. One category of behaviors is called crime because it threatens the public order. There is general agreement that acts such as homicide, arson, robbery, and larceny threaten the public order. About other categories of crime, however, there may be less agreement. For example, one such category grew out of the concern, religious beliefs, and moral outrage of our Puritan ancestors, who felt that in some areas of behavior, individuals must be protected from themselves. It is not so much that their behavior may be a threat to society as that their behavior may be a threat to themselves. Consequently, a number of acts, some of which we today call vices, were defined as crimes. These include gambling, prostitution, narcotics use, abortion, and a variety of sexual activities, including homosexuality. Opinions as to the "rightness" or "wrongness" of these acts vary from individual to individual and from group to group. However, there is in fact a substantial demand to engage in many of these acts.

Gambling is legal in some states and is widely sought and available throughout the country even where it is not legal. Prostitution is legal in some areas (in all Nevada counties but two, for example). Many young people as well as some of their elders wonder at the current laws punishing marijuana use. Studies show that marijuana is widely used and is relatively easy to obtain. Laws in many of our states outlaw sodomy, the "infamous crimes against nature," and yet Kinsey reported that many of these "crimes," which involve

various types of sexual behavior, are relatively common practices. People wonder why homosexual acts between consenting adults in private are not legal
if heterosexual acts under the same conditions are.

Edwin Schur has commented that because all parties involved in these actions (gambling, prostitution, abortion, and so on) are seeking them or at least
consenting to them, these are "crimes without victims."[12] If there is no victim,
Schur wonders whether anyone should be punished. He points out that when
these activities are called crimes, they are forced underground to a certain
extent and become more expensive to obtain. This leads to secondary crime—
theft, for example, to obtain funds to support the primary crime, such as narcotics or abortion. So, although some acts are defined as criminal by society,
this does not mean that there is universal agreement as to their "wrongness."

Juvenile delinquency refers to young people (under 21 in some states, under 18 in others) who commit criminal acts or who are wayward, disobedient,
uncontrollable, truant, or runaways. Delinquency is a confusing concept.
Young people *are* heavily involved in certain types of crimes, as can be seen
in Table 14.3. At the same time, many young people are defined as juvenile
delinquents for committing acts like disobedience or running away that are
ignored if committed by adults. American law is such that we are very strict
with juveniles: "Get ahold of the bad kids quick and change them so they
won't grow up to engage in worse activities." However, as we have seen in the
section on primary and secondary deviance, if the juvenile is thrown into a
reform school with other hardened delinquents, we almost guarantee the result we were trying to avoid. Most countries are more lenient than the United
States is in their treatment of juveniles.

Table 14.3 Arrests of Young People, 1992

	Percentage of Arrests for People:	
	Under Age 15	Under Age 18
Robbery	8	26
Burglary	14	34
Larceny-theft	14	31
Motor-vehicle theft	13	44
Arson	32	49
Vandalism	22	45

Uniform Crime Reports, 1992

SOCIAL CONTROL

Social control refers to the processes, planned or unplanned, by which people are made to conform to collective norms. Some amount of conformity seems to be essential in all societies. Predictability and order are necessary to the social organization of group behavior. If people could act in complete isolation, possibly they could ignore the existence of any norms. However, it is almost impossible to imagine such a situation because people do behave in groups, and although individuality and nonconformity are very popular, we should remember that they are acceptable only when most people, most of the time, conform.

Social control may be provided in a variety of ways. We could, for example, hope that people will naturally conform, that conformity is inherent in their nature. It seems, however, that people must be *trained* to conform. Basic social control is taught through the socialization process. Shortly after a child is born, the family begins telling him or her what to do and what not to do, what is right and what is wrong. Sanctions or punishments are applied if the child misbehaves. Children who fail to conform are told, in effect, that mother and father won't love them or will beat them, depending on the family's view of proper child-rearing practices. Later, influences outside the family continue the process. Peers, teachers in the school system, the church, the mass media all have a profound effect on inculcating group and societal norms and encouraging conformity to group expectations. Finally, if all else fails, the society can pass laws to ensure conformity. Laws force people to conform and are a response to breakdowns in other forms of social control. Again, sanctions are applied for failure. These sanctions may range from a fine to imprisonment to death.

Social control mechanisms vary depending on the type of group or society. In a primary group or small primitive society, rules are not written and social control is informal. Violators may be subjected to gossip or ridicule or possibly even ostracism; although these techniques do not have the dramatic effect of a raid by the FBI, they are surprisingly effective, even in modern mass society. Concern over what family and friends will think, potential disgrace and loss of reputation, and other informal sanctions are probably the first line of defense in producing conformity. Large societies add more formal types of social control—written laws, police, courts, prisons—making an exceedingly complicated legal system to guarantee an acceptable degree of conformity.

Social control mechanisms also vary depending on the type of norm violated. If the violated norm is a folkway regarded as insignificant or harmless to the group, informal and minor sanctions are elicited: whispering, giggling, or a little mild ridicule. If one of the mores of the group is being abused, however, more formal sanctions may be used.

The most extreme of the formal sanctions is the death penalty. Both the second and third research questions asked at the beginning of this chapter concern the death penalty. Question 2 concerns what happens when executions take place and are publicized: Do homicide rates go down or up? One of the arguments in favor of the death penalty is that its use will deter other people from committing serious crimes, which should result in a subsequent decline in the homicide rate. One of the arguments in opposition to the death penalty is that executions tend to brutalize society in that they suggest a continuance of violence and that the taking of revenge is appropriate. If this second argument were true, the homicide rate could be expected to go up after executions. A paper examined this issue by looking at executions, the amount of publicity about the execution (days of coverage, amount of TV air time), and the homicide rate in the United States. The idea was to see if more executions and more publicity had an effect on the homicide rate. What happened? In somewhat of an anticlimax, the researchers found that homicide rates were *unaffected* by executions or publicity. They neither went up nor down significantly. Maybe the two arguments (deterrence and brutalization) canceled each other out, or maybe neither is valid, or perhaps the number of executions performed is too small to be meaningful.[13]

The third research question from the beginning of this chapter wondered what might happen if death-row inmates—a group scheduled to be executed—were released, a frightening thought perhaps. This actually happened in 1972, when the U.S. Supreme Court ruled in *Furman v. Georgia* that the death penalty was unconstitutional. This meant that 600 death-row inmates nationwide went back into the general prison population. In 1972 there were 47 "Furman" inmates on death row in Texas; a study followed their behavior from 1973 to 1986. First they were transferred to the general prison population. In spite of what we might expect, they committed few serious rule violations, and most were model inmates during their time in prison. Thirty-one of the "Furman" inmates were eventually released on parole. Most of those released did not commit other crimes. Four of those released did commit new felonies: one murder, one rape, and two burglaries. Twenty-three inmates were successfully completing their parole at the time the article was written (1988). The success rate, in and out of prison, in a group that was seen as enough of a danger to be isolated from other prison inmates and enough of a threat to society to be executed, is surprising.[14]

Individual social control probably emerges from both internal and external sources. Internal constraints would be those aspects of the normative system that one internalizes through the socialization process. Internalized norms become the individual's *conscience*. External social control refers to the external

ANOTHER TYPE OF PUNISHMENT

The Reformed Mennonite Church imposed its "doctrine of avoidance" on Robert Bear of Carlisle, Pennsylvania. Commonly called shunning, this meant no one in the church including his family could speak or have anything to do with him. His wife and six children left him when the sanction was imposed. His parents died shortly after the church action and, according to Bear, "To the moment they died they looked at me with disgust. I was already burning in hell." Five years later the punishment remained in effect.

mechanisms—*rules* and *laws*—applied by society. It would be fortunate if the two were in balance, but as we saw in the previous section on deviant behavior, this is frequently not the case. Take, for example, the boy whose peers happen to be little hoods; his most important primary group is a delinquent gang. He internalizes a set of values and norms from them. He obeys the norms of the gang or he is ridiculed and ostracized. Yet in following gang social control he runs afoul of external societal social control, which in many cases makes opposing demands. To complete the dilemma, it is obvious that the social control system of the primary group, the gang, is much more important for one's behavior than is any external social control system. This type of gap occurs when the values of one segment of society are vastly different from the values of the rule makers of society.

Another type of gap or difference between internal and external social control occurs when people's attitudes, values, and norms change more rapidly than laws change. A cultural lag results. For example, there are sharp differences of opinion regarding whether to and how to enforce existing gambling and prostitution laws. Laws traditionally have been slow to change, and perhaps they should be, so as not to respond to every whim or alteration in public opinion. We might predict a pattern through which change in laws occurs: (1) Attitudes change; (2) laws relating to the changed attitudes are not enforced; and (3) laws change. It is likely, however, that there is a large lapse of time between Step Two and Step Three. Consequently, when attitudes change, laws tend to be kept on the books, but they are just not enforced as vigorously as they might be. Law-enforcement agencies and individual citizens often find themselves in a difficult position because of this situation. The marijuana laws in California—first strictly enforced with heavy penalties, then less strictly enforced, now changing—are a good example of this sequence.

White-collar crime and organized crime provide examples of another type of conflict between society's laws and people's attitudes. The public favors and encourages prosecution of offenses such as burglary, assault, larceny, homicide, and sex offenses. They are indignant with the small-time thieves and quick to lock them up for their sins. On the other hand, white-collar crime is seldom punished. White-collar crime refers to the illegal acts committed by middle- and upper-class people during the course of their regular business activities and includes such examples as the bank vice-president who embezzles funds, the physician who splits fees or performs unnecessary operations, the executive who cheats on his or her income tax, and the disk jockey who accepts payola.

Individual white-collar crime is only part of the picture. We know that some corporations are engaged in unethical business practices, such as price fixing by computer, steel, and electrical companies; false advertising of drugs and food; collusion between government regulatory agencies and the companies they are supposed to be regulating; campaign payoffs in return for favors; bribery; cost overruns on defense contracts; land and computer fraud; and on and on. Take one case: General Electric, Westinghouse, and some 20 other electrical companies cooperated to fix prices of electrical equipment. Bids, supposed to be secret and competitive, were neither. The companies worked out ahead of time which company would bid lowest on a contract and what the other bids should be to make it look good. The resulting lowest bid was far higher than it should have been, and the companies made good money—in the neighborhood of $2 billion more than they should have in one seven-year period. But (surprise) they were caught. An outraged judge levied $2 million in fines (a drop in the bucket compared to the take) and sentenced seven men to 30 days in jail. See whether the average thief who steals a car or burglarizes a house gets off with 30 days. However, people were amazed at the "harshness" of the sentences in the great electrical conspiracy, for corporate criminals almost never go to jail, even for 30 days.

Criminologist Edwin Sutherland has maintained that more money is lost yearly through white-collar crime than through all other forms of crime combined and that the damage to social relationships that results (no one can be trusted, everyone is on the take) is even more serious. Yet, as we said, the larceny of the white-collar criminal is seldom punished, and even when it is, the penalties are slight compared with those for other crimes. Why? Well, the activity may have involved substantial skill and ingenuity and it wasn't violent, so the public tends to identify with it or at least not object to it. Many times the victim, often a large organization or the public in general, is invisible or not easy to identify with. Often the culprit is a large corporation that has spent much money over the years to cultivate its image, and the public finds it difficult to believe that such an upstanding important company could be guilty of such dastardly deeds. And finally, the white-collar criminal is supposedly a "good" person—middle-class and respectable. So, the public prefers to look the other way.

Likewise, the activities of organized crime go largely unpunished, but the reasons for breakdown in social control in this case are more basic and easier to understand. Many people *want* what organized crime supplies: gambling, narcotics, tax-free cigarettes and alcohol, prostitution, and loan rackets. The conclusion is clear: The attitudes and values of society are reflected even more in punishment practices than they are in the written laws of society.

THEORY AND RESEARCH: A REVIEW

All three theoretical viewpoints—functional analysis theory, symbolic interactionist theory, and conflict theory—may be seen in this chapter. The comment

A TOMBSTONE EPITAPH—GIRARD, PENNSYLVANIA, 1870

In memory of
Ellen Shannon
Aged 26 years
Who was fatally burned
March 21st 1870
by the explosion of a lamp
filled with "R. E. Danforth's
Nonexplosive
Burning Fluid"

> *From* Folklore on the American Land
> *by Duncan Emrich (Boston: Little,
> Brown, 1972).*

that deviant behavior is not necessarily harmful to group life but is actually helpful in defining group boundaries and thus helps preserve group stability represents the *functional analysis* view. Labeling theory is *symbolic interactionist* in that it focuses on the definitions of people's behavior and sees the definitions as the product of a type of interaction between two or more parties. Conflict theorists would be interested in the question, Who gets labeled? From the *conflict theory* perspective, the criminal justice system—law makers, police, courts, prisons—is in the hands of the ruling class and is used to protect existing patterns of power and privilege. Conflict theorists see conflict among classes as producing unemployment, poor working conditions, and family breakdown, which are the important pressures that often lead to crime.

The major source of the data in this chapter is records: *reactive* and *nonreactive surveys* of information collected by police departments, the United Nations, census agencies, and organizations dealing with drugs, alcohol, and mental illness. Self-reports of people's activities are also used. The problem with this sort of information is that people, agencies, and governments don't always report accurately for a series of reasons ranging from embarrassment to self-incrimination to a desire to "look good" (encourage tourism) or "look bad" (increase the budget). Statistics on crime and deviant behavior may therefore get manipulated.

SUMMARY

The previous chapters in this section have dealt with social change, social disorganization, and collective behavior. In this chapter we returned to the topic of norms, which was first discussed in Chapter Three. A *norm* is defined as the accepted or required behavior in a specific situation. Most behavior is predictable because it is in accordance with generally accepted norms. However, some behavior is contrary, and this is called *deviant behavior.* Societies attempt to restrict the amount of deviant behavior through *social control* measures. This chapter addressed itself to these concepts: deviance and social control.

Individual variation and individual differences are a basic part of the human condition. When differences from group norms reach a particular point, however, social differentiation becomes social deviance. Many sociologists believe that deviance is best understood as a *labeling* process. Behaviors and/or characteristics of individuals are deviant when they are defined as such by members of society. Deviance is neither a natural nor a statistical phenomenon but is instead a product of group viewpoints and definitions.

Criminal and delinquent behaviors are also seen as products of a defining or labeling process. If society is especially concerned about specific types of behavior, it may define these behaviors as not only deviant but criminal as well. Frequently differences of opinion develop about the "rightness" or "wrongness" of particular acts. Such differences are more likely to occur in large, complex societies in which a variety of viewpoints exist about what is deviant and what is normal. These variations in definitions make it possible within a society for a given act to be viewed as criminal and deviant, as criminal but not deviant, as deviant but not criminal, or as "normal."

We briefly examined several categories of deviance—mental illness, mental deficiency, alcoholism, drug use, and suicide—as well as some descriptions of and explanations for criminal and delinquent behavior. Finally, we discussed social control, which refers to the processes by which people are made to conform to collective norms. Social control mechanisms are practiced by all societies and include norms internalized through the socialization process as well as laws enforced by the police and courts of a society.

Two readings follow, both of which deal with social control. Who goes to prison and who stays in prison? You might be surprised. The first reading looks at some of the problems confronting the California prison system and some of the possible consequences of the "three strikes" legislation. The second reading describes a different type of social control. It sounds like vigilante justice out of the Old West, but this article is about modern-day Northern Ireland.

TERMS FOR STUDY

altruistic suicide (436)	labeling theory (428)
anomic suicide (436)	misdemeanor (438)
criminal behavior (438)	neurosis (433)
deviant behavior (426)	organic disorder (433)
differential association (428)	primary deviance (431)
egoistic suicide (436)	psychosis (433)
felony (438)	secondary deviance (431)
functional disorder (433)	social control (441)
juvenile delinquency (440)	

Reading 14.1

CALIFORNIA PRISONS AS GERIATRIC WARDS

Miles Corwin

In the 1990s a belief has spread across the country that a good way to solve the crime problem is to build more prisons and lock up more people. Politicians are increasingly running on "get tough" platforms, and several states have already passed so-called three strikes and you're out laws. These laws aspire to put a person convicted of three major felonies in prison for life. Does this approach make sense? What would it cost? What would be the consequences to our penal system? What kinds of people are in our prisons now? Are we locking up (and turning loose) the people we should? Criminologists, judges, lawyers, and other people concerned with our criminal justice system are debating these questions. Los Angeles Times reporter Miles Corwin gives us insight to some of the issues in this look inside the California penal system.

Every time S. M. Cohen takes a pill, visits a doctor, has a blood test, receives a shot of insulin or undergoes surgery, Californians pay for it.

Cohen, 67, an inmate at the California Institution for Men, has cancer and diabetes and takes 11 pills a day. He has blood tests twice a month and sees three specialists. His medical bills are more than $125,000 a year. His recent heart bypass cost $76,000.

Cohen, who is doing nine years for grand theft, is housed in a unit inmates call "Old Man's Dorm," a cavernous hall filled with grizzled prisoners who mill about using canes, walkers, wheelchairs and crutches. The highlight of their week is the Wednesday night bingo game.

Soon the state's population of geriatric prisoners will increase dramatically because of the new "three strikes" law, which is aimed at putting three-time felons behind bars for life. And there will be many more facilities such as Old Man's Dorm—the only unit in the state designed for low-security, geriatric prisoners.

Californians will have to decide how long elderly offenders should be kept behind bars. This decision will highlight a philosophical divide between those who believe society's need for justice is paramount, and others who argue that many elderly prisoners should be given early release because they no longer are dangerous and it is expensive to imprison them.

The predicted wave of elderly "lifers" will "create an economic disaster for the state," said Dr. Armand Start, a national expert on prison medical care. Funding eventually will have to be siphoned off from other state programs such as education, he said.

California will not be able to build prisons fast enough to keep pace with the influx of elderly prisoners, said Start, a professor at the University of Wisconsin Medical School.

Dangerous first-time offenders eventually will have to be paroled early, he said, to make room for older, less violent prisoners serving mandatory life terms.

Only about 2 percent of the state's prisoners are over 55. And Old Man's Dorm—officially known as Elm Hall—has long been viewed as a curiosity. But now Elm Hall, which is essentially a convalescent hospital behind bars, offers a glimpse into the future of the state prison system. There are some younger, disabled inmates, but most of the 150 prisoners are elderly men with bifocals who never miss the weekly meetings of their senior citizens' group.

There are no cellblocks because the prisoners need wheelchair and ambulance access. And despite chronic prison overcrowding, there are no bunk beds because most of the oldtimers do not have the strength to climb to the top bed. Many are recovering from strokes, heart attacks or crippling bouts of arthritis.

Elm Hall is a large cinderblock dormitory with rows of cots and lockers. A nurse is on duty until 10 p.m., to make sure no one slips in the shower or cannot get out of bed. Dorm living is better than the hard time of a cellblock, but most prisoners share one great fear.

"Everyone is terrified of dying here," said Cohen, who has a pencil-thin mustache and neatly trimmed gray hair. "That's what bothers us the most. We've seen people die in here, and it's the loneliest death in the world."

Elm Hall was designed in the early 1960s—with wheelchair ramps, bathtubs and a heating system—to house the state's small population of elderly prisoners. At that time the old-timers mingled easily with the mainline population. But during the 1980s, when young gangbangers began pouring into the prison system, the clash of generations created friction.

Many old-timers say they have only contempt for the young prisoners. Old-timers call themselves "convicts," and refer contemptuously to the young prisoners as "inmates." The inmates, the old-timers complain, "don't know how to do time." They snitch on each other; they steal out of lockers; they cut in front of people in chow lines. One elderly prisoner said that if, in his day, you cut in front of someone in line you would get a shiv in the belly.

William Ross, 81, a bald diminutive man with a faint resemblance to George Burns, is the dean of Old Man's Dorm. Ross, who is serving a two-year sentence for check kiting, looks out of place amid the young, tatooed prisoners pumping up at the weight pile.

One night last month, Ross was walking from the canteen, after buying cream, coffee and sugar. Two young inmates ran up behind him and tried to rip off his bag of supplies. But a few elderly prisoners, who were tailing him, began shouting for a guard. Ross held tight to the bag and the thieves were scared off.

"We old-timers stick together," said Ross, squinting through thick glasses. "When we head out to the yard we try to go in groups. If somebody has to go out alone, the others try to keep an eye out for him."

Years ago, in the days when it was not so smoggy and they could see Mt. Baldy, prisoners founded the Mountain View Senior Citizens Group. Members still collect aluminum cans and sometimes sell candy and doughnuts to raise money. If an old-timer is about to be released, but does not have enough money to get home or needs a new suit of clothes, group members will chip in with the cash. The family of one man who died in prison wanted to give him a proper burial, but they did not have the money. The group raised $1,000.

One Elm Hall inmate had such a long criminal history, officials say, he started out robbing stagecoaches. When he was released, he was 91 and still chain-smoking unfiltered Lucky Strikes.

Another renowned resident was Ronnie Fairbanks, whom officials listed as 90 when he last entered prison. But they could not be sure. Fairbanks had given authorities 21 birth dates since he was first arrested for larceny in 1928. He also had used 88 aliases during his criminal career, which included more than 40 arrests. Fairbanks often boasted that he never committed a violent crime. He was last arrested for stealing a suitcase at Los Angeles International Airport and was released from Chino in 1988. He has not been back. Yet.

"He may have been 90, but he was in great shape," said Diana Smith, program administrator for Elm Hall. "He jogged every day and kept himself in top physical condition. He was a very charming guy, very genteel, always polite."

Not every prisoner is eligible for Elm Hall, Smith said. When Charles Manson becomes a senior citizen, she said, he will not be welcome. Prison officials will not admit notorious criminals, child molesters or anyone with a violent history.

Building more facilities such as Elm Hall is not the best way to deal with aging inmates, many prison experts say. When criminals hit middle age they begin "burning out," said Start, who heads the National Center for Correctional Health Care Studies. Criminals in their 40s and 50s—like the rest of the population—are affected by aging. They become less impetuous, less hostile, less prone to violent behavior, he said.

A federal Bureau of Justice Statistics report found age to be the single most reliable indicator in predicting recidivism. About 22 percent of prisoners ages 18 to 24 return to prison within a year of their release; for prisoners over 55 the rate is less than 2 percent.

The 20 new prisons that are planned in California will not keep pace with the deluge of new inmates generated by the "three strikes" law, said Jonathan Turley, a law professor at George Washington University. In other states with "three strikes" measures, Turley said, officials had to release violent first-time offenders early because of overcrowding. They needed to make room for the growing population of elderly prisoners with life sentences. Turley predicted that California prison administrators eventually will have to do the same.

In Texas, which has a "three strikes" measure and a number of other mandatory sentencing provisions, inmates last year served an average of only 11 percent of their sentences. (California inmates serve about 50 percent of their sentences.) The Texas law recently has been changed, requiring inmates to serve a higher proportion of their terms. But young violent offenders will still have to be paroled early to make room for older prisoners serving mandatory life sentences, Texas prison experts say.

"This system makes no sense," said Turley, who heads a Washington, D.C., project that evaluates the parole status of elderly prisoners. "We'll be keeping in prison the inmates who are at a point in their life when they are least likely to commit another crime. And we'll be releasing those young offenders who are most dangerous. This system works against everything we've learned about recidivism."

Geriatric prisoners cost institutions about $60,000 a year—three times more than younger inmates, according to recent studies. On average, inmates over 55 suffer three chronic illnesses while incarcerated. Some do most of their time in prison infirmaries. Others run up huge bills at private hospitals because most prisons contract with outside facilities for major medical work.

But David Beatty, a spokesman for the National Victim Center, said victims' right advocates are more concerned with the "human cost" of releasing repeat offenders than "the financial cost" of keeping them in prison. Even if the odds are slim that a geriatric prisoner, if released, would commit another crime, Beatty said, "most victims aren't willing to take that chance. They feel more comfortable with these kinds of people locked up forever."

No one in the Department of Corrections can predict how many elderly prisoners to expect in coming years. Prison officials say they have been preoccupied with other problems such as AIDS, tuberculosis and gangs. They plan to research the issue this year.

California's "three strikes" measure calls for a minimum of 25 years to life for any felon who has committed two prior violent or serious felonies—crimes that range from murder to home burglary. In today's political climate 25 years to life often is the equivalent of life without possibility of parole, said Heather MacKay, a staff attorney for the Prison Law Office, a Bay Area prison reform group. The parole board, she said, "because of the prevailing political climate, often is reluctant to grant any paroles at all."

While California gears up for a deluge of geriatric prisoners, many states are moving in the opposite direction. They are involved in "compassionate release" programs for elderly prisoners who do not appear to be a threat to society.

Law professor Turley heads The Project for Older Prisoners, which lobbies parole boards in dozens of states on the issue of early release for elderly prisoners. Of all the prisoners evaluated, the project has only recommended 10 percent—about 60—for early release. Not a single one the project has helped free has returned to prison, he said.

He founded the project five years ago when he represented Quenton Brown, a Louisiana man with an IQ of 51. Brown had robbed a bread store of $117 and a cherry pie. When police arrested him he was crouched beneath a house across the street, eating a piece of pie.

Turley heard about Brown's case after he had served 18 years of a 30-year sentence. Brown, then 67, was suffering from emphysema and bleeding ulcers and was released after Turley argued his case before the parole board. It did not make sense to continue to incarcerate inmates like Brown, Turley argued, when young, violent prisoners were being released under a court order to ease the overcrowding in Louisiana prisons.

Working with law student volunteers, Turley evaluates the criminal histories of prisoners and rejects those who were convicted of sex crimes or violent offenses. Many elderly prisoners, he said, belong behind bars for the rest of their lives. But how much of a threat is an inmate who is recovering from a stroke, has only a few months to live and wants to spend his final days with his family?

About 40 percent of Medicare dollars are spent in the last two weeks of a patient's life. Because some prisoners' families have insurance, Turley said, it makes more sense to release some of them than to stick taxpayers with exorbitant medical bills.

An elderly prisoner who costs the state $60,000 a year could be housed in a private nursing home for much less. Yet California will be building 20 prisons, at a cost of about $100,000 per cell, to accommodate a new wave of prisoners, many of them senior citizens.

Turley said he does not recommend "wholesale release of older prisoners." What he does advocate is legislation that will allow states, through special parole laws, to give consideration to low-risk, high-cost elderly inmates. But few California politicians are willing to support any measures calling for early release of prisoners—regardless of their age.

"Politicians are now running the prison system by sound bite," Turley said. "The truth of the matter is, by the time we interview inmates who are in their 60s, 70s and 80s, most of them are statistically less dangerous than the law students I drive to the prison with."

QUESTIONS 14.1

1. In what ways do the demographics of the prison system mirror the demographics of the population in general?

2. What is the single most reliable indicator in predicting recidivism? Explain why this is true.

3. Critique the "three strikes" laws— what's good about them, and what's bad about them?

4. If you were redesigning our criminal justice system, especially the incarceration aspect of it, what changes would you make?

Reading 14.2

HARSH BRAND OF JUSTICE IN BELFAST

Shawn Pogatchnik

Formal social control means laws, police, courts, and prisons and includes concepts like testimony, evidence, rights, and due process. But what if the formal system of social control doesn't work? As this article from the Los Angeles Times *shows, other methods of gaining conformity may emerge.*

Like so many wild-eyed young men of Catholic west Belfast, Seamus Clark liked to drink hard and speed recklessly about town in stolen cars. Worse, his friends and family say, was the fact that "Shamy" had a proud mouth around the wrong people.

That combination made Clark a prime candidate for punishment—not by Northern

Ireland's police and legal system, but at the Byzantine hands of local Irish Republican Army gun squads.

In December 1987, a month after neighborhood vigilantes pulled him from a stolen car and beat him unconscious, five masked IRA men came knocking on his door. They ordered his parents and siblings upstairs, turned up the volume on the television, pinned him to the floor and fired five bullets into his knees and ankles.

Last April the faceless men returned, clubbed Clark severely with bats and gave him 48 hours to get out of Belfast. Today, 22-year-old Clark is hiding out in a small town south of the city and faces amputation of his failing right leg.

Trial by Gun

Clark's story, which is not unusual, illustrates the quip popular in west Belfast circles that "there's no law, but there's order."

In a bizarre outgrowth of Northern Ireland's two decades of civil unrest, paramilitary gun squads throughout the province have shot more than 1,500 people and in other ways assaulted at least another 350 as punishment for "antisocial behavior"—the local umbrella term for car stealing, shoplifting, drug trafficking, rape, prostitution and a range of other crimes.

Such trial-by-gun-squad is not confined to IRA-dominated Roman Catholic neighborhoods. Police spokesmen say that local "hoods," as well as political opponents, are regularly dealt with on the Protestant, working-class side of town by members of the Ulster Defense Association, Ulster Volunteer Force, Red Hand Commandos and other loyalist gangs.

But the IRA's prominent role in punishment shootings was singled out for criticism last month by Peter Brooke, Britain's secretary of state for Northern Ireland. He said that British security forces wanted to get the IRA "off the backs of the people."

"There is no trial, no legal process," he said. "Instead there is the knock on the door at night, the cold fear of the victim and the weeks and months in the hospital. Or the final and enforced departure from home and family ties into exile . . . or the hood, the bullet in the back of the head and the sad funeral."

The province's official police force, the Royal Ulster Constabulary, has recorded more than 100 punishment shootings so far this year, plus another 65 punishment beatings. This form of punishment can involve the use of "hurley bats," used in Gaelic hockey, or "breeze blocking," in which a cinder-block is dropped onto a person's legs until they are heard to break.

"Six Pack" Punishment

Punishment shootings vary in intensity—first-time offenders often being shot once or twice through the fleshy thigh, while others get a "six-pack" of bullets through the elbows, knees and ankles.

Compared to the carnage of a typical American city, the numbers of such shootings may seem relatively minor. But in Northern Ireland, with a total population of 1.58 million, the numbers are significant.

Nearly everyone in the insular world of west Belfast knows someone—often several people—who have been "done by the Rah" (slang for IRA).

Sinn Fein, the IRA's legal political wing that commands majority support within the high-unemployment Catholic ghettos, maintains that such "kneecapping" and other punishments serve as a useful and locally accepted deterrent.

"Punishment shootings and beatings are the end of a very long process, which begins essentially with complaints being made by concerned citizens," said Richard Mac-Auley, a Sinn Fein spokesman, seated inside one of the party's six public advice centers. "There's an expectation, in the absence of confidence in the police force, that republicans will fill the gap."

He cited one highly publicized example recently, when a woman was gang-raped by seven men in the Divis Flats, a Catholic area of high-rises just west of downtown Belfast.

"About a hundred local women organized a picket to demand that something be done," MacAuley said. "They didn't go to the RUC (constabulary) barracks in the area. They came to this building to demand that the republican movement do something."

The perpetrators fled the country before they could be shot, MacAuley said, noting that "if they come back to Belfast, one would expect very stern and harsh action to be taken against them."

De Facto Police

How IRA gunmen gradually gained their role as de facto police in many nationalist areas of Northern Ireland, and particularly in west Belfast, is a matter of debate. But it followed the Protestant-Catholic riots and introduction of British troops onto the streets in 1969.

Since then, the Royal Ulster Constabulary has remained more than 90 percent Protestant and has been viewed by many Catholics here as the enemy.

Police bases in west Belfast resemble modern-day Fort Apaches, bolstered by corrugated iron walls behind which bristle surveillance cameras and listening equipment.

When constabulary members patrol the area, they do so in heavily armored Land Rovers, rarely leaving the relative protection of the vehicles at night. If on foot, they wear flak jackets and always are escorted by several heavily armed British army soldiers who guard against possible ambush.

Stopping "ordinary" crime in such hostile territory, police spokesmen concede, has been low on the constabulary's list of priorities in comparison with rooting out the terrorist threat. To some extent, IRA leaders appear to have opportunistically filled a perceived vacuum in those communities.

A sampling of local opinion suggests that a majority of residents within republican areas condone, if not openly support, the IRA's brand of frontier justice. One mother of three—a nurse at the local hospital—was blunt: "Kneecapping's too good for those wee hoods," she said, rocking her daughter in her arms.

However, a growing movement composed of victims of punishment shootings and their families is making a stand against the gunmen.

After her own 24-year-old son Paddy was kneecapped by local IRA men last July, Nancy Gracey founded Families Against Intimidation and Terror. The group is trying to protect would-be victims by getting them out of Belfast, or Northern Ireland if necessary, and bringing together kneecap victims and their families.

"We've a 21-year-old boy in our group who's had nine bullets in him since he was 17 years of age. The IRA took a hacksaw to the arm of another wee fella," Gracey said.

Community activists note that west Belfast teenagers and young adults have been weaned on a violent culture in which authority figures encourage their participation in politically motivated crime, such as throwing bricks at army patrols and hijacking vehicles to make road barricades and car bombs.

"If kneecapping somebody worked, solved the problem, maybe I could understand," said Carmel McCavana, a director of the west Belfast Parent Youth Support Group, which coordinates programs for "at risk" youth, many of whom already bear scars from IRA punishment attacks.

"OK, in the short term kneecapping put somebody in the hospital for a few weeks," she said. "But in the long term, we know several young people who've gotten right behind the wheel of a lifted (stolen) car again, before their wounds have even healed.

"One young lad I know of actually was driving with his crutch on the accelerator," she said. "The young people here have very little fear . . . and very little hope. They need help."

Help for a repeat criminal offender is hard to come by in this community, whose sense of siege traditionally is relieved through violent means.

Even Seamus Clark's family has mixed feelings about the role of IRA strongmen in their neighborhood.

"The IRA shooting its own, to me, is wrong. But someone's got to keep order," said his mother, 62-year-old Alice Clark. "The kids run wild on the streets here.

"My son was doing something he shouldn't have done. Everyone normally gets for their first offense one shot, no matter what they do. I'd have accepted that, so I would," she said.

One of the bullets used in the first attack still sits on her mantle, flattened from ricocheting off the floor and into a sacred-heart ornament on the wall.

"But the Rah shot him five times. That," she emphasized, "that wasn't fair."

QUESTIONS 14.2

1. Which type of social control—formal or informal—is likely to work better in most situations? Why?

2. We used the terms *deterrence* and *brutalization* in this chapter as arguments that are used in connection with capital punishment. Do those terms have any relevance to the type of informal law enforcement used in Northern Ireland?

3. Do you think the law enforcement as described in this article is working? What are its advantages and disadvantages?

4. Construct for a place like Belfast a system of social control that you would be comfortable with—that would be acceptable to you. Describe it.

NOTES

1. This discussion of the labeling aspect of deviation is drawn from Howard S. Becker, *Outsiders* (New York: Free Press, 1963), ch. 1. Also see his chapter 8 for a discussion of

rule creators and rule enforcers. The case of the long-haired basketball player was described in *Sports Illustrated*, 16 February 1976.

2. For the complete report, see "On Being Sane in Insane Places," by D. L. Rosenhan, in *Science*, 19 January 1973, pp. 250–258. *Saturday Review* also has a good summary of the research in its 24 February 1973 issue, pp. 55–56.

3. The idea of primary and secondary deviance was first developed by Edwin Lemert. See his *Human Deviance, Social Problems and Social Control* (Englewood Cliffs, N.J.: Prentice-Hall, 1967), especially pp. 17–18 and 40–64. Also see the discussion in Marshall Clinard and Robert Meier, *Sociology of Deviant Behavior*, 5th ed. (New York: Holt, Rinehart & Winston, 1979), pp. 75–76.

4. Kai Erikson deals with these ideas in detail in his *Wayward Puritans* (New York: John Wiley, 1966), ch. 1, and in "Notes on the Sociology of Deviance," in *The Other Side*, edited by Howard S. Becker (New York: Free Press, 1964), especially pp. 9–15.

5. This is summarized from *Psychology 73/74 Encyclopedia* (Guilford, Conn.: Dushkin Publishing Group, 1973), pp. 166–169.

6. A more detailed discussion of mental illness, as well as of alcoholism, narcotics, and suicide, appears in Clinard and Meier, *Sociology of Deviant Behavior* (see Note 3).

7. Some of the data in this paragraph come from the 30 November 1987 issue of *Time* on alcoholism. See pp. 80–87.

8. These figures on suicides come from *Demographic Yearbook* (a United Nations publication) and *Vital Statistics of the United States* (from the Department of Health, Education, and Welfare).

9. Emile Durkheim, *Suicide*, translated by John A. Spaulding and George Simpson (New York: Free Press, 1951).

10. See "The British Gas Suicide Story and Its Criminological Implications," by Ronald Clarke and Pat Mayhew, in Michael Tonry and Norval Morris, *Crime and Justice: An Annual Review of Research* (Chicago: University of Chicago Press, 1988), pp. 79–114.

11. The conflict perspective starts, of course, with Marx. Today, the major spokesman is Richard Quinney. He has written a number of books, including *Critique of Legal Order: Crime Control in Capitalist Society* (Boston: Little, Brown, 1974) and *Criminology*, 2d ed. (Boston: Little, Brown, 1979).

12. In his book, *Crimes Without Victims* (Englewood Cliffs, N.J.: Prentice-Hall, 1965), Edwin Schur discusses abortion, narcotics use, and homosexuality as victimless crimes.

13. See "Murder, Capital Punishment, and Television: Execution Publicity and Homicide Rates," by William Bailey, *American Sociological Review* 55 (October 1990), pp. 628–633.

14. See "Institutional and Post Release Behavior of Furman-Commuted Inmates in Texas," by James Marquart and Jonathan Sorensen, *Criminology* 26 (no. 4, 1988), pp. 677–693.

CHAPTER FIFTEEN

Sociology: Another Perspective

THIS BOOK HAS attempted to provide you with an understanding of sociology. We looked at theories, at methods, and especially at the concepts sociologists use to gain knowledge and understanding of the human condition. At this point, we should have a reasonably good idea of what sociologists *do.*

But there is another perspective to examine. To me, an important aspect of a profession or discipline is what the occupants *think* about what they do. This is a more informal (an insider's) view and examines such issues as: How do these people see themselves? What do they feel they should be accomplishing? What are their internal conflicts and arguments? What is their level of job satisfaction and morale? Think of doctors and police officers again. They interest me partly because of what they do, but I am fascinated about what they *think* about what they do: job satisfaction, attitude toward patients or clients, morale, amount of stress and anxiety related to the job, the type of internal issues and conflicts that they argue over, and so on. The cumulative effect of these matters must affect one's personality and give one a particular perspective, a way of looking at the world.

This last chapter will examine the insider's view: What do sociologists think about what they do? How do sociologists see themselves, and what are some of the issues on which they disagree? How does this affect the sociologists' perspectives?

A SOCIOLOGICAL PERSPECTIVE

Members of a discipline develop a particular way of looking at the world. They share these views as individuals, although the similarities of their perspectives are most readily seen when a group of people from the same discipline come together. Becoming a sociologist involves learning the meaning and use of a series of concepts and techniques and acquiring a body of knowledge. Going through this learning—a socialization process—provides one with a particular viewpoint. Members of any discipline may go through this process, and it affects the way they think, the way they talk, the way they behave.

The sociological perspective has a number of aspects. One is the tendency to see human behavior in the context of a group. To put it another way, sociologists understand people's behavior to be a product of their group affiliations. Consequently, it is on these group affiliations that sociologists focus in their studies. For example, let's look at Joe Blow as some of his friends and colleagues might look at him. What stands out to his physician is Joe's height, weight, pulse rate, blood pressure, and medical history. What his banker sees is his investment portfolio, cash reserves, and loan collateral. His minister may see him as a soul to be saved, a moral state to be improved. His professor may see him as a mind to be molded, a vessel to be filled with knowledge, or an intellect to be challenged. A sociologist looks at him and sees a white, middle-class, 21-year-old man who is a member of the Democratic Party.

Sociologists are typically more interested in characteristics of categories and groups than in specific individuals, and in this sociologists differ from psychologists, who are more interested in the *individual* and his or her personality, motivations, emotions, and behavior. For the sociologist, the focus in Joe's case would probably be on the characteristics of middle-class whites, of 21-year-old men, or of Democrats in general rather than on Joe as an individual. The sociologist usually studies numbers of people (groups, categories, and societies) and their patterns of behavior (norms, roles, and institutions).

Other aspects of the sociological perspective are discussed by Peter Berger in his book *Invitation to Sociology.* Berger feels that there is a "sociological consciousness" that has several facets.[1] A *debunking* tendency of sociologists is reflected in their study of things taken for granted. Berger suggests that in analyzing the structure and institutions of societies, it is often necessary to "unmask the pretensions and the propaganda with which men cloak their actions with each other." There is a constant attempt, then, to find the underlying explanations for phenomena instead of just accepting the handy or traditional explanations that people readily give. For example, sociologists would view it as important to put commonsense explanations aside and determine what the real relationship is between narcotics legislation and drug use or between use of the death penalty and the homicide rate.

A part of sociology is devoted to the study of the *unrespectable*. Early sociologists were reformers and muckrakers. They believed that sociologists should intervene and attempt to ameliorate problem conditions to ensure social progress. This preoccupation with social problems and with the unrespectable has abated somewhat today, but its influence in sociology is still strong. We see, for example, a variety of studies of underdogs, deviants, and outsiders. In addition to several studies of drug users and hippies, books have appeared dealing with such topics as shoplifting, pool playing, bar behavior, behavior of racetrack gamblers, prison behavior, death and dying, and mental illness, to mention a few.

This is not to imply that so-called *respectable* aspects of society are outside of the scope of sociology. On the contrary, much sociological study is of the respectable. (In fact, there has been criticism that sociology has become too interested in describing and supporting the *status quo* and less an agent for social change.) At any rate, we could conclude that although much of sociology deals with the respectable, a prominent and growing segment of the discipline deals with the unrespectable, with problems and deviant aspects of society.

Berger suggests that sociologists tend to be *relativist* and *cosmopolitan* in their outlook. Sociologists are involved in studying patterns of human interaction in a variety of cultures under many different conditions. They learn that behavior, ideas, and institutions are relative, relevant to specific cultures and specific locations. To make judgments about or evaluate certain behaviors, sociologists must, then, apply a broader, more cosmopolitan perspective; they must have a worldview, and this becomes part of their sociological consciousness.

Any similarities in outlook that are shared by sociologists emerge, to a great extent, as a result of their training. There are also many differences among them; some are evident when we examine how sociologists look at sociology.

SOCIOLOGICAL PERSPECTIVES OF SOCIOLOGY

The discipline of sociology encompasses specific themes or traditions. In combination, these traditions constitute a heritage from which all sociologists draw. A particular sociologist, however, will be influenced more strongly by one tradition or perspective than by others. Just as a physician specializes in particular aspects of medicine and a lawyer concentrates in particular aspects of law, so the sociologist specializes in specific areas of sociology (race, social class, or deviant behavior, for example) and follows a particular theme or tradition that he or she personally feels to be most relevant.

There has always been in sociology, as in other disciplines, a debate over who we are and what we should be doing. This is a continual self-analysis, which attracts the interest and attention of many sociologists. It seems appropriate to describe this as "sociological perspectives on sociology." It is an interesting debate and probably reflects the growing pains of an area of study that is still trying to define itself and its place in the world of knowledge.

As we mentioned in Chapter One, science is often divided into **pure science** and **applied science**. Into which of these categories does sociology properly fall? Sociology is seen by some as a *pure* science in that its practitioners seek knowledge for knowledge's sake alone, without regard for the possible uses the knowledge may have. In other words, sociology is a science, not a vehicle for social action. In recent years, however, a growing number of sociologists have encouraged the development of an *applied* sociology. In a variety of ways, sociologists are putting their concepts and techniques to practical use. In general, these efforts involve attempts to deal with the problems of society or with the individual segments of society: industry, education, labor, government, and community. Applied sociology is often designed to encourage desired social change or to forestall undesired social change. Applied sociology has been referred to as **action sociology,** and those involved are sometimes called *social engineers*. Philip Hauser points out that with the proliferation of economic and social planning and the increase in the welfare functions of government, the need for sociologists will continue to increase for the performance of both scientific and social engineering tasks.[2]

Some sociologists take a purely **scientific perspective.** For them, sociology is a social science, and they emphasize the procedures and techniques of science. Such sociologists see themselves as white-coated individuals who gain knowledge by applying scientific methods to social phenomena. They are objective, detached, free of any values or biases that might affect their work. They are, perhaps, bundles of sensory neurons moving in space, observing and recording the facts of social interaction. Sociologists today are developing sophisticated mathematical models and statistical techniques and using ever larger computers to aid them in the scientific analysis of human behavior.[3]

These individuals are often called **positivists.** They adhere pretty much to the classical view of science as outlined in Chapter One. Their role as scientist is based on the model that has worked so well for those in the physical and biological sciences, and its adoption by sociologists probably reflects the hope that our success may be as great as theirs. At the same time, the positivist model in sociology is criticized by some for several reasons. First, there is the criticism of science in general. There is concern that blind faith in and uncritical acceptance of science has led people to ignore the faults of science. Questions are raised about a series of ethical considerations: Shouldn't scientists be responsible for the possible harmful uses of their discoveries? Should scientists manipulate people without their knowledge, or even *with* their knowledge? Whom do the scientists represent—themselves, science in general, society, or those people (government agencies, large organizations, private concerns) who are paying their research bills? Failure to receive satisfactory answers to these questions has led many to be disillusioned with science.[4]

A second form of criticism of the positivist model is more specific to sociology. The complaint here is that the information sociologists collect is not valid, not accurate. This is a serious criticism. Derek Phillips states in his book *Knowledge From What?* that 90 percent of the data collected by sociologists is produced by means of interviews and questionnaires. Sociologists have failed to take into account, he maintains, that the questionnaire or interview situation is a *social* situation, that statements of the people studied are affected by the questionnaire, by their response to the interviewer and the interviewer's characteristics (sex, age, race, social class, demeanor, and so on), by the nature of the interaction between the interviewer and the interviewee, and by numerous other factors that have been left uncontrolled. The basic problem is that most sociological research is based on people's *reports* of their own behavior rather than on *actual observation* of their behavior. This is the issue of reactive versus nonreactive research, which we discussed in Chapter One. Phillips suggests that sociologists should instead concentrate on participant-observation and hidden-observation techniques. **Participant observation,** as we mentioned earlier, involves the investigator joining or living with the subjects being studied; the researcher becomes part of the group. Studies of primitive tribes, delinquent gangs, large organizations, and class and ethnic structure of communities, as well as many other topics, have been made using this technique. Participant observation allows the study of some events that could not otherwise be studied, and it may provide many insights that escape the more formal interview and questionnaire methods. **Hidden observation** refers to the use of records or traces left behind by people, such as checking the nose prints on the glass or the wear on the floor tile to determine the popularity of a museum exhibit, or to the actual observation of people without their knowing.[5] The key element in both participant and hidden observation is that we *watch* people's behavior rather than *ask them* about it. Concerns about the *accuracy* of our information, then, have led to disillusionment with some of the scientific aspects of sociology.

These problems, as well as other beliefs regarding the most appropriate path to knowledge, have led some sociologists to see the discipline not as science but as *social philosophy*. For them, sociologists are basically theorists who create grand and often highly complex and abstract generalizations about the nature of the universe. They frequently have remarkable insights about human behavior. However, from the scientist's viewpoint, the social philosopher's statements are often too general and abstract to be put to empirical test. Others see sociologists-as-social-philosophers as true scholars rather than as mere scientific technicians; they are widely read masters of many disciplines, and students may gather adoringly at their feet.

Some sociologists take a **humanistic perspective.** For them, the sociologist is interested in and concerned about human welfare, values, and conduct; he or she wants to improve the lot of people in general. An ultimate goal for the humanist is the self-realization and full development of the cultivated person. There may occasionally be a tendency for the humanistic and scientific perspectives to be at odds with each other. The humanist is interested in bettering the condition of people and in developing the individual to the fullest. The objective of science is the gaining of empirical knowledge about the world, without regard for the possible uses of such knowledge. Don Martindale suggests that humanism is a system of values describing "what ought to be" and modes of conduct designed to secure them; science is the value-free pursuit of knowledge, of "what is," renouncing all concern with what ought to be. It has also been suggested that the scientist is more interested in the *means*—gaining knowledge; the humanist, in the *ends*—improving the lot of people.[6]

Is sociology basically concerned with gathering empirical data about people's behavior in a scientific sense, or with the study of those conditions that will enable individuals to realize their fullest potential in a humanistic sense? It is impossible to put all of sociology into one camp or the other. Most sociologists probably use a combination of scientific and humanistic viewpoints. Peter Berger expresses the view of many, however, when he argues that sociology must be used for humanity's sake. Social science, like other sciences, can be and sometimes is dehumanizing and even inhuman. It should not be. When sociologists pursue their tasks with insight, sensitivity, empathy, humility, and a desire to *understand* the human condition, rather than with a cold and humorless scientism, then indeed, the "sociological perspective helps illuminate man's social existence."[7]

SOCIOLOGISTS—WHAT ARE THEY AND WHAT DO THEY DO?

What makes one a sociologist? From what we have said in this chapter, a simple answer would be, One who has a sociological perspective. Also, a sociologist is a person who has earned an advanced degree in sociology. Most professional sociologists have master's degrees or doctorates, and many students who receive bachelor's degrees in sociology go into other fields, so the exact number of sociologists practicing sociology for a living is difficult to determine.

Most sociologists spend most of their time teaching. Opportunities for teaching sociology are found predominantly at the college and university level, but increasing numbers of high schools are adding sociology courses to their curricula. Some sociology courses, especially at the high-school level, have more of a practical, how-to-do-it focus. These courses might deal with dating, marriage and family, or personal adjustment.

Many sociologists are involved in research, either full-time or in combination with their teaching. Full-time researchers may run their own research agencies, but more often they work for state or federal agencies. They may, for example, be responsible for collecting and analyzing population and census data. They may collect police arrest statistics and analyze and combine these into crime reports. Sociologists, especially at the larger universities, are about equally involved in research and teaching.

Sociologists work in a number of capacities for a variety of private, state, and federal agencies. Criminologists are hired by institutions, such as prisons and reformatories, and by departments of correction. Population experts are valuable in a policy-making capacity for a number of agencies. Sociologists act as consultants to cities for recreation programs, city planning, urban renewal, mass transit, and so on. The services of specialists in marriage and the family are used by schools, churches, and social agencies, and some sociologists are involved in private marriage counseling. Additional areas in which sociologists are involved in applied research include evaluation research (analyzing whether a project is working the way it is supposed to), legislative consulting, housing and environmental planning, military research (problems of military installations as communities, patterns of leadership), health services, consumer and advertising research, and so on. Applied sociology is likely to continue to grow as we become more and more an informational society.

SOCIOLOGY AND YOU

What good has this time spent in sociology done for you? Admittedly it is difficult to measure the knowledge you have acquired. If the subject were typing and you had progressed from two words a minute to 80, there would be tangible evidence of success. Because the subject is sociology, if you were to go out and solve eight social problems during summer vacation, that might prove that the course had taught you something. The problem, however, is that we are not teaching a skill or a technique but a viewpoint. And it is sometimes difficult to measure the immediate usefulness of a new viewpoint or a new way of looking at things.

After a semester of introductory sociology, you should know more about the world you live in. Principally, you should have discovered from the course a new way of looking at the world and at yourself. As C. Wright Mills suggested in *The Sociological Imagination*, the study of sociology should show you a way of evaluating yourself as more than an isolated individual in a sea of humanity. The study of sociology should help you place yourself and history in perspective so you can more accurately identify and evaluate the factors that affect your behavior and the behavior of others. You should be more critical, now, and more able to evaluate aspects of the world that heretofore you may have taken for granted.

"Who am I?" and "How did I get this way?" are questions that are difficult to answer to anyone's satisfaction. Perhaps a sociological viewpoint will help deal with these questions; perhaps it won't. If the sociological viewpoint only helps to make you more aware of your social environment and of some of the forces at work there, the time spent has been worthwhile, for this will provide you with a greater understanding of and compassion for others.

TERMS FOR STUDY

action sociology (459)

applied science (459)

hidden observation (460)

humanistic perspective (461)

participant observation (460)

positivist (460)

pure science (459)

scientific perspective (459)

NOTES

1. The discussion in this and the following paragraphs is drawn from Peter L. Berger's *Invitation to Sociology* (Garden City, N.Y.: Doubleday, 1963), ch. 2 ("Sociology as a Form of Consciousness").

2. Philip M. Hauser, "On Actionism in the Craft of Sociology," *Sociology Inquiry* 39 (Spring 1969), pp. 139–147. For more on applied sociology, see Arthur Shostak, ed., *Sociology in Action* (Homewood, Ill.: Dorsey

Press, 1966), and Alvin Gouldner and S. M. Miller, eds., *Applied Sociology* (New York: Free Press, 1965).

3. A thorough and interesting discussion of several sociological perspectives, including "abstracted empiricism," is included in C. Wright Mills's excellent book, *The Sociological Imagination* (New York: Oxford University Press, 1959).

4. A number of works deal with these issues. See, for example, Gideon Sjoberg, ed., *Ethics, Politics, and Social Research* (Cambridge, Mass.: Schenkman, 1967); Ralph Beals, *Politics of Social Research* (Chicago: Aldine, 1968); and Irving L. Horowitz, ed., *The Rise and Fall of Project Camelot* (Cambridge, Mass.: M.I.T. Press, l967). Most methodology texts contain a discussion of scientific ethics.

5. Derek Phillips, *Knowledge From What?* (Chicago: Rand McNally, 1971). Also see John Lofland, *Analyzing Social Settings* (Belmont, Calif.: Wadsworth, 1971) and Eugene Webb, Donald Campbell, Richard Schwartz, and Lee Sechrest, *Unobtrusive Measures: Nonreactive Research in the Social Sciences* (Chicago: Rand McNally, 1966).

6. This discussion of humanistic and scientific thought follows from Don Martindale's analysis in his *Social Life and Cultural Change* (Princeton, N.J.: Van Nostrand, 1962), pp. 443–462.

7. See Peter Berger's chapter, "Sociology as a Humanistic Discipline," in *Invitation to Sociology* (see Note 1). Also see R. P. Cuzzort's discussion of Berger in Cuzzort's book, *Humanity and Modern Sociological Thought* (New York: Holt, Rinehart & Winston, 1969), ch. 10.

GLOSSARY

Acculturation The process of assimilating, blending in, and taking on the characteristics of another culture.

Achieved statuses Positions in society that are earned or achieved in some way.

Action sociology The practical use of sociological concepts and techniques to deal with the problems of society; also called *applied sociology*.

Aggregate A number of people clustered together in one place.

Alienation A feeling of isolation, meaninglessness, and powerlessness; may occur when workers see themselves as only a small and insignificant part of a large, impersonal bureaucracy.

Altruistic suicide The result of strong group attachments—the individual commits suicide to benefit the group and to follow group norms.

Amalgamation Refers to biological (rather than cultural) mixing.

Annihilation Elimination of one group by another.

Anomic suicide Occurs in situations in which norms are confused or are breaking down, as in economic depressions or periods of rapid social change.

Anomie A state of normlessness; a condition in which the norms and rules governing people's aspirations and moral conduct have disintegrated.

Anticipatory socialization Occurs when a person adopts the values, behavior, or viewpoints of a group he or she would like to, but does not yet, belong to.

Apollonian ceremony Sober, restrained, dignified, self-controlled religious ceremony; contrasts with a Dionysian ceremony.

Applied science Concerned with the practical and utilitarian uses of knowledge, with making knowledge useful to people.

Ascribed statuses Positions automatically conferred on individuals through no choice of their own; statuses people are born with.

Assimilation Mixing and merging of unlike cultures so that the two groups come to have a common culture.

Audience A collection of people at (watching, listening to) a public event; a passive crowd.

Authoritarian personality A type of personality especially prone to prejudice; tends to be ethnocentric, to be rigidly conformist, and to worship authority and strength.

Authority Legitimate, socially approved power.

Birth rate Number of live births per 1,000 people in the population.

Bureaucracy (formal organization) An organization in which the activities of some people are systematically planned by other people to achieve some purpose; usually involves such characteristics as division of labor, hierarchy of authority, system of rules, impersonality, and technical efficiency.

Capitalism An economic system that emphasizes the ideas of private property, the profit motive, and free enterprise and free competition.

Caste system A system of stratification in which lines or boundaries between levels (castes) are firmly drawn; mobility between castes is difficult if not impossible.

Category A number of people who have a particular characteristic in common.

Cause-effect relationship Describes a situation in which an event B (effect) always follows a previous event A (cause); and in the absence of A, B does not appear.

Centralization The tendency of people to gather around some central or pivotal point in a city.

Charisma A quality in an individual that sets him or her apart from others—the person is viewed as superhuman and capable of exceptional acts.

Charismatic authority Power stemming from the unique personality of a particular person.

Chronological age Age based on time from date of birth, for example, 8 years, 3 months or 15 years, 6 months, and so on. (Compare with *developmental age*.)

Church A large and highly organized religious organization that represents and supports the *status quo*, is respectable, and in which membership is automatic; one is born into a church.

Class consciousness An awareness of one's position or class in society and of the circumstances, interests, and concerns that that class has in common; from Karl Marx.

Class model A theory describing the distribution of power that holds that economic activities lead to the development of two classes: the bourgeoisie, or capitalists, who own the means of production, and the proletariat, who own only their own labor; from Karl Marx.

Closed group A group into which it is difficult if not impossible to gain membership.

Collective behavior Spontaneous, unstructured, and unstable group behavior.

Complementary needs theory A mate selection theory that holds that people select partners who make up for, balance, or supply characteristics they themselves don't have.

Compound family The family resulting from polygamy; marriages involving multiple spouses—several wives and/or husbands—at the same time.

Conflict theory A sociological theory that holds that the most appropriate way to understand society is to focus on the consequences of the conflict, competition, and discord that are common in social interaction; also used to explain social change; from Karl Marx.

Control (1) A product of knowledge—if information about an event is available, then regulation (control) of the consequences of the event are possible. (2) The attempt by scientists to manipulate (control) factors that may have an effect on what they are studying.

Counterculture A subculture that has norms and values that are in opposition to or in conflict with norms and values of the dominant culture.

Crazes The relatively short-term obsessions that members of a society or members of groups have toward specific behaviors or objects.

Criminal behavior Behavior prohibited by law and subject to formal punishment.

Crowd A temporary collection of people in close physical contact reacting together to a common stimulus.

Cult A religious organization that is a small, short-lived, often local group that is frequently built around a dominant leader. The cult is smaller, less organized, and more transitory than the sect.

Cultural lag Phenomenon that occurs when related segments of society change at different rates.

Cultural pluralism A pattern of interaction in which unlike cultures maintain their own identity and yet interact with each other relatively peacefully.

Cultural relativism Opposite of ethnocentrism; suggests that each culture be judged from its own viewpoint without imposing outside standards of judgment.

Culture The complex set of learned and shared beliefs, customs, skills, habits, traditions, and knowledge common to the members of a society; the social heritage of a society.

Cyclic theories   Theories of social change that hold that societies repeatedly pass back and forth through a series of stages (for example, order and disorder or prosperity and decline).

Decentralization   The tendency of people to move outward and away from the center of the city.

Definition of the situation   The suggestion, first offered by W. I. Thomas, that reality is socially structured, with the result that people respond more to the meaning a situation has for them than to the objective features of that situation.

Democratic model   Reflects the functionalist view that the institution of education provides equal opportunity for all students and rewards ability and hard work regardless of the student's origin.

Democratic system   A system of government in which the power of the state is limited by the wishes of the people, who are seen as having direct input into the decision-making process.

Demographic transition   The movement of a population from high birth and death rates to high birth and low death rates (and rapid population growth) to low birth and death rates.

Demography   The study of human population—its distribution, composition, and change.

Denomination   A religious organization that has less universality than the church and appeals to a smaller category of people; a racial, ethnic, or social-class grouping that is conventional and respectable.

Dependent variable   The "effect"; that quantity or aspect of a study whose change the researcher wants to understand.

Developmental age   Age based on a student's level of maturation or behavior, which would suggest a "readiness" to be placed at a particular grade level. (Compare with *chronological age*.)

Deviant behavior   Behavior contrary to generally accepted norms; its limits are often established by custom or public opinion, sometimes by law.

Differential association   Sutherland's theory of crime, which says that behavior, normal and deviant alike, is learned during interaction with others.

Diffusion   The spread of objects or ideas from one society to another or from one group to another within the same society, resulting in changes in society.

Dionysian ceremony   Ecstatic, emotional, nonrational religious ceremony characterized by loss of self-control; contrasts with an Apollonian ceremony.

Disaster   A situation in which there is a basic disruption of the social context within which individuals and groups function; a radical departure from the pattern of normal expectations.

Discrimination   Actual behavior resulting in unfavorable and unequal treatment of individuals or groups.

Dysfunction   An act that leads to change or destruction of some unit (organism, social group or institution, and so on).

Education   One of the five main institutional areas, the major focus of which is the process of socialization—learning the knowledge, skills, and appropriate ways of behaving to exist in a larger social world.

Ego   In Freud's view, the acting self; the mediator between the id and superego.

Egoistic suicide   Occurs when personal relationships are secondary, distant, and not group-oriented; when problems appear, the absence of emotional support from others makes suicide more likely.

Elite model   A theory describing the distribution of power that holds that power is controlled by an elite who rule because of their superior organization, their personal attributes, or their control of valued resources such as wealth, government authority, and communication facilities.

Elite-pluralism A theory describing the distribution of power that holds that although power is spread across society, the impossibility of mass political participation means that small elite groups develop and make major decisions.

Empiricism Knowledge gained through experience and sense observation.

Endogamy Marriage within a certain group.

Ethnicity Refers to people bound together by cultural ties.

Ethnocentrism A type of prejudice that maintains that one's own culture's ways are right and other cultures' ways, if different, are wrong.

Evolutionary theories Theories of social change that see change as linear—as going in one direction, and that direction is usually toward progress and greater complexity.

Exchange theory A theory of mate selection that proposes the idea of a marketplace in which the seeker wants to maximize his chances for a happy marriage and looks for a good deal in which his or her own assets and liabilities are compared with those of the potential partner; the relationship is continued on the assumption that one will get more out of it than it will cost.

Exogamy Marriage outside a certain group.

Experiment Type of research in which the researcher introduces a stimulus to a subject or group and then observes the response; the researcher manipulates the independent variable (exercise, for example) and watches for change in the dependent variable (heart disease, for example).

Expulsion Removal of a group from the territory in which it resides.

Extended family More than two generations of the same family living together in close association or under the same roof.

Fads The relatively short-term obsessions that members of a society or members of specific groups have toward specific behaviors or objects.

Family of orientation The family into which one is born.

Family of procreation The family of which one is a parent.

Fashions Temporary attachments to specific behaviors, styles, or objects; similar to fads but more widespread and of longer duration.

Fecundity The physical or biological ability to reproduce.

Felony Serious crime that is punished by a year or more in a state prison or by death.

Fertility Refers to the number of children born.

Folk society Robert Redfield's term used to describe isolated villages in nonindustrialized countries that are small, homogeneous, relatively self-contained, and largely based on subsistence activities.

Folkways Norms that are less obligatory than mores, the "shoulds" of society; sanctions for violation are mild.

Function An act that contributes to the existence of a unit, such as breathing and eating for the life of the organism.

Functional analysis A sociological theory that focuses on the structures that emerge in society and on the functions that these structures perform in the operation of society as a whole.

Functional disorders Illnesses caused by environmental factors—stress, rejection, unhappy family, and so on.

Functional theories Theories of social change that hold that there are tremendous forces resisting change but that these forces may perhaps be overcome slowly and adaptively because of changes outside the system, growth by differentiation, and internal innovations.

Gemeinschaft Primary, closely knit society in which relationships are personal and informal and there is a commitment to or identification with the community.

General fertility rate The number of births per 1,000 women age 15 through 44.

Generalized other The sum of the viewpoints and expectations of the social group or community to which an individual belongs.

Genocide The intentional extermination of a whole ethnic or racial group.

Gesellschaft Secondary society based on contractual arrangements, bargaining, a well-developed division of labor, and rational thought rather than emotion.

Group A number of people who have shared or patterned interaction and who feel bound together by a "consciousness of kind" or a "we" feeling.

Group marriage The marriage of two or more men to two or more women at the same time.

Hawthorne effect The fact that people respond to the process of being observed.

Hidden curriculum Refers to the subtle, unintended activities that occur in the process of schooling whereby students are taught the values, attitudes, manners, and personality characteristics appropriate to their future place in society.

Hidden observation The use of records or traces left behind by people, or the actual observation of people without their knowing.

Horizontal groups Groups whose members come predominantly from one social-class level, such as associations of doctors or carpenters.

Horizontal social mobility Movement from one occupation to another within the same social class; also used to refer to spatial or geographical mobility.

Human ecology The adaptation of people to their physical environment, their location in space.

Humanistic perspective A perspective in which the individual (sociologist) is interested in and concerned about human welfare, values, and conduct; there is a desire to improve the lot of people in general, with the ultimate goal being the self-realization and full development of the cultivated person.

Hypothesis A testable statement or proposition used to guide an investigation.

I Mead's concept of the subjective, acting self that initiates spontaneous and original behavior; is contrasted with the "me."

Id In Freud's view, the primitive part of the personality made up of inborn, instinctual, antisocial drives.

Idealistic culture Sorokin's term describing a culture midway between the sensate and ideational; the culture emphasizes logic and rationality.

Ideal norms How a person should behave in a particular situation. (Compare with *statistical norms*.)

Ideational culture Sorokin's term describing a culture that emphasizes feelings and emotions and is subjective, expressive, and religious.

Incest taboo A prohibition that forbids or limits sexual contact between certain family members.

Independent variable The "cause"; that quantity or aspect of the study that seems to produce change in another variable.

Influence Subtle, informal, indirect power based on persuasion rather than coercion and force.

Institutionalization The process in which values, beliefs, norms, and roles regarding a society's basic problems or functions become organized, formalized, and generally understood.

Interaction process analysis (IPA) A technique for group study in which interaction between group members is broken down into small parts or acts. The acts are then classified into a set of predetermined categories (shows agreement, gives information, asks for opinion, and so on).

Interest group A number of people who organize to influence decisions on issues important to them.

Invasion The penetration of one group or function into a territory dominated by another group or function.

Invention Change introduced through the creation of a new object or idea.

Involuntary groups Groups in which membership is automatic and the participant has no choice regarding joining, such as one's family or the army platoon to which one is assigned.

Juvenile delinquency Refers to young people (under 21 in some states, under 18 in others) who commit criminal acts or who are wayward, truant, uncontrollable, or runaways.

Keynoting The process of focusing the behavior of an ambivalent crowd on a particular activity or issue.

Labeling The public stamping, typing, or categorizing of a person as deviant.

Labeling theory A theory that sees deviant behavior as a product of group definitions; becoming deviant involves a labeling process, and one is deviant because a particular group labels one as such.

Latent function The unintended and less obvious, often unrecognized function.

Legal-rational authority Power attached to positions rather than to people; positions are defined by a formal set of rules and procedures; typical of governments and businesses in modern societies.

Life expectancy A person's expectation at birth of length of life, based on risks of death for people born in that year.

Lobbyists People in the business of knowing, influencing, and persuading the right people in government.

Looking-glass self A person's attitudes toward self derived from the person's interpretations of how others evaluate that person.

Manifest function The intended and recognized function of something.

Marginal person A person who is caught between two antagonistic cultures; a product of both but a true member of neither.

Mass hysteria A type of collective behavior in which a particular behavior, fear, or belief sweeps through a large number of people such as a crowd, a city, or a nation.

Material culture The real, tangible things that a society creates and uses: screwdriver, house, classroom, supersonic jet, computer.

Matriarchal society Society in which decisions are made by females.

Me Mead's concept that describes the objective part of self, represents others' attitudes and viewpoints, and judges and evaluates; is contrasted with the "I."

Mechanical solidarity Durkheim's description of the type of social cohesion likely to exist in small, homogeneous, preindustrial societies: uniformity among people and a lack of differentiation or specialization.

Migration A permanent change of residence with the consequent relocation of one's interests and activities.

Milling The excited, restless, physical movement of the people in a crowd through which communication occurs, leading to a definition of the situation and possible collective action.

Minority status Social position in which people are treated as lower in social ranking and are subject to domination by other segments of the population.

Misdemeanors Crimes that are less serious than felonies, punishable by up to a year in a county jail.

Mob A focused, acting crowd that is emotionally aroused and intent on taking aggressive action.

Modern bureaucracy In Weber's view, a type of large organization in which specialization, a system of rules, formality, and a hierarchy of authority are highly developed.

Modern professional organization A type of large organization in which responsibility is decentralized; less bureaucratic in Weber's view than the modern bureaucracy.

Monogamy   Marriage with one person at a time.

Mores   Obligatory norms, the "musts" of society; sanctions are harsh if mores are violated.

Mortality   The number of deaths occurring in a population.

Multinational   A large corporation whose influence goes beyond national boundaries, often with connections throughout the world.

Multiple causation   The situation in which an event is caused by more than one factor.

Mysticism   Information or knowledge gained by mystical means: intuition, revelation, inspiration, magic, visions, spells; tends to be private knowledge experienced only by the mystic.

Nativism   The attempt to improve a people's existence by eliminating all foreign persons, objects, and customs.

Neoevolutionary theories   Theories of social change that see societies as moving through a series of evolutionary stages but in a multilinear fashion; societies are influenced by varied factors, and not all cultures change at the same direction or at the same speed.

Neuroses   The mildest and most common of mental illnesses, in which the person is anxious, nervous, compulsive but generally able to function adequately in society.

Nonmaterial culture   The abstract creations of a society, such as customs, laws, ideas, values, beliefs.

Nonreactive research   Research in which records, physical traces, and signs left by people are studied; observation of subjects without direct intrusion, subject is not aware of being studied. (Compare with *reactive research*.)

Norms   The accepted or required behavior for a person in a particular situation.

Nuclear family   A married couple and their children.

Oligarchy   Rule by a few; Robert Michels's "Iron Law of Oligarchy" describes the tendency in large organizations for an increasingly conservative elite to emerge and take over, with abuse of power a likely consequence.

Open group   A group to which membership is easy to gain.

Operational definition   Precise definitions of terms or conditions, also known as "working definitions," that allow others to know exactly the dimensions of what was studied.

Organic disorders   Illnesses caused by brain injury, hereditary factors, or physiological deterioration.

Organic solidarity   Durkheim's description of the type of social cohesion likely to exist in modern industrialized societies: a high degree of specialization and division of labor, and increasing individualism.

Panic   Nonadaptive or nonrational flight resulting from extreme fear and loss of self-control.

Participant observation   Research in which the researcher is or appears to be a participant in the activity or group that is being studied.

Patriarchal society   Society in which decisions are made by males.

Peer group   Consists of people of relatively the same age, interests, and social position with whom one has reasonably close association and contact.

Personality   The sum total of the physical, mental, emotional, social, and behavioral characteristics of an individual.

Pluralistic model   A theory describing the distribution of power that holds that power is diffuse, or spread across many diverse interest groups.

Political party   A number of people who organize to gain legitimate governmental power.

Polyandry   A form of polygamy in which one woman has several husbands at a time.

Polygamy   Plural marriage; the practice of having more than one husband or wife at a time.

Polygyny A form of polygamy in which one man has several wives at a time.

Population pyramid A pictorial or graphed profile of the age and gender characteristics of a population.

Positivist A term describing a perspective in which sociologists gain knowledge by closely following the scientific perspective; apply scientific methods to social phenomena; are objective and detached and try to be free of any values or biases that might affect their work.

Power The ability of one party (either an individual or group) to affect the behavior of another party.

Prejudice Favorable or unfavorable attitudes toward a person or group that are not based on actual experience.

Prestige One's distinction in the eyes of others; one's reputation.

Primary deviance Beginning, often minor, acts of deviant behavior that do not affect the individual's self-concept—the individual does not see himself or herself as a deviant. (Compare with *secondary deviance*.)

Primary group A group in which contacts between members are intimate, personal, and face to face.

Primary socialization The first socialization an individual undergoes in childhood; one must undergo this to become a member of society; it ends with the establishment of the generalized other.

Process theory A theory of mate selection that proposes that through a series of social and psychological processes, the field of potential mates is progressively narrowed down by physical attraction, religious and racial differences, role and value similarities, psychological influences, and so on.

Propaganda Attempts to influence and change the public's viewpoint on an issue.

Pseudocharisma Refers to public figures who appear to have charisma, but in fact have a packaged and carefully created image manufactured by public-relations techniques.

Psychoses More severe of the mental illnesses, in which the person loses touch with reality, may have hallucinations and delusions, or may withdraw into a dream world.

Public A number of people who have an interest in, and difference of opinion about, a common issue.

Public opinion The ideas and beliefs held by a public on a given issue.

Pure science Attempts to discover facts and principles about the universe without regard for possible uses the knowledge may have.

Race People related by common descent or heredity; usually identified by hereditary physical features.

Racism The belief that people are divided into distinct hereditary groups that are innately different in their social behavior and mental capacities and that because of these differences can be ranked as superior and inferior; this ranking leads to differential treatment in terms of access to society's resources such as jobs, living conditions, wealth, and power.

Random sample A sample of subjects in which every element in the total population being studied has an equal chance of being selected; the small sample of subjects is then theoretically representative of the larger population from which it was selected.

Rationalism A method of gaining knowledge through the use of reason and logic.

Reactive research Research in which subjects react or respond to a questionnaire or an interviewer. (Compare with *nonreactive research*.)

Reference groups Groups that serve as models for our behavior; groups whose perspectives we take as our own and use to mold our behavior.

Reform movement A movement that seeks modifications in certain aspects of society.

Religion A unified system of beliefs, feelings, and behaviors related to things defined as sacred.

Replacement level The average number of births necessary per woman over her lifetime for the population eventually to reach zero growth.

Replication The repetition of studies (research efforts) in similar and unique circumstances to see if the results are consistent with earlier results.

Resocialization A process in which major modifications or reconstructions are made of a person's patterns of behavior, personality, and self-concept.

Resource control A type of power generated by the possession or control of valued items, such as an energy source, major communication facilities, and so on.

Revolutionary movement A movement that seeks a complete change in the social order.

Riot A situation in which mob behavior becomes increasingly widespread and destructive.

Role The behavior of one who occupies a particular status or position in society.

Role and value theories Theories of mate selection that hold that people who share common values and common definitions about roles are more likely to select each other.

Role conflict Conflict that occurs when a person occupies several statuses or positions that have contradictory role requirements.

Role requirements Norms specifying the expected behavior of people holding a particular position in society.

Role strain Strain that occurs when there are differing and conflicting expectations regarding one's status or position.

Rumor Unconfirmed, although not necessarily false, person-to-person communications.

Sanction The punishment one receives for violation of a norm (or the reward granted for compliance with a norm).

Schooling Refers to that part of education involving formal instruction in a classroom setting.

Science The gaining of knowledge through sense observation; understanding through description by means of measurement, making possible prediction and thus adjustment to or control of the environment.

Scientific perspective A perspective in which the individual (sociologist) emphasizes the procedures and techniques of science; the view that knowledge is properly gained by applying scientific methods to social phenomena.

Secession The formal withdrawal of a group of people from a political, religious, or national union.

Secondary deviance Deviant acts that represent a defense, attack, or adjustment to the problems created by reactions to primary deviation; the self-concept changes so that the individual begins to see and define self as deviant. (Compare with *primary deviance*.)

Secondary group Group in which contacts between members are more impersonal than in a primary group; interaction is more superficial and is probably based on utilitarian goals.

Secondary socialization Takes over where primary socialization leaves off; involves internalizing knowledge of new facets and sectors of life and of special skills and techniques.

Sect Small religious organization that is less organized than the church and in which membership is voluntary. Members of sects usually show greater depth and fervor in their religious commitment than do members of churches.

Secularization Becoming more worldly or unspiritual, as opposed to becoming more religious or spiritual.

Segregation (1) Group conflict context: the setting apart of one group. (2) Ecological context: the clustering together of similar people.

Self One's awareness of and ideas and attitudes about one's own personal and social identity.

Self-fulfilling prophecy Occurs when a false definition of a situation evokes a new behavior that makes the originally false conception come true.

Sensate culture Sorokin's term describing a culture that emphasizes the senses and is objective, scientific, materialistic, profane, and instrumental.

Sex ratio Number of males per 100 females.

Significant others People (mother, father, older brother, and so on) whose viewpoints and expectations the individual considers particularly important.

Social change Significant alterations in the social relationships and cultural ideas of a society.

Social control The processes, planned or unplanned, by which people are made to conform to collective norms.

Social differentiation The process of defining, describing, and distinguishing among different categories of people.

Social disorganization The condition resulting when norms and roles break down and customary ways of behaving no longer operate.

Social inequality A consequence of social differentiation in which categories of people are ranked or valued at different levels, based on such factors as wealth, race, ethnicity, age, or gender.

Social institutions Organized systems of social relationships that embody some common values and procedures and meet some basic needs of society.

Social interaction The process of being aware of others when we act and of modifying our behavior in accordance with others' responses.

Socialism An economic system that emphasizes the ideas that property should be collectively owned; that goods and services should be produced according to peoples' needs, regardless of profitability; that centralized planning should determine what is produced; and that wages and benefits should be equally distributed.

Socialization The social process whereby one learns the expectations, habits, skills, values, beliefs, and other requirements necessary for effective participation in social groups.

Social mobility Refers to movement (up, down, or sideways) within the social-class structure.

Social movement A group of people acting with some continuity to promote or to resist a change in their society or group.

Social organization The social fabric of society; the integrated set of norms, roles, cultural values, and beliefs through which people interact with each other, individually and in groups.

Social reproduction model Reflects the conflict view that the institution of education works to reproduce the social class of the student; that is, schooling functions in such a way that the social class of the student replicates the social class of the family from which the student comes.

Social stratification The division of people in society into layers or ranks. The source of the ranking may be one or a combination of factors: wealth, power, prestige, race, sex, age, religion, and so on.

Social structure The network of norms, roles, statuses, groups, and institutions through which people relate to each other in society.

Society A number of people living in a specific area who are relatively organized, self-sufficient, and independent, and who share a common culture.

Sociobiology The systematic study of the biological basis of social behavior in all kinds of organisms, including humans.

Sociology The science or study of the origin, development, organization, and functioning of human society, including social behavior, groups, institutions, and social organization.

Sociometry A technique for group study in which group members are asked to designate the person or people in the group whom they "like most," "would most like to work with," "would like expelled from the group," and so on. The resulting responses for all members of the group are then plotted in a figure called a sociogram, which shows the patterns and directions of interaction in the group.

State The political body organized to govern within society; the center of legitimate power.

Statistical norms How a person actually behaves in a particular situation. (Compare with *ideal norms*.)

Status A position in society or in a group.

Status inconsistency Situation that occurs when the factors that determine an individual's rank in society are not consistent with each other—for example, a college-educated carpenter or a black court justice.

Status symbols Objects (often material possessions) or behaviors that one uses to show (or exaggerate) one's place on the social-class ladder.

Stereotyping Using a simplified and standardized image or label to describe a group or category of people.

Strain for consistency William Graham Sumner's view that there is a fundamental resistance to change in all societies in that people are basically conservative and seek stability.

Subcultures Groups or segments of society that share many of the characteristics of the dominant culture, but that have some of their own specific customs or ways that tend to separate them from the rest of society.

Succession The complete displacement or removal of an established group; the end product of invasion.

Superego In Freud's view, that part of the personality made up of an internalized set of rules and regulations forming the conscience.

Survey research Research that involves the collection of information from subjects through use of questionnaires or interviews; may be descriptive or causal-explanatory.

Symbolic interaction A sociological theory that focuses on the *process* of person-to-person interaction, on how people come to develop viewpoints about themselves and others, and on forms of interpersonal communication.

Total fertility rate An estimate of the number of children born per 1,000 women.

Totalitarian system A system of government in which the state has absolute and total control, power is highly centralized in one party or group, and dissent is not allowed.

Tracking A procedure whereby students are placed in certain curricular or ability patterns based on test scores and on input from teachers, parents, and school administrators.

Traditional authority Power that is granted according to custom; leaders are selected by inheritance.

Urban society Robert Redfield's term to describe modern Western industrialized nations; opposite of folk society.

Values General opinions and beliefs that people have about which ways of behaving are proper and acceptable, and about which ways are improper and unacceptable.

Variable A condition or trait that changes or has different values (such as I.Q., temperature, or weight).

Vertical group A group whose members come from a variety of social classes.

Vertical social mobility Movement up or down the social-class ladder.

Voluntary groups Groups that have open membership; people may join or not as they wish.

Wealthfare system According to Turner and Starnes, through the use of power and influence a tax system has been constructed that allows the upper classes to maintain their privileged position by permitting them to avoid taxation and by giving them and the businesses they own direct cash payments and subsidies.

Welfare system Ostensibly, according to Turner and Starnes, a humanitarian effort to keep people from starving by giving monetary support; in fact, it perpetuates poverty by keeping payments low, by forcing people to work in any available job, and by diffusing any collective power the poor may have.

Index

477